D0208154

ADVANCES IN CHEMISTRY SERIES **228**

CSC SYMPOSIUM SERIES 2

Electron Transfer in Inorganic, Organic, and Biological Systems

James R. Bolton, EDITOR
University of Western Ontario

Noboru Mataga, EDITOR
Osaka University

George McLendon, EDITOR
University of Rochester

Developed from a symposium sponsored
by the International Chemical Congress
of Pacific Basin Societies, Honolulu, Hawaii,
December 17–22, 1989

American Chemical Society, Washington, DC 1991
Canadian Society for Chemistry, Ottawa, Canada

Library of Congress Cataloging-in-Publication Data

Electron transfer in inorganic, organic, and biological systems / James R. Bolton, Noboru Mataga, George McLendon, editors.

   p.   cm.—(Advances in chemistry series; 228) (CSC symposium series; v. 2)

   "Developed from a symposium sponsored by the International Chemical Congress of Pacific Basin Societies, Honolulu, Hawaii, December 17–22, 1989."

   Includes bibliographical references and indexes.

   ISBN 0–8412–1846–3

   1. Oxidation–reduction reaction—Congresses.

   I. Bolton, James R., 1937–   . II. Mataga, Noboru, 1927–   . III. McLendon, George, 1952–   . IV. International Chemical Congress of Pacific Basin Societies (1989: Honolulu, Hawaii) V. Series. VI. Series: CSC symposium series; v. 2.

QD63.O9E34 1991
541.3′93—dc20

QD
1
.A355
no.228

91–12841
CIP

The paper used in this publication meets the minimum requirements of American National Standard for Information Sciences—Permanence of Paper for Printed Library Materials, ANSI Z39.48–1984.

Advances in Chemistry Series

M. Joan Comstock, *Series Editor*

1991 ACS Books Advisory Board

FOREWORD

The ADVANCES IN CHEMISTRY SERIES was founded in 1949 by the American Chemical Society as an outlet for symposia and collections of data in special areas of topical interest that could not be accommodated in the Society's journals. It provides a medium for symposia that would otherwise be fragmented because their papers would be distributed among several journals or not published at all. Papers are reviewed critically according to ACS editorial standards and receive the careful attention and processing characteristic of ACS publications. Volumes in the ADVANCES IN CHEMISTRY SERIES maintain the integrity of the symposia on which they are based; however, verbatim reproductions of previously published papers are not accepted. Papers may include reports of research as well as reviews, because symposia may embrace both types of presentation.

ABOUT THE EDITORS

JAMES R. BOLTON has been active in the field of photoinduced electron transfer (PET) reactions for almost 25 years. He was involved in some of the early work on PET reactions in photosynthesis using the technique of electron paramagnetic resonance spectroscopy. More recently, he and his group have synthesized donor–acceptor model compounds, involving a porphyrin linked to a p-benzoquinone via amide and other bridge units. The study of PET rates in compounds of this type has provided some important information on the mechanism of intramolecular PET and has provided some insights as to how PET occurs in photosynthesis. He is a Fellow of the Chemical Institute of Canada and the American Association for the Advancement of Science. In 1978 he received the Noranda Award of the Chemical Institute of Canada.

NOBORU MATAGA has been active for more than 35 years in the study of excited molecules, including electron transfer, energy transfer, and related phenomena. His 1955 work on excited-state solvation was seminal, and his formula of fluorescence solvatochromism led to the study of dynamic solvent effects on charge-transfer phenomena. He and his co-workers made pioneering investigation of intramolecular exciplexes with conjugate π-electronic donors and acceptors and extended it to demonstrate charge separation by two-step electron transfer in photosynthetic models. They used femtosecond-picosecond laser spectroscopy to elucidate molecular interactions in the excited state, especially electron-transfer mechanisms. His professional contributions were acknowledged with the Award of the Chemical Society of Japan in 1986.

GEORGE MCLENDON received a B.S. from the University of Texas—El Paso and a Ph. D. (with Arthur Martell) from Texas A&M University. In 1976 he joined the faculty of the University of Rochester, where he became Tracy Harris Professor of Natural Science in 1990. He has held concurrent visiting scientist positions at AT&T Bell Laboratories, Argonne National Laboratory, Caltech, and the University of Oxford. His teaching and research reflect a special interest in electron-transfer reactions in proteins and the solid state. His awards include a Sloan Fellowship, Dreyfus Teacher–Scholar Award, Guggenheim Fellowship, ACS Eli Lilly Award for Fundamental Research in Biochemistry, the ACS Award in Pure Chemistry, and several teaching awards. He consults widely for government and industry, serving on the NSF Advisory Board, DOE Visiting Committee, and NIH study sections.

CONTENTS

INDEXES

Introduction to Electron Transfer in Inorganic, Organic, and Biological Systems

James R. Bolton[1], Noboru Mataga[2], and George McLendon[3]

[1]Photochemistry Unit, Department of Chemistry, University of Western Ontario, London, Ontario N6A 5B7, Canada
[2]Department of Chemistry, Faculty of Engineering Science, Osaka University, Toyonaka, Osaka 560 Japan
[3]Department of Chemistry, University of Rochester, Rochester, NY 14627

This chapter provides an overview of the current developments in the field of electron transfer in theory and inorganic, organic, and biological systems. It shows how the various chapters of this volume contribute to the efforts to find common solutions to common problems.

THE CURRENT IMPORTANCE OF ELECTRON-TRANSFER REACTIONS is illustrated by the fact that four of the symposia and 81 of the papers presented at the Conference of the Pacific Basin Chemical Societies in Hawaii in December 1989, had the words "electron transfer" in the title.

This intensity of activity has been building since the mid-1950s, when the basic elements of electron-transfer theory were introduced by Marcus (1–5), with later contributions from Hush (6), Levich and Dogonadze (7, 8), and others. Recent progress in the elucidation of its mechanisms in various fields of photochemistry and photobiology is partly due to remarkable advances in experimental methods such as ultrafast laser spectroscopy (9, 10). These techniques have made possible more direct and detailed observations of electron-transfer processes. A brief review of basic electron-transfer theory is provided in this volume (11).

Interest in electron-transfer processes can be divided into four general

0065–2393/91/0228–0001$06.00/0

areas: theory, inorganic systems, organic systems, and biological systems. Although there has been considerable interaction between theory and applications, the three experimental fields have tended to develop separately, even to the point of using different nomenclature. The key purposes of this volume are to bring these fields together, to increase interactions among them, and to seek common solutions to common problems.

Much of the current interest in electron-transfer processes stems from the exciting advances made in the past decade in our understanding of the primary processes involved in photosynthesis (12). The determination of the crystal structure of the reaction center protein in photosynthetic bacteria (13, 14), together with the results of ultrafast laser photolysis and related investigations on photoinduced electron-transfer processes in these systems, has provided not only a vivid picture of how photoinduced electron transfer occurs in photosynthesis, but also an insight into how nature has optimized the efficiency of the system. Attention is now shifting to the reaction centers of green-plant systems, particularly Photosystem I (15).

Elegant work on natural photosynthesis has stimulated studies of model donor–acceptor molecules, joined by a spacer or directly by a single bond (16–18), as well as of modified protein systems. These studies attempted to define the important factors that control electron transfer from a donor to an acceptor.

On the other hand, studies of the electron-transfer mechanism in the fluorescence quenching reaction between uncombined donor and acceptor systems, including various dyes and aromatic molecules as fluorescers and various organic and inorganic molecules as quenchers in solution, have a long history dating from the 1930s (19). This subject has been one of the most important aspects in elucidation of the mechanisms underlying photoinduced electron transfer in solution. Nevertheless, there are still some problems, such as the energy gap dependence of the electron-transfer rate constant in the fluorescence quenching reaction, that cannot be interpreted satisfactorily on the basis of standard electron-transfer theories (10, 20).

Electron-transfer theory has continued to develop since the pioneering work mentioned in the foregoing discussion. Current interest is focused on explaining the dependence of function and dynamics on the structure of the reaction center protein and on examining the interplay between electronic and nuclear factors (21). This information can provide a better understanding of the role of electron tunneling and of solvation (20) and solvent dynamics (22, 23). These electron-transfer factors have been examined in many systems, including solutions, molecular assemblies, and various model systems.

Most of the model systems, with donor and acceptor groups separated by a spacer, are composed of organic or metal ion entities. It is a tribute to the synthetic chemist that we have so many model compounds today. The studies on these molecules, many of which are discussed in this volume, have provided a rich harvest of information on the various important factors that control intramolecular electron transfer. Among these factors are energy of the excited state, exergonicity ($-\Delta G^0$), distance between the donor (D) and acceptor (A), orientation of D with respect to A, nature of the bridge, nature of the solvent, and temperature.

Work in this area can be broadly subdivided into

1. photoinduced electron-transfer reactions in which both forward (charge separation) and reverse (charge recombination) intramolecular electron-transfer rate constants have been measured;

2. charge-shift reactions, in which an electron is introduced (e.g., by pulse radiolysis) into one entity in the molecule and then undergoes intramolecular electron transfer to another entity; and

3. bimolecular studies using uncombined donor and acceptor entities, in which it has been possible to infer the unimolecular rate constants for charge recombination within the geminate radical ion pairs formed by charge separation at encounter between an excited molecule and a quencher.

Studies on combined donor–acceptor model systems have expanded to include multiple donors or acceptors, up to four interacting entities. Some of these molecules have also provided excellent model systems for the study of triplet excitation transfer.

Studies of organic donor–acceptor adducts have been complemented and extended by synthesis and detailed investigation of inorganic analogs. The variability of oxidation states and reorganization energies afforded by metal complexes has been used to particular advantage in studies ranging from the classic Creutz–Taube compound (24) to the elegant "intervalence" electron-transfer peptide spacer studies by Isied (25, 26) and others [e.g., Ohno et al. (27)].

The photochemical properties of transition metal complexes have been strongly linked to their electron-transfer properties. Most recently, keen interest in photochemical "water splitting" catalyzed by tris(4,4'-bipyridyl)Ru(II) and its homologs has spawned a deeper appreciation of the fundamental relation between photochemical radiationless transitions and electron transfer, as summarized by the energy gap law in the inverted region (10).

Heme proteins have also provided a very useful system for the study of electron-transfer reactions across a fixed distance. In these studies, inorganic complexes (e.g., Ru complexes) have been attached to peripheral amino acids, as exemplified by the work of Therien et al. (28), Isied (25, 26), and others. Photoinduced electron transfer is then initiated by excitation of either the Ru complex or the heme group. This subject has been helped considerably by theoretical work such as that by Beratan and Onuchic (29).

A complementary approach has been pursued by Natan and Hoffman (30), McLendon et al. (31), and others, who studied electron transfer between two proteins. In general, these protein pairs are physiological partners. In such studies, some uncertainty may exist about the precise structure of the protein–protein complex. Indeed, recent work has suggested that such protein–protein complexes can be highly dynamic, with relative motion along the interacting protein surfaces. It is difficult in such systems to discuss the "distance dependence" or "pathway dependence" of electron-transfer rates, because the system investigated describes not just one structure, but a family of structures. Conversely, such studies do provide key information on aspects of biological design. This information leads not only to adequate reaction rates but also to excellent reaction specificities. It is hoped that such insights may ultimately lead to the study and understanding of in vivo electron-transport systems.

Researchers continue to study electron transfer in photosynthetic systems. By carrying out specific modifications to the structure of the reaction center, either by extracting and replacing components or by genetic modifications, it is possible to obtain information on the importance of certain components and structural features in controlling the rate of electron transfer in the reaction center. In addition, many sophisticated techniques (e.g., Stark spectroscopy and quantitative femtosecond spectroscopy) are being used to elucidate the nature of the electron-transfer process. Some of this work is described in this book (32).

Finally, Miller (33) has provided an excellent summary of the puzzles of electron-transfer processes, and Marcus (34) has summarized the important points arising from the symposium upon which this volume was based.

In the future there is likely to be an expansion and better understanding of the studies summarized in this introduction, as well as expansion into other areas. The results will be a better understanding of the mechanisms of electron transfer in various molecular assemblies and of interfacial electron transfer. Applications of this work are likely to lead to the design of more efficient photoconverters of solar energy, photosensing devices, information storage devices, and many other developments. It is an exciting field!

References

1. Marcus, R. A. *J. Chem. Phys.* **1956**, *24*, 966.
2. Marcus, R. A. *Faraday Discuss. Chem. Soc.* **1960**, *29*, 21.
3. Marcus, R. A. *Annu. Rev. Phys. Chem.* **1964**, *15*, 155.
4. Marcus, R. A. *J. Chem. Phys.* **1965**, *43*, 679.
5. Marcus, R. A. *Faraday Discuss. Chem. Soc.* **1982**, *74*, 7.
6. Hush, N. S. *Trans. Faraday Soc.* **1961**, *57*, 557.

7. Levich, V. G.; Dogonadze, R. R. *Dokl. Akad. Nauk. SSSR* **1959**, *124*, 123; *Dokl. Phys. Chem. (Engl. Transl.)* **1959**, *124*, 9.
8. Levich, V. G. *Adv. Electrochem. Electrochem. Eng.* **1966**, *4*, 249.
9. Mataga, N.; Miyasaka, H.; Asahi, T.; Ojima, S.; Okada, T. In *Ultrafast Phenomena VI*; Springer Verlag: Berlin, 1988; pp 511–516.
10. Mataga, N. In *Electron Transfer in Inorganic, Organic, and Biological Systems*; Bolton, J. R.; Mataga, N.; McLendon, G., Eds.; Advances in Chemistry Series 228; American Chemical Society: Washington, DC, 1991; Chapter 6 and references cited therein.
11. Bolton, J. R.; Archer, M. D. In *Electron Transfer in Inorganic, Organic, and Biological Systems*; Bolton, J. R.; Mataga, N.; McLendon, G., Eds.; Advances in Chemistry Series 228; American Chemical Society: Washington, DC, 1991; Chapter 2.
12. Deisenhofer, J.; Michel, H. *Science (Washington, D.C.)* **1989**, *245*, 1463.
13. Deisenhofer, J.; Epp, O.; Miki, K.; Huber, R.; Michel, H. *J. Mol. Biol.* **1984**, *80*, 385.
14. Chang, C. H.; Tiede, D.; Tang, J.; Smith, U.; Norris, J. R.; Schiffer, M. *FEBS Lett.* **1986**, *205*, 82.
15. Iwaki, M.; Itoh, S. In *Electron Transfer in Inorganic, Organic, and Biological Systems*; Bolton, J. R.; Mataga, N.; McLendon, G., Eds.; Advances in Chemistry Series 228; American Chemical Society: Washington, DC, 1991; Chapter 10.
16. Connolly, J. S.; Bolton, J. R. In *Photoinduced Electron Transfer*; Fox, M. A.; Chanon, M., Eds.; Elsevier: New York, 1989; Vol 4, pp 303–393.
17. Wasielewski, M. R.; Johnson, D. G.; Niemczyk, M. P.; Gains, G. L.; O'Neil, M. P.; Svec, W. A. In *Electron Transfer in Inorganic, Organic, and Biological Systems*; Bolton, J. R.; Mataga, N.; McLendon, G., Eds.; Advances in Chemistry Series 228; American Chemical Society: Washington, DC, 1991; Chapter 8.
18. Bolton, J. R.; Schmidt, J. A.; Ho, T.-F.; Liu, J.-Y.; Roach, K. J.; Weedon, A. C.; Archer, M. D.; Wilford, J. H.; Gadzekpo, V. P. Y. In *Electron Transfer in Inorganic, Organic, and Biological Systems*; Bolton, J. R.; Mataga, N.; McLendon, G., Eds.; Advances in Chemistry Series 228; American Chemical Society: Washington, DC, 1991; Chapter 7.
19. See for example Förster, Th. *Fluoreszenz Organischer Verbindungen*; Vandenhoeck & Ruprecht: Göttingen, 1951; p 219.
20. Kakitani, T.; Yoshimori, A.; Mataga, N. In *Electron Transfer in Inorganic, Organic, and Biological Systems*; Bolton, J. R.; Mataga, N.; McLendon, G., Eds.; Advances in Chemistry Series 228; American Chemical Society: Washington, DC, 1991; Chapter 4.
21. Sutin, N. In *Electron Transfer in Inorganic, Organic, and Biological Systems*; Bolton, J. R.; Mataga, N.; McLendon, G., Eds.; Advances in Chemistry Series 228; American Chemical Society: Washington, DC, 1991; Chapter 3.
22. Maroncelli, M.; McInnis, J.; Fleming, G. R. *Science (Washington, D.C.)* **1989**, *243*, 1674.
23. Zhang, X.; Kozik, M.; Sutin, N.; Winkler, J. R. In *Electron Transfer in Inorganic, Organic, and Biological Systems*; Bolton, J. R.; Mataga, N.; McLendon, G., Eds.; Advances in Chemistry Series 228; American Chemical Society: Washington, DC, 1991; Chapter 16.
24. Creutz, C.; Taube, H. *J. Am. Chem. Soc.* **1969**, *91*, 3988.
25. Isied, S. S. *Prog. Inorg. Chem.* **1984**, *32*, 443.
26. Isied, S. S. In *Electron Transfer in Inorganic, Organic, and Biological Systems*; Bolton, J. R.; Mataga, N.; McLendon, G., Eds.; Advances in Chemistry Series 228; American Chemical Society: Washington, DC, 1991; Chapter 15.
27. Ohno, T.; Yoshimura, A.; Ikeda, N.; Haga, M.-A. In *Electron Transfer in In-*

organic, Organic, and Biological Systems; Bolton, J. R.; Mataga, N.; McLendon, G., Eds.; Advances in Chemistry Series 228; American Chemical Society: Washington, DC, 1991; Chapter 14.

28. Therien, M. J.; Bowler, B. E.; Selman, M. A.; Gray, H. B.; Chang, I-J.; Winkler, J. R. In *Electron Transfer in Inorganic, Organic, and Biological Systems*; Bolton, J. R.; Mataga, N.; McLendon, G., Eds.; Advances in Chemistry Series 228; American Chemical Society: Washington, DC, 1991; Chapter 12.

29. Beratan, D. N.; Onuchic, J. N. In *Electron Transfer in Inorganic, Organic, and Biological Systems*; Bolton, J. R.; Mataga, N.; McLendon, G., Eds.; Advances in Chemistry Series 228; American Chemical Society: Washington, DC, 1991; Chapter 5.

30. Natan, M. J.; Hoffman, B. M. In *Electron Transfer in Inorganic, Organic, and Biological Systems*; Bolton, J. R.; Mataga, N.; McLendon, G., Eds.; Advances in Chemistry Series 228; American Chemical Society: Washington, DC, 1991; Chapter 13.

31. McLendon, G. L.; Hickey, D.; Sherman, F.; Brayer, G. In *Electron Transfer in Inorganic, Organic, and Biological Systems*; Bolton, J. R.; Mataga, N.; McLendon, G., Eds.; Advances in Chemistry Series 228; American Chemical Society: Washington, DC, 1991; Chapter 11.

32. Franzen, S.; Boxer, S. G. In *Electron Transfer in Inorganic, Organic, and Biological Systems*; Bolton, J. R.; Mataga, N.; McLendon, G., Eds.; Advances in Chemistry Series 228; American Chemical Society: Washington, DC, 1991; Chapter 9.

33. Miller, J. R. In *Electron Transfer in Inorganic, Organic, and Biological Systems*; Bolton, J. R.; Mataga, N.; McLendon, G., Eds.; Advances in Chemistry Series 228; American Chemical Society: Washington, DC, 1991; Chapter 17.

34. Marcus, R. A. In *Electron Transfer in Inorganic, Organic, and Biological Systems*; Bolton, J. R.; Mataga, N.; McLendon, G., Eds.; Advances in Chemistry Series 228; American Chemical Society: Washington, DC, 1991; Epilogue.

RECEIVED for review April 27, 1990. ACCEPTED revised manuscript July 19, 1990.

Basic Electron-Transfer Theory

James R. Bolton[1] and Mary D. Archer[2]

[1]Photochemistry Unit, Department of Chemistry, University of Western Ontario, London, Ontario N6A 5B7, Canada
[2]Newnham College, Cambridge, England CB3 9DF

This chapter provides an introduction to basic electron-transfer theory. The classical Marcus theory is developed, and the reorganization energy is defined. The difference between adiabatic and nonadiabatic electron-transfer reactions is explained. Quantum mechanical theories of electron transfer are outlined for nonadiabatic reactions with particular application to the Marcus inverted region. Finally, the effect of solvent dynamics is examined.

THE BASICS OF ELECTRON-TRANSFER THEORY are presented in this chapter so that the authors of subsequent chapters can refer to it for the fundamental equations and nomenclature. It should also serve as a tutorial for those who are not familiar with the basic theory, although this is a only a brief outline.

By far the most successful theory of electron transfer (ET) is that introduced and developed by Marcus (1–5); thus, this outline will deal almost exclusively with a summary of that theory and the important equations derived therefrom. Hush (6) developed a theory similar to that of Marcus, based on concepts involved in ET at electrode surfaces; however, Hush's theory does not predict the inverted region (vide infra). Comprehensive reviews by Newton and Sutin (7) and Marcus and Sutin (8, 9) offer a thorough development of the Marcus theory of electron transfer.

Usually Marcus theory is used for outer-sphere ET reactions between a donor D and an acceptor A. (For convenience, we assume that D and A are neutral molecules. The case of charged reactants introduces only the possibility of electrostatic effects that can be incorporated with little difficulty into the theory.) Either D or A may be in an excited state (D* or A*), in which case the process is called photoinduced electron transfer (PET). How-

0065–2393/91/0228–0007$06.00/0

ever, other than a change in the starting-state energies, the principles of electron-transfer theory apply equally well to photoinduced and to ground-state electron-transfer reactions.

For second-order reactions the ET reaction can be divided into three steps. In the first step D and A diffuse together to form an outer-sphere *precursor complex* $D|A$ (rate constant k_a in eq 1a usually approaches the diffusion-controlled limit).

$$D + A \underset{k_d}{\overset{k_a}{\rightleftharpoons}} D|A \tag{1a}$$

In the second step, the precursor complex $D|A$ undergoes a reorganization toward a transition state in which electron transfer takes place to form a *successor complex* $D^+|A^-$.

$$D|A \underset{k_{-ET}}{\overset{k_{ET}}{\rightleftharpoons}} D^+|A^- \tag{1b}$$

Because of the Franck–Condon principle, the nuclear configuration of the precursor and successor complexes at the transition state must be the same.

Finally, the successor complex dissociates to form the product ions D^+ and A^-.

$$D^+|A^- \overset{k_s}{\longrightarrow} D^+ + A^- \tag{1c}$$

If D and A are covalently linked or held together in a matrix (e.g., a protein), then only step 1b can occur; k_{ET} and (in principle) k_{-ET} are then directly measurable.

A steady-state analysis of steps 1a to 1c yields the following expression for the observed bimolecular ET rate constant

$$k_{obs} = \frac{k_a}{1 + \dfrac{k_d}{k_{ET}} + \dfrac{k_d k_{-ET}}{k_s k_{ET}}} \tag{2a}$$

or

$$\frac{1}{k_{obs}} = \frac{1}{k_a} + \frac{1}{K_A k_{ET}}\left[1 + \frac{k_{-ET}}{k_s}\right] \tag{2b}$$

where $K_A = k_a/k_d$. If $k_s \gg k_{-ET}$, eq 2b reduces to

$$\frac{1}{k_{obs}} = \frac{1}{k_a} + \frac{1}{K_A k_{ET}} = \frac{1}{k_a}\left[1 + \frac{k_d}{k_{ET}}\right] \tag{2c}$$

If, in addition, $k_d \gg k_{ET}$, then

$$k_{obs} \cong K_A k_{ET} \tag{2d}$$

On the other hand, if $k_d \ll k_{ET}$

$$k_{obs} \cong k_a \tag{2e}$$

and the observed second-order ET rate constant will not contain any information about k_{ET}.

Knowledge of the various state energies with respect to the ground state of D–A is very important for the interpretation of kinetics in Marcus theory. This is particularly true in PET, where excited-state energies must be known.

The energy of the first excited singlet state, S_1, is usually estimated from the wavelength at which the normalized chromophore (D or A) absorption and fluorescence spectra cross. The energy of the lowest triplet state, T_1, is not as easily measured. It is usually estimated from the blue edge of the low-temperature phosphorescence spectrum of the triplet chromophore, making allowance, if possible, for any perturbation produced by the D–A linkage and for the Stokes shift.

The energies of D^+ and A^- are usually obtained from redox potentials, which are often measured by using cyclic voltammetry or differential pulse voltammetry (10). The difference in Gibbs energy ΔG^0 between the D + A states and the $D^+ + A^-$ states may be approximated as

$$\Delta G^0 = e(E^0_{D^+/D} - E^0_{A/A^-}) + w^P - w^R \tag{3}$$

$E^0_{D^+/D}$ and E^0_{A/A^-} are the standard reduction potentials for D^+ and A^-; w^P and w^R (usually negative) correct for the work of bringing D^+ and A^- and D and A, respectively, together; and e is the electronic charge.

The relative placement of the state energies may be solvent-dependent, but not always in a simple calculable way. Therefore, direct measurement [e.g., by measuring redox potentials in each solvent (11)] is preferable to indirect estimates or extrapolations.

The Classical Marcus Theory

The important electron-transfer step in the mechanism of eqs 1a–1c is step 1b, in which the forward and reverse reactions are unimolecular. Of course, in the case of linked D–A systems, this is the only step.

The initial reactant state or precursor complex D|A will have a potential energy that is a function of many nuclear coordinates (including solvent coordinates), which results in a multidimensional potential-energy surface.

There will exist a similar surface for the product state or successor complex $D^+|A^-$. In transition-state theory, a *reaction coordinate* is introduced so that the potential-energy surface can be reduced to a one-dimensional profile, as shown in Figure 1 for reactions of zero ΔG^0. Curve R represents the potential energy of the reactant state $D|A$ and curve P that of the product state $D^+|A^-$. [The curves in Figure 1 are "symmetrized" with an effective reduced force constant $k = 2k^r k^p/(k^r + k^p)$, rather than assuming that $k^r = k^p$ (4).] For ET to occur, the reactant state must normally distort along the reaction coordinate from its equilibrium precursor position A to position B, the transition state, which has the same nuclear configuration as the initially formed product state. ET occurs at this position, and the resulting product state then relaxes to its equilibrium successor position C.

When all internal and external (solvent) modes are considered, the potential-energy profiles of the reactant and product states along the reaction coordinate are found to be markedly nonparabolic. This is primarily because solvent motion, which occurs at low frequencies where deviations from harmonic motion are large, plays an important role in ET. If, however, the system is represented in a Gibbs (free) energy space, then it can be shown (8, 9) that the Gibbs energy profiles along the reaction coordinate can be well approximated as parabolas. Hence, in further discussion we shall use Gibbs energy profiles rather than potential-energy profiles.

For purposes of presenting the theory, we consider first a ground-state

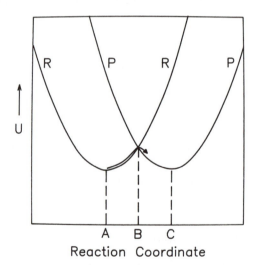

Reaction Coordinate

Figure 1. Section cut along the reaction coordinate through the multidimensional potential surface of the reactant state (R) and that of the product state (P). A and C denote the equilibrium nuclear configurations of R and P, respectively; B denotes the configuration at the intersection (transition state) of the two potential-energy surfaces. The diagram represents the case in which the equilibrium potential energies of R and P are the same.

reaction where $\Delta G^0 < 0$. According to classical transition-state theory, the first-order rate constant k_{ET} is given by

$$k_{ET} = \kappa_{el}\nu_n\exp\left[\frac{-\Delta G^{\ddagger}}{k_B T}\right] \tag{4}$$

where ν_n is the frequency of passage (nuclear motion) through the transition state $(D|A)^{\ddagger}$ $(\nu_n \sim 10^{13}\ s^{-1})$, $\Delta G^{\ddagger}$ is the Gibbs energy of activation for the ET process, κ_{el} is the electronic transmission coefficient, k_B is the Boltzmann constant, and T is temperature. In the classical treatment κ_{el} is usually taken to be unity.

Figure 2 illustrates the parabolic Gibbs energy surfaces as a function of reaction coordinate for a variety of conditions. In Marcus theory, the curvature of the reactant and product surfaces is assumed to be the same. The important quantities in this diagram are λ, the *reorganization energy*, de-

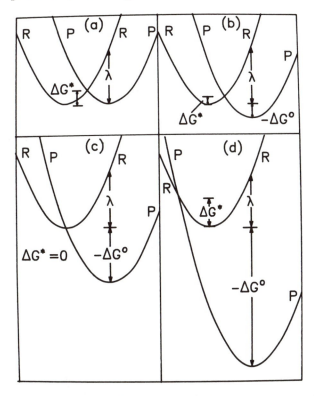

Figure 2. Diagrams showing the intersection of the Gibbs energy surfaces for the reactant (R) state (D|A) and the product (P) state (D$^+$|A$^-$): (a), an isoergonic reaction with $\Delta G^0 = 0$; (b), the normal region where $0 \le -\Delta G^0 \le \lambda$; (c), the condition for maximum rate constant where $-\Delta G^0 = \lambda$; (d), the "inverted region" where $-\Delta G^0 > \lambda$.

fined as the change in Gibbs energy if the reactant state $(D|A)$ were to distort to the equilibrium configuration of the product state $(D^+|A^-)$ without transfer of the electron; ΔG^*, the Gibbs energy of activation for forward ET; and ΔG^0, the difference in Gibbs energy between the equilibrium configurations of the product and reactant states. We assume that ΔG^0 represents the Gibbs energy of reaction when the donor and acceptor are at a distance r_{DA} apart in the prevailing medium; hence the need for the w term in eq 3. In addition, we distinguish between the Gibbs activation energy ΔG^*, obtained from the intersection of the parabolas in diagrams such as those in Figure 2, and the experimental Gibbs activation energy $\Delta G^{\ddagger}$ obtained from the thermodynamic treatment of the transition-state theory.

Figure 2a represents the situation for ET between like species where $\Delta G^0 = 0$. Here the two parabolic surfaces are identical, except that the product surface is displaced along the reaction coordinate with respect to the reactant surface. There is a significant Gibbs energy of activation, even though $\Delta G^0 = 0$. It follows from the properties of parabolas that for a self-exchange reaction or other reaction of zero ΔG^0, ignoring the effect of any work terms

$$\Delta G^* = \frac{\lambda}{4} \tag{5a}$$

For reactions where $\Delta G^0 \neq 0$ (Figure 2b), it is usually assumed that the $D^+|A^-$ surface simply shifts vertically by ΔG^0 with respect to the $D|A$ surface. Again it follows from the analytical geometry of intersecting parabolas that

$$\Delta G^* = \frac{(\lambda + \Delta G^0)^2}{4\lambda} \tag{5b}$$

where the zero of Gibbs energy is taken as that of the precursor complex $D|A$ at the reaction distance r_{DA}. Inserting eq 5b into eq 4 yields the classical Marcus equation

$$k_{ET} = \kappa_{el}\nu_n \exp\left[-\frac{(\lambda + \Delta G^0)^2}{4\lambda k_B T}\right] \tag{6}$$

Equations 5b and 6 and Figure 2 indicate that for moderately exergonic reactions ΔG^* will decrease and k_{ET} will consequently increase, as ΔG^0 becomes more negative. When $-\Delta G^0 = \lambda$ (Figure 2c), $\Delta G^* = 0$ and k_{ET} reaches its maximum value of $\kappa_{el}\nu_n$. However, as ΔG^0 becomes ever more negative in a highly exergonic reaction, the intersection point of the R and P surfaces moves to the left of the center of the R surface, as shown in Figure 2d. This shift indicates that ΔG^* should increase again and thus the striking prediction of eq 6 is that k_{ET} will decrease as the reaction becomes highly exergonic in what has been called the Marcus *inverted region*. Physically, this means that the product must initially be formed in an increasingly distorted and high-energy state.

Reorganization Energy. The reorganization energy, λ, is usually divided into two contributions

$$\lambda = \lambda_{in} + \lambda_{out} \tag{7}$$

The solvent-independent inner term λ_{in} arises from structural differences between the equilibrium configurations of the reactant and product states. λ_{in} is usually treated harmonically so that

$$\lambda_{in} = \frac{1}{2} \sum_i \bar{f}_i (r_R^{eq} - r_P^{eq})^2 \tag{8}$$

where r_R^{eq} and r_P^{eq} are the equilibrium bond lengths in the reactant and product states, respectively; $\bar{f}_i$ is a reduced force constant for the ith vibration, and the sum is taken over all significant intramolecular vibrations. In the few cases where λ_{in} values have been calculated, they have been found to be fairly small [0.1–0.3 eV; *see* Brunschwig et al. (*12*)]; however, in some inorganic complexes [e.g., $Co(NH_3)_6^{2+/3+}$] λ_{in} can be quite large.

The outer term λ_{out} is called the *solvent reorganization energy* because it arises from differences between the orientation and polarization of solvent molecules around $D|A$ and $D^+|A^-$. If the surrounding solvent is treated as a dielectric continuum, then it can be shown (*1, 13*) that

$$\lambda_{out} = \frac{1}{2} \epsilon_0 \left[\frac{1}{\epsilon_{op}} - \frac{1}{\epsilon_s} \right] \int (E^R - E^P)^2 \, dV \tag{9}$$

where E^R and E^P are the electric fields exerted in vacuo at a distance r from the centers of the reactant and product states, respectively; ϵ_{op} and ϵ_s are the optical and static dielectric constants, respectively, of the surrounding solvent medium ($\epsilon_{op} = n^2$, where n is the refractive index of the medium); ϵ_0 is the permittivity of vacuum; and the cyclic integration is carried out over the volume V. The term $(1/\epsilon_{op} - 1/\epsilon_s)$ arises because λ_{out} is the energy of reorganizing the solvent molecules around the equilibrium $D|A$ complex until they are in the orientation of the solvent molecules around the equilibrium $D^+|A^-$ complex but without transfer of the electron. This corresponds to changing the orientation polarization but not the nuclear and electronic polarization.

The computation of the integral of eq 9 requires a specific model so that appropriate boundary conditions can be set. Most authors have chosen a spherical reagent model, which gives on integration (14)

$$\lambda_{out} = \frac{(\Delta e)^2}{4\pi\epsilon_0} \left[\frac{1}{2a_D} + \frac{1}{2a_A} - \frac{1}{r_{DA}} \right] \left[\frac{1}{\epsilon_{op}} - \frac{1}{\epsilon_s} \right] \tag{10}$$

where Δe is the charge transferred in the reaction (almost always one electronic charge); a_D and a_A are the radii of the donor and acceptor, respectively; and r_{DA} is the center-to-center distance between the donor and acceptor.

Cannon (13) and Marcus (15) have also considered a more realistic ellipsoidal model, but it generates rather complex equations for λ_{out}. Nevertheless, irrespective of the model chosen, it is usually possible to approximate λ_{out} by

$$\lambda_{out} \cong B \left[\frac{1}{\epsilon_{op}} - \frac{1}{\epsilon_s} \right] \tag{11}$$

where B is a solvent-independent parameter whose value depends on the model and the molecular dimensions. The value of λ_{out} varies from near zero for very nonpolar solvents (for which $\epsilon_{op} \approx \epsilon_s$) to 1.0–1.5 eV for polar solvents; thus λ_{out} is usually the dominant term in eq 7.

λ_{out} is a function of distance because B in eq 11 is a function of r_{DA} [see eq 10]. Also, λ_{out} is slightly temperature-dependent, as both ϵ_{op} and ϵ_s vary with temperature. For most solvents it is possible to express $\lambda_{out} = \lambda_H - T\lambda_S$, where λ_H and λ_S are enthalpic and entropic components of λ_{out}, respectively; this emphasizes that λ is a Gibbs energy term. For most liquid solvents λ_{out} does not vary by more than 5% over a 100 K temperature range.

Adiabatic vs. Nonadiabatic Electron-Transfer Reactions. Two types of ET reactions can be distinguished according to the magnitude of the *electronic coupling energy* H_{rp} between the reactant and product states (some authors use the symbol V for this term), defined by

$$H_{rp} = \langle \psi_R^\circ | \hat{\mathcal{H}}_{el} | \psi_P^\circ \rangle \tag{12}$$

where ψ_R° and ψ_P° are the electronic wave functions of the equilibrium reactant and product states, respectively; and $\hat{\mathcal{H}}_{el}$ is the Born–Oppenheimer (rigid nuclei) electronic Hamiltonian for the system.

The ET reaction is said to be *adiabatic* if H_{rp} is moderately large, so that the Gibbs energy surfaces interact as shown in Figure 3. Because the surfaces are separated in the intersection region, the reaction always remains on the lower surface as it proceeds through the transition state and the transmission coefficient $\kappa_{el} \approx 1$ in eq 6. When H_{rp} becomes so small that

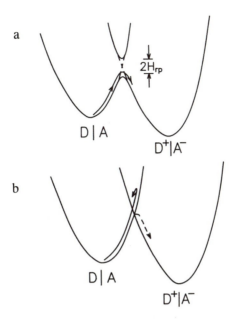

a

b

Figure 3. Adiabatic (a) and nonadiabatic (b) electron transfer. H_{rp} is the electronic coupling energy defined by eq 12.

the R and P surfaces no longer interact significantly, the ET reaction is said to be *nonadiabatic* (this corresponds to $\kappa_{el} \ll 1$ in eq 6). As indicated in Figure 3b, the system will then usually remain on the D|A surface as it passes through the intersection region and will return to the equilibrium state of the reactant. Only occasionally does it cross over to the P surface, bringing about the ET reaction.

The point at which a reaction is to be regarded as adiabatic or nonadiabatic varies with the system. However, Newton and Sutin (7) indicate that for typical transition metal redox reactions, the point of demarcation is $H_{rp} \approx 0.025$ eV. H_{rp} falls off exponentially with distance between D and A. Thus, adiabatic reactions are generally found in those cases in which D and A are relatively close together. In practice, this means either van der Waals contact of D and A in the reactant state or close coupling of D and A in an intramolecular entity.

Quantum Mechanical Electron-Transfer Theories for Nonadiabatic Reactions

Classical Marcus theory generally works well for ET reactions where $\kappa_{el} = 1$, corresponding to a unit probability of electron transfer at the transition state (i.e., most adiabatic ET reactions). However, to explain ET reactions where $\kappa_{el} \ll 1$ (i.e., most nonadiabatic ET reactions), a quantum mechanical approach to nuclear motion is required. We must introduce the concepts of both electron tunneling from D to A and nuclear tunneling from the reactant to the product states. This development dates from the work of Levich and Dogonadze (*16*) and Levich (*17*), followed by several others (e.g., *18–22*).

In the quantum mechanical model, the electronic coupling energy H_{rp} (eq 12) (which involves the overlap of the electronic wave functions of the D and A moieties) and the overlap between the vibrational wave functions of the reactant D|A state and the product $D^+|A^-$ state are of key importance. The latter point is illustrated in Figure 4. The quantum mechanical interpretation of thermal activation is that it permits the population of levels near the intersection of the surfaces where vibrational overlap is significant. An additional feature in the quantum model is that nuclear tunneling, accompanied by electron tunneling from D to A, is possible below the intersection point. Because nuclei are relatively massive, nuclear tunneling must involve only small displacements (~0.1 Å) of nuclei from the reactant to the product surface.

ET may occur nonadiabatically in three ways:

1. *Electron tunneling at the transition state.* When the reactant and product states have the same nuclear configuration at the intersection point, even though H_{rp} may be very small, there is a finite probability of electron tunneling from D to A. This is illustrated by wave function a in Figure 4. Even though the probability of electron tunneling is generally temperature-independent, the overall reaction rate will be temperature-

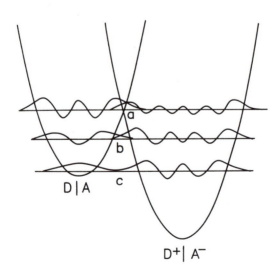

Figure 4. Potential-energy diagram for the quantum mechanical model of electron transfer. The vibrational wave functions are shown symbolically to illustrate the importance of vibrational overlap.

dependent by virtue of the activation required to reach the intersection point.

2. *Activated nuclear tunneling.* Even though the system may not be at the intersection point, the surfaces are close enough for nuclear tunneling from the R to the P surface to be possible. If this takes place from a thermally activated state (e.g., wave function b in Figure 4), the reaction rate will again be temperature-dependent.

3. *Temperature-independent nuclear tunneling.* At very low temperatures, all activated processes are very slow. There may remain a perceptible temperature-independent ET rate arising from nuclear tunneling from the lowest vibrational state of the reactant state to the product surface, as indicated by wave function c in Figure 4.

The quantum model treats the whole donor–acceptor entity as one system. Time-dependent perturbation theory gives the transition rate constant w_j from an initial level j in the reactant state to a set of levels i in the product state as the Fermi "Golden Rule" expression

$$w_j = \frac{2\pi}{\hbar} H_{rp}^2 \sum_i \langle \chi_{Pi}^\circ \mid \chi_{Rj}^\circ \rangle^2 \, \delta(\epsilon_{Pi} - \epsilon_{Rj}) \tag{13}$$

where χ_{Pi}° and χ_{Rj}° are the vibrational wave functions for the equilibrium product state at level i and the equilibrium reactant state at level j, respectively; ϵ_{Pi} and ϵ_{Rj} are the vibrational energies of level i in the product state and level j in the reactant state, respectively; the sum is taken over all vibronic states (including solvent oscillators) in the product state; H_{rp} is the electronic coupling energy defined in eq 12; δ is a Dirac delta function that ensures energy conservation (i.e., $\delta = 0$ if $\epsilon_{Pi} \neq \epsilon_{Rj}$; $\delta = 1$ if $\epsilon_{Pi} = \epsilon_{Rj}$). $\langle \chi_{Pi}^\circ \mid \chi_{Rj}^\circ \rangle^2$ is sometimes written as $(FC)_{ij}$ and called the Franck–Condon factor for levels i and j.

The overall first-order rate constant is then obtained by summing w_j over all the vibrational levels j of the reactant state, each weighted by the Boltzmann probability $P(\epsilon_{Rj})$

$$P(\epsilon_{Rj}) = \frac{\exp\left[-\epsilon_{Rj}/k_B T\right]}{\sum_j \exp\left[-\epsilon_{Rj}/k_B T\right]} \tag{14}$$

of finding the reactant state in the level j with energy ϵ_{Rj}, so that

$$k_{ET} = \sum_j w_j \, P(\epsilon_{Rj}) \tag{15a}$$

$$k_{ET} = \frac{2\pi}{\hbar} H_{rp}^{2} \sum_{j} \sum_{i} \langle \chi_{Pi}^{\circ} \mid \chi_{Rj}^{\circ} \rangle^{2} P(\epsilon_{Rj}) \delta(\epsilon_{Pi} - \epsilon_{Rj}) \qquad (15b)$$

$$k_{ET} = \frac{2\pi}{\hbar} H_{rp}^{2} \, \text{FCWD} \qquad (15c)$$

where FCWD stands for Franck–Condon weighted density-of-states. As ϵ_{Rj} increases, $\langle \chi_{Pi}^{\circ} \mid \chi_{Rj}^{\circ} \rangle$ increases but $P(\epsilon_{Rj})$ decreases.

In a full quantum analysis, the sum in eq 15b is taken over all internal and solvent vibrational modes. However, the solvent vibrations usually occur at low frequencies and are often treated classically, leaving only the relatively high-frequency internal modes to be treated quantum mechanically. In this analysis, eq 15b can be rewritten as the semiclassical Marcus equations (8, 9)

$$\begin{aligned} k_{ET} = {} & \frac{2\pi}{\hbar} H_{rp}^{2} \, (4\pi\lambda_{out} k_B T)^{-1/2} \\ & \times \sum_{j} \sum_{i} \langle \chi_{Pi}^{\circ} \mid \chi_{Rj}^{\circ} \rangle^{2} \\ & \times P(\epsilon_{Rj}) \exp\left[-\frac{(\Delta G^{0} + \epsilon_{Pi} - \epsilon_{Rj} + \lambda_{out})^{2}}{4\lambda_{out} k_B T} \right] \end{aligned} \qquad (16)$$

where λ_{out} is the solvent reorganization energy.

In some treatments [e.g., that of Jortner (20); see also Meyer (23), Miller et al. (24), and Brunschwig and Sutin (25)] it is assumed that the relevant high-frequency vibrations in the reactant state can be replaced by one averaged mode with a frequency ν. Equation 16 then reduces to

$$\begin{aligned} k_{ET} = {} & \frac{2\pi}{\hbar} H_{rp}^{2} \, (4\pi\lambda_{out} k_B T)^{-1/2} \\ & \times \sum_{m=0}^{\infty} \left[\frac{e^{-S} S^{m}}{m!} \right] \exp\left[-\frac{(\lambda_{out} + \Delta G^{0} + mh\nu)^{2}}{4\lambda_{out} k_B T} \right] \end{aligned} \qquad (17a)$$

where m is an integer and

$$S = \frac{\lambda_{in}}{h\nu} \qquad (17b)$$

Eq 17a maximizes at the point where $-\Delta G^{0} \sim \lambda_{out} + \lambda_{in} = \lambda$. In the normal region ($-\Delta G^{0} < \lambda$), eq 17a is represented very well (see Figure 6) by the simplified expression

$$k_{ET} = \frac{2\pi}{\hbar} H_{rp}^{2} \, (4\pi\lambda k_B T)^{-1/2} \exp\left[-\frac{(\lambda + \Delta G^{0})^{2}}{4\lambda k_B T} \right] \qquad (18a)$$

This is called the "high-temperature limit" of the semiclassical Marcus expressions. Equations 17a and 18a track each other very well in the normal region, and thus eq 18a is often used in that region even though the conditions do not justify the high-temperature limit. Equation 18a is equivalent to eq 6 with

$$\kappa_{el}\nu_n = \frac{2\pi}{\hbar} H_{rp}^2 (4\pi\lambda k_B T)^{-1/2} \tag{18b}$$

Thus, the classical Gibbs energy diagrams can be retained, provided that $\kappa_{el}\nu_n$ is reinterpreted according to eq 18b.

The Inverted Region. The situation depicted in Figure 2d and in more detail in Figure 5 is the Marcus inverted region where $-\Delta G^0 > \lambda$. This is an important concept because it indicates a strategy for slowing down a highly exergonic charge recombination reaction following an energy-storing charge separation reaction. According to classical Marcus theory (eq 6) log k_{ET} should first increase as the reaction becomes exergonic, reach a maximum when $-\Delta G^0 = \lambda$, and then decrease quadratically with $-\Delta G^0$ in the inverted region. However, we have seen that classically ET can occur only at the intersection point of the two surfaces; there may be a more effective route via quantum mechanical nuclear tunneling from the reactant surface to the product surface, in which case eq 15a, 16, or 17a may apply. This condition is likely to be particularly important in the inverted region, where the vibrational wave functions of the reactant and product states are embedded

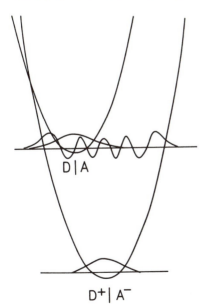

D | A

D⁺ | A⁻

Figure 5. Embedded reactant and product vibrational wave functions in the inverted region.

(as shown in Figure 5), so the Franck–Condon factors are much larger than in the normal region. Siders and Marcus (26) have shown that, if nuclear tunneling is taken into account, a falloff in log k_{ET} is still expected in the inverted region, but it is predicted to be approximately linear in $-\Delta G^0$ rather than quadratic. Figure 6 illustrates this effect. Curve a corresponds to the classical Marcus expression (eq 6). Curves b, c, and d have been calculated by using the semiclassical expression (eq 17a), with three different vibrational frequencies ν. There is relatively little difference between the results for eq 17a and 18a in the normal region (this is the justification for using eq 18a in the normal region), but marked differences occur in the inverted region. Thus, eq 18a must not be used in the inverted region.

There have been many attempts to detect the inverted region predicted by Marcus theory, but until 1984 almost all failed to observe a significant decrease in rate constants at high exergonicities. However, Creutz and Sutin (27) did observe a few bimolecular examples showing a slight decrease in rate constant for $-\Delta G^0 > 1.5$ eV.

The first indication that the inverted region might exist came from a study by Beitz and Miller (28) of the reactions of electrons trapped in a glassy matrix. The first definitive observations were made in 1984 by Miller et al. (24) on intermolecular charge-shift reactions between the biphenyl radical anion and various acceptors in a rigid low-temperature glass, and in an elegant study by Miller et al. (29) of intramolecular ET from the biphenyl radical anion to various acceptors separated by a rigid hydrocarbon linkage.

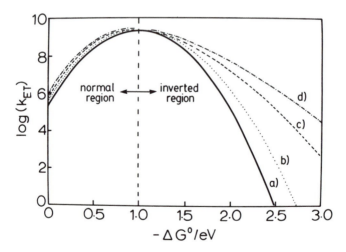

Figure 6. Variation of log k_{ET} with the exergonicity $(-\Delta G^0)$ of the reaction. Curve a was calculated by using the simplified semiclassical expression (eq 18a) with $\lambda = \lambda_{in} + \lambda_{out}$; curves b, c, and d were calculated by using a semiclassical expression (eq 17) with vibrational frequency $\bar{\nu} = 750$, 1500, and 2250 cm^{-1}, respectively. $\lambda_{in} = 0.20$ eV; $\lambda_{out} = 0.80$ eV; $H_{rp} = 0.0003$ eV; and $T = 300$ K.

Since then several other workers (*30–35*) have found strong evidence for the inverted region.

Kakitani and Mataga (*36*) have proposed a modification to Marcus theory. They suggest that because of dielectric saturation in the first solvent shell around ionic components, the curvature (i.e., λ) of the Gibbs energy surfaces should be much larger for ion pairs than for neutral or singly charged entities. From their analysis they predict that charge separation reactions should enter the inverted region only at very large values of $-\Delta G^0$; whereas the charge-recombination reactions should enter the inverted region at much smaller values of $-\Delta G^0$. Thus, the lifetime of the charge-separated products should be increased helpfully. This theory has recently been modified and extended (*37*) and may account for the nonobservance of the inverted region in the bimolecular fluorescence quenching reactions of Rehm and Weller (*38*).

Effect of Solvent Dynamics.   We have so far implicitly assumed that the molecular and solvent reorganization required as an ET reaction passes from the reactant state to the transition state can keep pace with the rate of electron transfer. But as k_{ET} becomes very large, k_{ET}^{-1} becomes comparable with the time required for solvent molecules to reorient, and the observed ET rate is then determined primarily by solvent dynamics. Zusman (*39*) first considered this effect, which has since been treated by several groups (*40–46*). In some instances, ET becomes nonexponential or multiexponential (*47*). In other cases, a single exponential decay is still found, which can be described by the following modification to eq 18a

$$k_{ET} = \frac{2\pi}{\hbar} H_{rp}^2 (4\pi\lambda k_B T)^{-1/2} \left[\frac{1}{1+g} \right] \exp\left[-\frac{(\lambda + \Delta G)^2}{4\lambda k_B T} \right] \quad (19)$$

where g is an "adiabaticity" factor defined by

$$g = k_{ET}^{NA} F \tau_L \quad (20)$$

where k_{ET}^{NA} is the (nonadiabatic) value of k_{ET} when $g = 0$, F is a function of the ratios $\lambda_{in}/\lambda_{out}$ and $\Delta G^*/k_B T$, and τ_L is the "constant charge" solvent dielectric relaxation time given by

$$\tau_L = \tau_D \frac{\epsilon_{op}}{\epsilon_s} \quad (21)$$

where τ_D is the usual "constant field" dielectric relaxation time (*48*). For typical solvents of moderately low viscosity, τ_L is in the range 1–500 ps. Marcus (49) notes four limiting cases reached from eq 19:

1. The slow reaction limit where the solvent relaxation rate is rapid compared with the intrinsic ET rate $(\tau_L^{-1} >> k_{ET}^{NA})$; in this limit $g << 1$ and eq 17a or 18a apply.

2. The fast reaction limit where the intrinsic ET rate is rapid compared with the solvent reorientation rate $(k_{ET}^{NA} >> \tau_L^{-1})$; in this limit $g >> 1$ and the observed $k_{ET} \propto \tau_L^{-1}$).

3. When $\Delta G^*/k_B T$ is small, k_{ET}^{NA} becomes fast. If, in addition, $\lambda_{in}/\lambda_{out}$ is small, F approaches unity and k_{ET} approaches τ_L^{-1}.

4. When $\lambda_{in}/\lambda_{out}$ is large enough, F tends to zero, even when $\Delta G^*/k_B T$ is small; k_{ET} then becomes independent of τ_L^{-1} and approaches k_{ET}^{NA}.

Acknowledgments

We thank R. A. Marcus, N. Mataga, N. Sutin, and T. Kakitani for their helpful comments. Financial support for this work was provided by the Natural Sciences and Engineering Research Council of Canada.

References

1. Marcus, R. A. *J. Chem. Phys.* **1956**, *24*, 966.
2. Marcus, R. A. *Faraday Discuss. Chem. Soc.* **1960**, *29*, 21.
3. Marcus, R. A. *Annu. Rev. Phys. Chem.* **1964**, *15*, 155.
4. Marcus, R. A. *J. Chem. Phys.* **1965**, *43*, 679.
5. Marcus, R. A. *Faraday Discuss. Chem. Soc.* **1982**, *74*, 7.
6. Hush, N. S. *Trans. Faraday Soc.* **1961**, *57*, 557.
7. Newton, M. D.; Sutin, N. *Annu. Rev. Phys. Chem.* **1984**, *35*, 437.
8. Marcus, R. A.; Sutin, N. *Biochim. Biophys. Acta* **1985**, *811*, 265.
9. Marcus, R. A.; Sutin, N. *Comments Inorg. Chem.* **1986**, *5*, 119.
10. Wilford, J. H.; Archer, M. D.; Bolton, J. R.; Ho, T.-F.; Schmidt, J. A.; Weedon, A. C. *J. Phys. Chem.* **1985**, *89*, 5395.
11. Schmidt, J. A.; Liu, J.-Y.; Bolton, J. R.; Archer, M. D.; Gadzekpo, V. P. Y. *J. Chem. Soc., Faraday Trans. 1* **1989**, *85*, 1027.
12. Brunschwig, B. S.; Ehrenson, S.; Sutin, N. *J. Phys. Chem.* **1986**, *90*, 3657.
13. Cannon, R. D. *Chem. Phys. Lett.* **1977**, *49*, 299.
14. Cannon, R. D. *Electron Transfer Reactions*; Butterworth: London, 1980.
15. Marcus, R. A. *J. Chem. Phys.* **1965**, *43*, 1261.
16. Levich, V. G.; Dogonadze, R. R. *Dokl. Akad. Nauk. SSSR* **1959**, *124*, 123; *Dokl. Phys. Chem. (Engl. Transl.)* **1959**, *124*, 9.
17. Levich, V. G. *Adv. Electrochem. Electrochem. Eng.* **1966**, *4*, 249.
18. Kestner, N. R.; Logan, J.; Jortner, J. *J. Phys. Chem.* **1974**, *78*, 2148.
19. Ulstrup, J.; Jortner, J. *J. Chem. Phys.* **1975**, *63*, 4358.
20. Jortner, J. *J. Chem. Phys.* **1976**, *64*, 4860.
21. Siders, P.; Marcus, R. A. *J. Am. Chem. Soc.* **1981**, *103*, 741.

22. Bixon, M.; Jortner, J. *Faraday Discuss. Chem. Soc.* **1982**, *74*, 17.
23. Meyer, T. J. *Prog. Inorg. Chem.* **1983**, *30*, 389.
24. Miller, J. R.; Beitz, J. V.; Huddleston, R. K. *J. Am. Chem. Soc.* **1984**, *106*, 5057.
25. Brunschwig, B. S.; Sutin, N. *Comments Inorg. Chem.* **1987**, *6*, 209.
26. Siders, P.; Marcus, R. A. *J. Am. Chem. Soc.* **1981**, *103*, 748.
27. Creutz, C.; Sutin, N. *J. Am. Chem. Soc.* **1977**, *99*, 241.
28. Beitz, J. V.; Miller, J. R. *J. Chem. Phys.* **1979**, *71*, 4579.
29. Miller, J. R.; Calcaterra, L. T.; Closs, G. L. *J. Am. Chem. Soc.* **1984**, *106*, 3047.
30. Gould, I. R.; Ege, D.; Mattes, S. L.; Farid, S. *J. Am. Chem. Soc.* **1987**, *109*, 3794.
31. Gould, I. R.; Moser, J. E.; Ege, D.; Farid, S. *J. Am. Chem. Soc.* **1988**, *110*, 1991.
32. Irvine, M. P.; Harrison, R. J.; Beddard, G. S.; Leighton, P.; Sanders, J. K. M. *Chem. Phys.* **1986**, *104*, 315.
33. Levin, P. P.; Pluzhnikov, P. F.; Kuzmin, V. A. *Chem. Phys. Lett.* **1988**, *147*, 283.
34. Ohno, T.; Yoshimura, A.; Mataga, N. *J. Phys. Chem.* **1986**, *90*, 3295.
35. Vauthey, E.; Suppan, P.; Haselbach, E. *Helv. Chim. Acta* **1988**, *71*, 93.
36. Kakitani, T.; Mataga, N. *J. Phys. Chem.* **1985**, *89*, 8 and 4752.
37. Yoshimori, A.; Kakitani, T.; Enomoto, Y.; Mataga, N. *J. Phys. Chem.* **1989**, *93*, 8316.
38. Rehm, D.; Weller, A. *Isr. J. Chem.* **1970**, *8*, 259.
39. Zusman, L. D. *Chem. Phys.* **1980**, *49*, 295.
40. Calef, D. F.; Wolynes, P. G. *J. Phys. Chem.* **1983**, *87*, 3387.
41. Calef, D. F.; Wolynes, P. G. *J. Chem. Phys.* **1983**, *78*, 470.
42. Sumi, H.; Marcus, R. A. *J. Chem. Phys.* **1986**, *84*, 4272.
43. Sumi, H.; Marcus, R. A. *J. Chem. Phys.* **1986**, *84*, 4894.
44. Marcus, R. A.; Sumi, H. *J. Electroanal. Chem.* **1986**, *204*, 59.
45. Onuchic, J. N.; Beratan, D. N.; Hopfield, J. J. *J. Phys. Chem.* **1986**, *90*, 3707.
46. Rips, I.; Jortner, J. *Chem. Phys. Lett.* **1987**, *133*, 411.
47. Mataga, N., Osaka University, private communication.
48. Frölich, H. *Theory of Dielectrics: Dielectric Constant and Dielectric Loss*, 2nd ed.; Oxford University Press: Oxford, England, 1958.
49. Marcus, R. A. In *Understanding Molecular Properties*; Avery, J. et al., Eds.; D. Reidel Publishing Co.: Dordrecht, Netherlands, 1987; pp 229–236.

RECEIVED for review April 27, 1990. ACCEPTED revised manuscript July 18, 1990.

Nuclear and Electronic Factors in Electron Transfer

Distance Dependence of Electron-Transfer Rates

Norman Sutin

Chemistry Department, Brookhaven National Laboratory, Upton, NY 11973

The factors that determine the distance dependence of electron-transfer rates are discussed in terms of current models. These models are used to analyze recent data on intramolecular electron-transfer rates in bridged systems. It is found that, in certain systems, the distance dependence of the nuclear factor is larger than that of the electronic factor, although the opposite is true in other systems. Theoretical models are available for calculating the dependence of the nuclear factor on separation distance, and a great deal of progress has been made in deriving expressions describing the distance dependence of the electronic factor. Despite these achievements, there is still considerable uncertainty regarding the values of certain of the key parameters to be used in calculating the magnitudes of the electronic coupling elements in complex systems.

RECENT STUDIES OF MODEL COMPOUNDS (1–8) and naturally occurring systems (9–21) have led to a deeper understanding of the factors determining the distance dependence of electron-transfer rates. These studies have shown that the nuclear and electronic factors both decrease with increasing separation of the redox sites. This finding is in accord with theoretical predictions (22–28). This chapter presents some recent results on the distance dependence of optical and thermal electron-transfer rates. It includes examples of how studies of model systems can provide information about the detailed electron-transfer pathways in complex, naturally occurring systems.

0065–2393/91/0228–0025$06.00/0
© 1991 American Chemical Society

Theoretical Framework

The focus of this chapter will be intramolecular electron transfer at room temperature. First, the formalism that will be used is outlined. Nuclear tunneling corrections will be neglected; such corrections are not large unless the temperature is low or the activation process involves changes in high-frequency modes. In the absence of such effects, the first-order rate constant for intramolecular electron transfer between a donor and an acceptor site can be expressed as the product of an electronic transmission coefficient κ_{el}, an effective nuclear vibration frequency ν_n that destroys the activated complex configuration, and a nuclear factor κ_n (eq 1) (22, 29–33)

$$k = \kappa_{el}\nu_n\kappa_n \tag{1}$$

The electron-transfer reaction is adiabatic ($\kappa_{el} \sim 1$) when the probability of electron transfer in the activated complex is high, and nonadiabatic ($\kappa_{el} \ll 1$) when the electron-transfer probability is low. Within the Landau–Zener treatment of avoided crossings, κ_{el} is given by

$$\kappa_{el} = \frac{2\left[1 - \exp\left(\frac{-\nu_{el}}{2\nu_n}\right)\right]}{2 - \exp\left(\frac{-\nu_{el}}{2\nu_n}\right)} \tag{2a}$$

$$\nu_{el} = \frac{H_{rp}^2}{\hbar}\left(\frac{\pi}{\lambda RT}\right)^{1/2} \tag{2b}$$

where ν_{el} is the electron-hopping frequency in the intersection region (Figure 1), H_{rp} is the electronic coupling matrix element, λ is the reorganization parameter, $\hbar = h/2\pi$, R is the universal gas constant (cal/deg), and T is absolute temperature. Equation 2 shows that $\kappa_{el} = 1$ (i.e., the reaction is adiabatic) when $\nu_{el} \gg 2\nu_n$ and that $\kappa_{el} = \nu_{el}/\nu_n \ll 1$ (i.e., the reaction is nonadiabatic) when $\nu_{el} \ll 2\nu_n$. For nonadiabatic reactions, the product $\kappa_{el}\nu_n$ is independent of the nuclear vibration frequency and is given by eq 2b. Because H_{rp} and λ have opposite distance dependences and because of the $\lambda^{1/2}$ in the denominator of eq 2b, the distance dependence of the electronic factor for a nonadiabatic reaction (ν_{el}) may be larger than that of H_{rp}^2. This difference may be appreciable when the distance dependence of λ is large, as is the case for electron transfers in polar media.

Provided that the free-energy surfaces representing the reactants and products are harmonic with identical "reduced" force constants, the classical nuclear factor κ_n is given by eqs 3–5. In these expressions ΔG^* is the

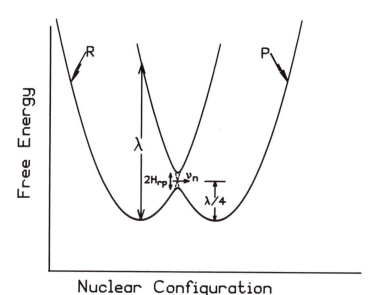

Figure 1. Free energy of the reactants (left parabola) and products (right parabola) as a function of nuclear configuration (reaction coordinate) for an electron-exchange reaction ($\Delta G^0 = 0$). The splitting at the intersection of the curves is equal to $2H_{rp}$, where H_{rp} is the electronic coupling matrix element. λ is the vertical difference between the free energies of the reactants and products at the equilibrium nuclear configuration of the reactants. Thermal electron transfer occurs at the nuclear configuration appropriate to the intersection of the reactant and product curves.

contribution to the free energy of activation from the nuclear reorganization, and ΔG^0 is the free-energy change for the electron transfer.

$$\kappa_n = \exp\left(\frac{-\Delta G^*}{RT}\right) \tag{3}$$

$$\Delta G^* = \frac{(\lambda + \Delta G^0)^2}{4\lambda} \tag{4}$$

$$\lambda = \lambda_{in} + \lambda_{out} \tag{5}$$

Reorganization Energy

The reorganization parameter λ is equal to the energy difference between the reactants' and products' free-energy surfaces at the reactants' equilibrium nuclear configuration when $\Delta G^0 = 0$ (Figure 1). λ contains a contribution from each mode that undergoes a displacement as a consequence of the electron transfer. Two broad classes of contributions to λ are generally distinguished: λ_{in} is associated with (generally fast) changes in the intramolecular (inner-shell) bond distances and angles; λ_{out} is associated with (slow) changes in the polarization of the surrounding medium (outer-shell). λ_{out} depends upon solvent polarity, on the separation of the redox sites, and (for a given separation) on the shape of the molecule (34–36). In the case of proteins or other macromolecules, conformational changes may be included in either λ_{in}, λ_{out}, or ΔG^0, depending upon the time scales for the structural changes (37, 38).

The magnitude of λ_{out} is generally calculated from a classical model in which the medium outside the inner-coordination shells of the reactants is treated as a dielectric continuum with a polarization made up of two parts, a relatively rapid electronic contribution and a slower vibrational–orientational response. For the case of spherical donors and acceptors of radii a_D and a_A with a center-to-center separation distance r, λ_{out} is given by eq 6 with ϵ_{op} and ϵ_s equal to the optical and static dielectric constants of the bulk solvent, respectively (29)

$$\lambda_{out} = \frac{(\Delta e)^2}{4\pi\epsilon_0} \left(\frac{1}{2a_D} + \frac{1}{2a_A} - \frac{1}{r} \right) \left(\frac{1}{\epsilon_{op}} - \frac{1}{\epsilon_s} \right) \tag{6}$$

where Δe is the charge transferred in the reaction and ϵ_0 is permittivity of vacuum. Expressions for λ_{out} are also available for the case in which the two redox sites are located in an ellipsoidally or spherically shaped molecule (29, 34, 35). In all cases λ_{out} increases with increasing separation of the redox sites. ΔG^0 may also depend on r, because $\Delta G^0(r) = \Delta G^0(r=\infty) + w_p - w_r$, where w_r and w_p are the work required to bring the two redox sites to their separation distance in the reactant and product states, respectively. Consequently, in addition to the distance dependence of λ, that of ΔG^0 also needs to be considered in interpreting measured electron-transfer rates.

The consequences of changes in ΔG^0 and λ depend upon whether the reaction is in the normal or the inverted regime and can be quite complex. In the normal regime ($|\Delta G^0| < \lambda$), ΔG^* decreases and the rate constant increases as the driving force for the electron transfer becomes more favorable (eq 4). When $|\Delta G^0| = \lambda$, $\Delta G^* = 0$, $\kappa_n = 1$, and the reaction is barrierless. If the driving force is increased even further ($|\Delta G^0| > \lambda$), ΔG^* will increase and the rate constant will decrease with increasing driving force. This is the inverted regime.

The difference between the normal and inverted regimes can be further illustrated by considering the variation of the rate constant with λ. Again, the rate is maximal when $\lambda = |\Delta G^0|$. When $\lambda > |\Delta G^0|$, increasing λ increases ΔG^* and decreases the rate. By contrast, when $|\Delta G^0| > \lambda$, increasing λ decreases ΔG^* and increases the rate. Thus, the rate responds oppositely to changes in λ in the normal and inverted regions. Because λ decreases as either reactant separation, solvent polarity, or structural differences diminish, such changes promote rapid electron transfer in the normal regime but lead to decreased rates in the inverted regime. These effects need to be kept in mind in interpreting the distance dependence of electron-transfer rates. For example, as previously discussed (36), a rate maximum as a function of reactant separation may be observed in the inverted region under suitable conditions.

The Electronic Factor

In terms of eqs 1–3, the rate constant for nonadiabatic electron transfer is given (22, 29–33) by

$$k = \frac{H_{rp}^2}{\hbar} \left(\frac{\pi}{\lambda RT} \right)^{1/2} e^{-\Delta G^*/RT} \tag{7}$$

The exact dependence of H_{rp} on the distance separating the redox sites remains an open question and will, in general, depend on the nature of the particular system. For many systems H_{rp} appears to decrease exponentially with increasing separation distance (eq 8)

$$H_{rp} = H_{rp}^0 \exp \left(\frac{-\beta(r - r_0)}{2} \right) \tag{8}$$

where H_{rp}^0 is the value of H_{rp} at $r = r_0$, r_0 is the close-contact separation of the redox sites, and β measures the rate of decrease of the electronic coupling with separation. If H_{rp}^0 is sufficiently large (>50 cm^{-1}) so that $\nu_{el} \sim \nu_n$ at r_0, then the electron transfer will be essentially adiabatic at $r = r_0$ (more exactly, the condition $\nu_{el} = \nu_n$ gives $\kappa_{el} = 0.57$) (38). Under these conditions the rate constant at separation distance r is given by

$$k = \nu_n e^{-\beta(r - r_0)} e^{-\Delta G^*/RT} \tag{9a}$$

where the factor 0.57 has been neglected. Alternatively, if the electron transfer is nonadiabatic at $r = r_0$, then the rate constant at separation distance r is given by

$$k = \frac{(H_{rp}^0)^2}{\hbar} \left(\frac{\pi}{\lambda RT} \right)^{1/2} e^{-\beta(r - r_0)} e^{-\Delta G^*/RT} \tag{9b}$$

Another parameter that is useful for discussing the distance dependence of the electronic coupling element is the dimensionless quantity α. In the case of two redox sites linked by a single chain consisting of n equivalent bridging units (spacers), α is defined by eqs 10a–10c

$$H_{rp} = H_{rp}^{0}\, \alpha^{n} \tag{10a}$$

$$n = \frac{r - r_0}{l} \tag{10b}$$

$$\ln \alpha = \frac{-\beta l}{2} \tag{10c}$$

where l is the length of the bridging unit. Essentially, α is the multiplicative factor by which H_{rp} changes (decays) per bridging unit, where the bridging unit may be a bond (e.g., a C–C or a C–N bond) or a group of atoms (e.g., an amino acid residue) (24, 28); in the latter case, α is the product of the decays for the individual C–C, C–N, and N–C bonds. In certain applications it is convenient to consider an average α for the bonds in the group (e.g., an average α for the C–C, C–N, and N–C bonds in the amino acid residue) (24).

Because of the decreased coupling with increasing separation, most reactions will be nonadiabatic at large separations of the redox sites. However, there is a caveat: Equation 2a shows that the relevant parameter determining the degree of adiabaticity is not H_{rp} but the ratio v_{el}/v_{n}. As a consequence, a reaction may remain adiabatic even up to large r if v_{n} is sufficiently small. For example, certain long-range electron transfers in metalloproteins might be adiabatic if coupled to a sufficiently slow protein-conformation change (38).

When the coupling of the redox sites is sufficiently weak (i.e., at large separations of the centers), H_{rp} may be calculated by use of perturbation theory. Consider the case of n sequential bridging units depicted in Scheme 1.

In this scheme the bridging units are either coupled to one another ($H_{i,i+1}$) or, in the case of the terminal units, to the redox sites (H_{D1} for the

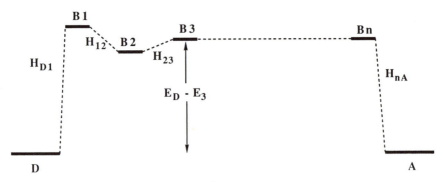

Scheme 1.

donor site and H_{nA} for the acceptor site), E_D is the donor (or acceptor) energy, and E_i is the energy of bridging unit i. For this case the perturbation method yields eq 11 for H_{rp} (39–45).

$$H_{rp} = \frac{H_{D1}H_{nA}}{E_D - E_1} \prod_{i=1}^{n-1} \left(\frac{H_{i,i+1}}{E_D - E_{i+1}} \right) \tag{11}$$

If all of the bridging units are identical ($H_{i,i+1} = H_{BB}$, $E_D - E_{i+1} = \Delta E_{DB}$), then H_{rp}, β, and α are given by eqs 12a–12c, where H_{BB} is the coupling between adjacent bridging units.

$$H_{rp} = \frac{H_{DB}H_{BB}{}^{n-1}H_{BA}}{(\Delta E_{DB})^n} \tag{12a}$$

$$\beta = \left(\frac{2}{l} \right) \ln \left(\frac{\Delta E_{DB}}{H_{BB}} \right) \tag{12b}$$

$$\alpha = \frac{H_{BB}}{\Delta E_{DB}} \tag{12c}$$

Although mediation via the LUMO (lowest unoccupied molecular orbitals) of the bridge (electron-type superexchange) is depicted in Scheme 1, mediation via the HOMO (highest occupied molecular orbitals) of the bridge (hole-type superexchange) can be treated in an analogous manner.

Alternative approaches are available for calculating H_{rp} in systems where the coupling element is fairly large (i.e., when $\alpha \rightarrow 1$) (40–43). In an approach based upon extended Hückel theory, $2H_{rp}$ is taken as equal to the difference between the energies of the lowest unoccupied and highest occupied molecular orbitals that span D and A (one molecular orbital is symmetric and the other antisymmetric with respect to D and A). The coupling can be calculated by use of a partitioning method (40–43). For two redox sites connected by a single bridge of n atoms, this procedure gives eq 13

$$H_{rp} = H_{D1}H_{nA} \sum_{\nu} \frac{a_{1\nu}a_{n\nu}}{E_D - E_{B\nu}} \tag{13}$$

where H_{D1} and H_{nA} are the coupling elements between the orbitals of D and A and the atomic orbitals of the adjacent bridge atoms 1 and n, respectively; $a_{1\nu}$ and $a_{n\nu}$ are the coefficients of the νth bridge molecular orbital at the atoms bonded to D and A; $E_{B\nu}$ is the energy of the νth molecular orbital of the bridge; and the summation is over the molecular orbitals of the bridge. Equation 13 shows that, for a bridge orbital to contribute significantly to the coupling pathway, the coefficients of the terminal atoms need to be relatively large and the energy of the bridge orbital needs to be fairly close to that of the donor (or acceptor) orbital.

Values of H_{ij} and of α per bond for substitution in these expressions can be obtained by use of the Wolfsberg–Helmholtz approximation (46)

$$H_{ij} = \frac{1.75(H_{ii} + H_{jj})S_{ij}}{2} \tag{14a}$$

$$\alpha_{ij} \sim \frac{(H_{ii} + H_{jj})S_{ij}}{[E_D - (H_{ii} + H_{jj})/2]} \tag{14b}$$

where H_{ii} and H_{jj} are the energies of the orbitals involved in the bond and S_{ij} is their overlap.

Note that α and β are not universal properties of the bridge, but also depend upon the energies of the donor and acceptor orbitals and upon their symmetries.

Relationship Between the Coupling Elements for Thermal and Optical Electron Transfer

Information regarding the magnitude of the electronic coupling element can be obtained from the intensities of the metal-to-metal charge-transfer (MMCT) transitions in mixed-valence systems. In symmetrical systems the effective coupling element for the optical electron transfer is related to the intensity of the MMCT transition by (47, 48)

$$H_{rp}(cm^{-1}) = \frac{2.06 \times 10^{-2}}{r} (\epsilon_{max}\bar{\nu}_{max}\Delta\bar{\nu}_{1/2})^{1/2} \tag{15}$$

where r is in angstroms, ϵ_{max} is the molar absorptivity coefficient, $\bar{\nu}_{max}$ is the transition maximum in reciprocal centimeters, and $\Delta\bar{\nu}_{1/2}$ is the full width of the (Gaussian-shaped) MMCT band, also in reciprocal centimeters.

The effective electronic coupling element determined from the intensity of the MMCT transition (eq 15) is not necessarily equal to the electronic coupling element determined from rate measurements (eq 7). Optical electron transfer occurs at the equilibrium nuclear configuration of the reactants; thermal electron transfer occurs at the nuclear configuration appropriate to the intersection of the reactant and product surfaces (Figure 1). Consequently, H_{rp} will be different for optical and thermal electron transfer to the extent that the electronic wave functions for the (zero-order) reactant and product states depend upon their nuclear configurations.

The differences between the coupling elements for optical and thermal electron transfer will be even larger when the coupling is dominated by superexchange interactions because the degree of mixing of the reactant and product states with the bridge state depends on their energy separation and therefore upon the nuclear configuration at which the transfer occurs (49–51). This situation can be illustrated by considering a symmetrical three-state system in which three energy surfaces—the reactant state (R), the product state (P), and the donor-to-bridge charge-transfer state (S)—have identical force constants. For such a system

$$\Delta E_{rs}(Q^*) = \Delta E_{ps}(Q^*) = E_{CT} - \frac{\lambda}{2} \tag{16a}$$

$$\Delta E_{rs}(Q_r^0) = E_{CT} \tag{16b}$$

$$\Delta E_{ps}(Q_r^0) = E_{CT} - \lambda \tag{16c}$$

where Q^* and Q_r^0 denote the nuclear configuration at the intersection region and the reactants' equilibrium configuration, respectively (Figure 2). The effective electronic coupling elements in the superexchange mechanism are proportional to the reciprocals of the relevant energy differences (eq 17a); the particular proportionality constants for the reciprocal energy differences in the optical case (eq 17b) are derived in ref. 49.

$$H_{rp}(th) \propto \frac{1}{\Delta E_{rs}(Q^*)} \tag{17a}$$

$$H_{rp}(op) \propto \frac{1}{2} \left(\frac{1}{\Delta E_{rs}(Q_r^0)} + \frac{1}{\Delta E_{ps}(Q_r^0)} \right) \tag{17b}$$

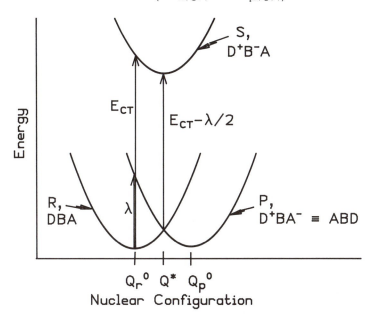

Figure 2. Energy surfaces for the zero-order reactant (R), product (P), and charge-transfer (S) states for electron transfer in a symmetrical bridged system. E_{CT} is the energy for the donor-to-bridge charge-transfer transition.

It therefore follows that, for $E_{CT} >> \lambda/2$, the ratio of the effective electronic coupling elements for thermal and optical electron transfer is approximately given by eq 18.

$$\frac{H_{rp}(\text{th})}{H_{rp}(\text{op})} = 1 - \left(\frac{\lambda/2}{E_{CT} - \lambda/2}\right)^2 \tag{18}$$

Although the foregoing treatment has been illustrated by considering mixing with a metal-to-bridge charge-transfer excited state (i.e., for electron superexchange), an analogous set of equations can be written for mixing with a bridge-to-metal charge-transfer excited state (i.e., for hole superexchange). Thus eq 18 is quite general and is applicable to symmetrical three-state systems with E_{CT} representing either the metal-to-bridge or the bridge-to-metal charge-transfer energy at the equilibrium configuration of the initial state.

Equation 18 shows that $H_{rp}(\text{th}) \sim H_{rp}(\text{op})$ when $E_{CT} >> \lambda/2$ and that $H_{rp}(\text{th}) << H_{rp}(\text{op})$ when λ is large or the bridge state is relatively low-lying. (However, the perturbation treatment may no longer be valid in the latter limit.) In a polar medium λ will increase with increasing separation of the redox sites. To the extent that this effect is dominant, $H_{rp}(\text{th})$ will be smaller, and increase faster with redox site separation, than $H_{rp}(\text{op})$.

Polyene-Bridged Systems

The energies and intensities of the metal-to-metal charge-transfer (MMCT) absorption bands in symmetric polyene-bridged diruthenium systems (eq 19) will be considered first.

$(NH_3)_5Ru^{II}py(CH=CH)_n pyRu^{III}(NH_3)_5$

$\xrightarrow{h\nu = E_{op}} [(NH_3)_5Ru^{III}py(CH=CH)_n pyRu^{II}(NH_3)_5]^*$ (19)

For such systems the energy of the MMCT band maximum is given by (47, 48)

$$E_{op} = \lambda_{out} + \lambda_{in} + \Delta E_{ex} \tag{20}$$

where ΔE_{ex} is the spin-orbit excitation energy of the product state [~0.25 eV for the diruthenium systems (2, 34)]. Data for the $n = 0$ and $n = 1$ complexes in D_2O have been available for a number of years (52). Spectral results for the $n = 2–4$ complexes in nitrobenzene have recently been reported (6), but the interpretation of the data is complicated by the fact that the MMCT absorption bands are not well resolved from the metal-to-ligand charge-transfer (MLCT) transitions. The absorption bands for the $n = 2–4$ complexes in nitrobenzene have been deconvoluted by assuming a Gaussian band-shape (53), and λ values in D_2O were then calculated from the λ values in nitrobenzene by correcting for the different $(1/\epsilon_s - 1/\epsilon_{op})$ values for the two solvents (30).

The λ values for the $n = 2\text{--}4$ complexes in D_2O estimated as described, together with the λ values for the $n = 0,1$ complexes calculated from the measurements in D_2O, are plotted vs. the metal–metal distance in Figure 3, which also includes λ values calculated from the two-sphere (dashed curve) and ellipsoidal (solid curve) models (34). In these calculations a value of 0.18 eV was used for λ_{in} and 4.05 Å was taken for the radii of the end spheres. The calculated λ values for all of the polyene-bridged complexes lie above the experimental values, possibly due to specific solvent interactions (54–57). Ion pairing may also be important for such highly charged complexes (55). Moreover, the λ_{out} calculations assume that a full unit of electronic charge is transferred in the optical excitation. This may be a poor assumption because, as a result of mixing with the MLCT state, some metal electron density resides on the pyridine ligand (57). This delocalization will decrease the dipole moment change and lower the λ_{out} value.

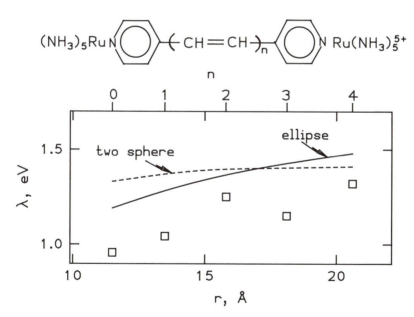

Figure 3. Plot of λ vs. separation distance for the polyene-bridged ($n = 0\text{--}4$) diruthenium complexes. The squares are the experimental data (refs. 6 and 52) and the results for the $n = 2\text{--}4$ complexes have been corrected as described in the text. The solid and dashed curves were constructed with λ_{out} values calculated from the ellipsoidal (ref. 34) and two-sphere models (eq 6), respectively. A value of 0.18 eV was used for λ_{in}, and 4.05 Å was taken for the radii of the end spheres.

The values of ln H_{rp} for the polyene-bridged systems, calculated from the intensities of the MMCT transitions (eqs 2 and 15) with the concentrations corrected by using $K = 10$ for the comproportionation constant (53), are plotted vs. r in Figure 4 (open circles). As discussed for λ_{out}, the data used to calculate H_{rp} for the $n = 2$–4 complexes were obtained by deconvoluting the published MMCT spectra in nitrobenzene (53); H_{rp} for the $n = 0,1$ complexes was calculated from spectral data in D_2O. Despite the approximations made, this analysis is justified because the H_{rp} value for the $n = 2$ complex derived here for nitrobenzene (270 ± 30 cm^{-1}) is very similar to the value obtained from recent measurements in D_2O (58). The value of β (eq 8) is only 0.2 Å^{-1}, and the value of α per bond is 0.9. A recent detailed calculation (51), in which allowance is made for changes in the coplanarity of the pyridine rings with the number of bridging groups, gives $\beta = 0.25$ Å^{-1}. These values are consistent with the interpretation that the polyene bridge is acting as a "molecular wire" in these π-bonded systems (6).

For purposes of comparison with other systems, values of ln κ_n (calculated from the energies of the MMCT transitions) for the polyene-bridged diruthenium systems and ln κ_{el} (calculated from the intensities of the MMCT transitions, eqs 2 and 15) are plotted vs. r in Figure 5. It is apparent from this figure that the κ_{el} values for all the complexes are close to unity, a result suggesting that the thermal electron transfers in these systems are likely to be adiabatic. This conclusion persists even after corrections are made for

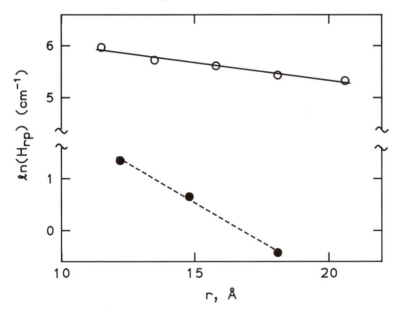

Figure 4. Plot of ln H_{rp} vs. separation distance for the polyene-bridged ruthenium(II)–ruthenium(III) complexes (open circles) and the polyproline-bridged osmium(II)–ruthenium(III) complexes (closed circles).

the difference between $H_{rp}(\text{th})$ and $H_{rp}(\text{op})$ (eq 18). Interestingly, thermal electron transfer from $Fe^{II}(CN)_5$ or $Ru^{II}(NH_3)_4(H_2O)$ to $Co^{III}(NH_3)_5$ via the same bridging ligands also appears to be adiabatic (7, 59, 60). Because of the large λ_{out} for the polyene-bridged complexes, $\kappa_n \ll 1$ for these systems and the distance dependence of $\ln \kappa_n$ is much steeper than that of $\ln \kappa_{el}$.

Not all extensively conjugated ruthenium centers are strongly coupled. For example, the metal–metal electronic coupling in the phenyl-bridged complex $(NH_3)_5Ru^{II}py(ph)_npyRu^{III}(NH_3)_5$ where $n = 1,2$ is relatively weak, possibly because the aromatic rings are rotated about 40° relative to one another (61); the lack of coplanarity of the rings presumably reduces the π-interaction of the two ruthenium centers.

Polyproline-Bridged Systems

Extensive data are available for thermal electron transfer within polyproline-bridged Os(II)–Ru(III) complexes (eq 21)

$(NH_3)_5Os^{II}-iso(\text{proline})_n-Ru^{III}(NH_3)_5$

$$\xrightarrow{k} (NH_3)_5Os^{III}-iso(\text{proline})_n-Ru^{II}(NH_3)_5 \quad (21)$$

where iso is the isonicotinyl residue (2). Values of the activation energy

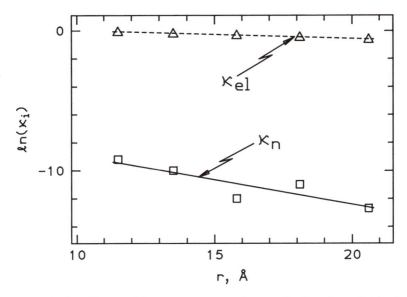

Figure 5. Plot of ln κ_{el} and ln κ_n vs. separation distance for the polyene-bridged (n = 0–4) ruthenium(II)–ruthenium(III) complexes. The results for the n = 2–4 complexes in ref. 6 have been corrected as described in the text. The parameters were derived from the energies and intensities of the metal-to-metal charge-transfer transitions.

(natural log of the nuclear factor) and entropy (natural log of the electronic factor) as a function of the distance separating the two metal centers were determined from the temperature dependence of the rate as a function of the number of proline units in the bridge. Unlike the polyene-bridged system already discussed, both the nuclear and electronic factors for the polyproline-bridged system are $<< 1$, with the distance dependence of the electronic factor (slope 0.68 Å^{-1}, Figure 6) being smaller than that of the nuclear factor (slope 0.91 Å^{-1}).

The values of ln H_{rp} for the polyproline-bridged systems are plotted vs. r in Figure 4 (closed circles). The value of β is 0.60 Å^{-1}, considerably larger than the value for the polyene-bridged system (0.2 Å^{-1}). The average value of α per bond is also smaller for proline than for ethylene. Thus the saturated polyproline bridge is a poorer electron mediator than the conjugated polyene bridge for this combination of donors and acceptors. In terms of eq 12c, this condition arises, in part, because the energy of the LUMO of the bridge is considerably higher for polyproline than for polyene, while the average of the redox potentials of the donor and acceptor sites is only about 0.3 eV more positive for the polyene than for the polyproline system. By use of the same approach that leads to eq 13, Siddarth and Marcus (62) have calculated that $\beta = 0.75$ Å^{-1} for the polyproline-bridged Os(II)–Ru(III) system, in fair agreement with the experimental value of 0.60 Å^{-1}. However, their calculated H_{rp} values are much larger than the experimental values.

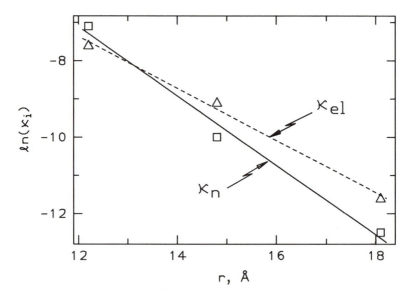

Figure 6. Plot of ln κ_{el} and ln κ_n vs. separation distance for the polyproline-bridged osmium(II)–ruthenium(III) complexes (2). The parameters were derived from measurements of the temperature dependence of the rate.

As discussed, β and α are energy-dependent quantities, and the proximity of the energy of the transferring electron [average of the donor and acceptor redox potentials, converted to an absolute energy scale (24)] to the energy of the HOMO or LUMO of the bridge is an important parameter. Consistent with this finding, the value of β for electron transfer in the proline-bridged (bpy$^-$)Ru^{II}–Co^{III} system (Table I) is a factor of 2 smaller than β for the proline-bridged Os(II)–Ru(III) system considered. The smaller β for the former system could arise from the lower average redox potential of its redox sites; this smaller β would result in more efficient mixing of the donor levels with the unoccupied molecular orbitals of the proline bridge.

Electron transfer in the totally organic tyrosine(proline)$_n$tryptophan system (64) affords an interesting comparison with the mixed-metal complexes. The electron transfer in the bridged tyrosine–tryptophan system is from the tyrosine to the neutral indolyl radical, formed by oxidation of the tryptophan side chain by a pulse radiolysis method, and is accompanied by proton transfer (eq 22). The latter may or may not occur in concert with the actual electron transfer.

$$(HO)Tyr(Pro)_n(Trp\cdot) \rightarrow (O\cdot)Tyr(Pro)_nTrpH \qquad (22)$$

The distance dependence of the electron-transfer rate constant in the tyrosine–tryptophan system (slope of $\ln k$ vs. r is ~ 0.3 Å^{-1} for $n = 1$–3), and consequently also the distance dependence of the electronic factor is much smaller than that observed for the Os(II)–Ru(III) system and similar to that found for the (bpy$^-$)Ru^{II}–Co^{III} system. At first sight the latter result is surprising because the redox potentials for the indolyl and phenol side chains of histidine and tyrosine residues are quite positive (64). However, hole transfer (rather than electron transfer) may be more important in this system than in the mixed-metal complexes, where the redox potentials of the redox sites are lower. The increased importance of hole transfer could account for the relatively low β value for the organic system.

Table I. Distance Dependence of the Electronic Coupling Element

System	Method[a]	β, Å^{-1}	α_g	α_b	Ref.
(NH$_3$)$_5$Ru^{II}py(CH=CH)$_n$pyRu^{III}(NH$_3$)$_5$	Op	0.2	0.8	0.9	6
(NH$_3$)$_5$Os^{II}iso(Pro)$_n$Ru^{III}(NH$_3$)$_5$[b]	Th	0.60	0.41	0.74	2
(bpy)$_2$Ri^{II}L(Pro)$_n$Co^{III}(NH$_3$)$_5$[c]	Th	0.30	0.64	0.86	63
Tyr(Pro)$_n$Trp[d]	Th	≤ 0.3	≥ 0.6	≥ 0.8	64

NOTE: H_{rp} decays by the multiplicative factor α_g per group (or residue) or a factor α_b per bond; $\beta l = -2 \ln \alpha$, where l is the length of the bridging unit.
[a]Op and Th denote whether the parameters are based on measurements of optically or thermally induced electron transfer.
[b]iso denotes isonicotinyl residue, Pro denotes proline; each proline residue extends the bridge by three σ bonds.
[c]The electron donor is the bipyridine radical anion (bpy$^-$). The proline residues are attached by an amide linkage to bis(2,2′-bipyridine)(4-carboxy-4′-methyl-2,2′-bipyridine)ruthenium(II).
[d]The electron donor is tyrosine (Tyr) and the acceptor is oxidized tryptophan (Trp).

As these examples show, the distance dependence of the nuclear and electronic factors can be quite large—and different. The dependence of the nuclear factor on separation distance will be small when the medium is nonpolar (unless ΔG^0 has a large distance dependence) or when $|\Delta G^0| \sim \lambda$, that is, when the reaction is barrierless. The distance dependence of the electronic factor κ_{el}, but not necessarily that of the electronic coupling element H_{rp}, will be small when the reaction is adiabatic and large at large separations of the redox sites. The distance dependence of the coupling element depends on the nature of the bridging group and, for a given bridging group, on the potentials of the redox sites.

Electron Transfer in Metalloproteins

This section illustrates how the results obtained for model systems can be used to evaluate possible electron-transfer pathways in more complex systems. Specifically, the distance dependence of the electronic coupling element for the polyproline-bridged Os(II)–Ru(III) system is used to compare the magnitude of the coupling provided by different pathways for the electron transfer from the $Ru^{II}(NH_3)_5$ attached to histidine-33 of cytochrome c to the heme iron(III).

The value of H_{rp} for the derivatized cytochrome c system, derived from rate measurements and from comparisons with related systems, is 0.03 cm^{-1} (20). As a basis for discussion of the electron-transfer pathway in the derivatized cytochrome c, the factor of 0.41 for the decay of H_{rp} per amino acid residue in the Os(II)–Ru(III) polyproline-bridged system is assumed to be applicable also to the Ru(II)–Fe(III) cytochrome c derivative. The average of the redox potentials of the donor and acceptor sites is not very different in the Os(II)–Ru(III) polyproline-bridged and Ru(II)–Fe(III) cytochrome c systems. Thus this approach is justified to the extent that the energetics for mediation by the polyproline and cytochrome c polypeptide bridge residues are similar; it would break down, for example, if the polypeptide chain contained many aromatic residues. The latter is not the case for cytochrome c.

In a through-bond pathway, the electron transfer proceeds exclusively via the tortuous polypeptide chain of the protein. In other words, the electron transfer from the ruthenium(II) bound to the histidine-33 proceeds via the 14 intervening amino acid residues to the histidine-18 attached to the heme iron(III). If $H_{rp}{}^0$ is assumed to be ~ 10 cm^{-1}, then the value of H_{rp} for mediation by 14 amino acid residues is estimated to be $10(0.41)^{14}$ or $\sim 5 \times 10^{-5}$ cm^{-1}, almost 3 orders of magnitude smaller than the experimental value (20) of 0.03 cm^{-1}. This disparity strongly suggests that the electron transfer in the ruthenium-derivatized ferricytochrome c does not proceed "along" the protein backbone.

The experimental H_{rp} of 0.03 cm^{-1}, with $H_{rp}{}^0 \sim 10$ cm^{-1} and $\alpha = 0.41$ per amino acid residue, suggests that fewer than seven amino acid residues mediate the coupling. Consequently, paths that jump or short-circuit the polypeptide chain are invoked. Such paths could involve electron transfer via hydrogen bonds formed across bends in the polypeptide chain, or could be "through space", possibly via aromatic side chains of amino acids suitably positioned along the protein backbone (24). The couplings provided by such short-circuit pathways are likely to be very distance- and orientation-dependent. On the basis of an examination of the structure of cytochrome c, Meade et al. (20) recently proposed a hydrogen-bonded pathway. In this mechanism the "electron transfer" proceeds from the ruthenium bound to the histidine-33 via the polypeptide chain to proline-30. The proline-30 is coupled to histidine-18 via the hydrogen bond formed between the carbonyl group of the proline-30 and the NH of histidine-18.

$$\text{Ru-(His-33)} \sim \text{(Pro-30)--H--(His-18)-Fe} \tag{23}$$

Unless the electronic coupling provided by the H bond is very weak (and in ref. 28 it is argued that it is not), the electronic interaction through the three amino acids and the hydrogen bond should be sufficient to account for the observed rate.

To conclude, the distance dependence of electron-transfer rates is determined by a number of factors. In certain systems the distance dependence of the nuclear factor is larger than that of the electronic factor, although the opposite is true in other systems. Theoretical models are available for calculating the distance dependence of the nuclear factor and the variation of the electronic factor with distance. However, there is still considerable uncertainty regarding the values of certain of the key parameters to be used in calculating the coupling elements in complex systems. It is hoped that, with the increasing ingenuity being used to synthesize model systems to selectively modify amino acids in metalloproteins, many of these questions will be resolved in the near future.

Acknowledgment

I acknowledge very helpful discussions with Carol Creutz, Bruce Brunschwig, Noel Hush, Marshall Newton, and Jay Winkler. I also thank Jean-Pierre Launay for making available unpublished data on the spectra of the $n = 2$ polyene-bridged diruthenium complexes in D_2O. This research was carried out at Brookhaven National Laboratory under contract DE–AC02–76CH00016 with the U.S. Department of Energy and supported by its Division of Chemical Sciences, Office of Basic Energy Sciences.

References

1. Isied, S. S.; Vassilian, A.; Magnuson, R. H.; Schwarz, H. A. *J. Am. Chem. Soc.* **1985,** *107,* 7432.
2. Isied, S. S.; Vassilian, A.; Wishart, J. F.; Creutz, C.; Schwarz, H. A.; Sutin, N. *J. Am. Chem. Soc.* **1988,** *110,* 635.
3. Oevering, H.; Paddon-Row, M. N.; Heppener, M.; Oliver, A. M.; Cotsaris, E.; Verhoeven, J. W.; Hush, N. S. *J. Am. Chem. Soc.* **1987,** *109,* 3258.
4. Penfield, K. W.; Miller, J. R.; Paddon-Row, M. N.; Cotsaris, E.; Oliver, A. M.; Hush, N. S. *J. Am. Chem. Soc.* **1987,** *109,* 5061.
5. Paddon-Row, M. N.; Oliver, A. M.; Warman, J. M.; Smit, K. J.; de Haas, M. P.; Oevering, H.; Verhoeven, J. W. *J. Phys. Chem.* **1988,** *92,* 6958.
6. Woitellier, S.; Launay, J. P.; Spangler, C. W. *Inorg. Chem.* **1989,** *28,* 758.
7. Lee, G.-H.; Della Ciana, L.; Haim, A. *J. Am. Chem. Soc.* **1989,** *111,* 2535.
8. Wasielewski, M. R.; Niemczyk, M. P.; Svec, W. A.; Pewitt, E. B. *J. Am. Chem. Soc.* **1985,** *107,* 5562.
9. Ho, P. S.; Sutoris, C.; Liang, N.; Margoliash, E.; Hoffman, B. M. *J. Am. Chem. Soc.* **1985,** *107,* 1070.
10. Petterson-Kennedy, S. E.; McGourty, J. L.; Kalweit, J. A.; Hoffman, B. M. *J. Am. Chem. Soc.* **1986,** *108,* 1739.
11. Mayo, S. L.; Ellis, W. R.; Crutchley, R. J.; Gray, H. B. *Science (Washington, D.C.)* **1986,** *233,* 948.
12. Winkler, J.; Nocera, D. G.; Yocom, K. M.; Bordignon, E.; Gray H. B. *J. Am. Chem. Soc.* **1982,** *104,* 5782.
13. Nocera, D. G.; Winkler, J. R.; Yocom, K. M.; Bordignon, E.; Gray, H. B. *J. Am. Chem. Soc.* **1984,** *106,* 5145.
14. Isied, S. S.; Kuehn, C.; Worosila, G. *J. Am. Chem. Soc.* **1984,** *106,* 1722.
15. Conklin, K. T.; McLendon, G. *J. Am. Chem. Soc.* **1988,** *110,* 3345.
16. Brunschwig, B. S.; DeLaive, P. J.; English, A. M.; Goldberg, M.; Gray, H. B.; Mayo, S. L.; Sutin, N. *Inorg. Chem.* **1985,** *24,* 3473.
17. Elias, H.; Chou, M. H.; Winkler, J. R. *J. Am. Chem. Soc.* **1988,** *110,* 429.
18. Axup, A. W.; Albin, M.; Mayo, S. L.; Crutchley, R. J.; Gray, H. B. *J. Am. Chem. Soc.* **1988,** *110,* 435.
19. Karas, J. L.; Lieber, C. M.; Gray, H. B. *J. Am. Chem. Soc.* **1988,** *110,* 559.
20. Meade, T. J.; Gray, H. B.; Winkler, J. R. *J. Am. Chem. Soc.* **1989,** *111,* 4353.
21. Bowler, B. E.; Meade, T. J.; Mayo, S. L.; Richards, J. H.; Gray, H. B. *J. Am. Chem. Soc.* **1989,** *111,* 8757.
22. Marcus, R. A.; Sutin, N. *Biochim. Biophys. Acta* **1985,** *811,* 265.
23. Sutin, N.; Brunschwig, B. S.; Creutz, C.; Winkler, J. R. *Pure Appl. Chem.* **1988,** *60,* 1817.
24. Beratan, D. N.; Onuchic, J. N.; Hopfield, J. J. *J. Chem. Phys.* **1987,** *86,* 4488.
25. Closs, G. L.; Miller, J. R. *Science (Washington, D.C.)* **1988,** *240,* 440.
26. McLendon, G. *Acc. Chem. Res.* **1988,** *21,* 160.
27. Larsson, S. *Chem. Scr.* **1988,** *28A,* 15.
28. Cowan, J. A.; Upmacis, R. K.; Beratan, D. N.; Onuchic, J. N.; Gray, H. B. *Ann. N.Y. Acad. Sci.* **1989,** *550,* 68.
29. Marcus, R. A. *Annu. Rev. Phys. Chem.* **1964,** *15,* 155.
30. Marcus, R. A. *J. Chem. Phys.* **1965,** *43,* 679.
31. Sutin, N. *Acc. Chem. Res.* **1982,** *15,* 275.
32. Sutin, N. *Prog. Inorg. Chem.* **1983,** *30,* 441.
33. Newton, M. D.; Sutin, N. *Annu. Rev. Phys. Chem.* **1984,** *35,* 437.
34. Brunschwig, B. S.; Ehrenson, S.; Sutin, N. *J. Phys. Chem.* **1986,** *90,* 3657.
35. Brunschwig, B. S.; Ehrenson, S.; Sutin, N. *J. Phys. Chem.* **1987,** *91,* 4714.

36. Brunschwig, B. S.; Ehrenson, S.; Sutin, N. *J. Am. Chem Soc.* **1984**, *106*, 6858.
37. Brunschwig, B. S.; Sutin, N. *J. Am. Chem. Soc.* **1989**, *111*, 7454.
38. Sutin, N.; Brunschwig, B. S. In *Electron Transfer in Biology and the Solid State: Inorganic Compounds with Unusual Properties*; Johnson, M. K.; King, R. B.; Kurtz, D. M.; Kutal, C.; Norton, M. L.; Scott, R. A., Eds.; Advances in Chemistry 226; American Chemical Society: Washington, DC, 1990; pp 65–88.
39. McConnell, H. M. *J. Chem. Phys.* **1961**, *35*, 308.
40. Larsson, S. *J. Am. Chem. Soc.* **1981**, *103*, 4034.
41. Larsson, S. *Chem. Phys. Lett.* **1982**, *90*, 136.
42. Larsson, S. *J. Phys. Chem.* **1984**, *88*, 1321.
43. Larsson, S. *J. Chem. Soc., Faraday Trans. 2* **1983**, *79*, 1375.
44. Onuchic, J. N.; Beratan, D. N. *J. Am. Chem. Soc.* **1987**, *109*, 6771.
45. Ratner, M. A. *J. Phys. Chem.* **1990**, *94*, 4877.
46. Wolfsberg, M.; Helmholtz, L. *J. Chem. Phys.* **1952**, *20*, 837.
47. Hush, N. S. *Prog. Inorg. Chem.* **1967**, *8*, 391.
48. Creutz, C. *Prog. Inorg. Chem.* **1983**, *30*, 1.
49. Newton, M. D. *Chem. Rev.*, in press.
50. Reimers, J. R.; Hush, N. S. *Chem. Phys.* **1989**, *134*, 323.
51. Reimers, J. R.; Hush, N. S. *Inorg. Chem.* **1990**, *29*, 3686.
52. Sutton, J. E.; Taube, H. *Inorg. Chem.* **1981**, *20*, 3125.
53. Creutz, C., unpublished work.
54. Curtis, J. C.; Sullivan, B. P.; Meyer, T. J. *Inorg. Chem.* **1983**, *22*, 224.
55. Chang, J. P.; Fung, E. Y.; Curtis, J. C. *Inorg. Chem.* **1986**, *25*, 4233.
56. Hupp, J. T.; Weaver, M. J. *J. Phys. Chem.* **1985**, *89*, 1601.
57. Creutz, C.; Chou, M. H. *Inorg. Chem.* **1985**, *89*, 1601.
58. Launay, J. P., Centre National de Recherche Scientifique, personal communication.
59. Szecsy, A. P.; Haim, A. *J. Am. Chem. Soc.* **1981**, *103*, 1679.
60. Geselowitz, D. *Inorg. Chem.* **1987**, *26*, 4135.
61. Kim, Y.; Lieber, C. M. *Inorg. Chem.* **1989**, *28*, 3990.
62. Siddarth, P.; Marcus, R. A. *J. Phys. Chem.* **1990**, *94*, 2985.
63. Ogawa, M.; Wishart, J. F.; Isied, S. S., to be published.
64. Faraggi, M.; DeFillippis, M. R.; Klapper, M. H. *J. Am. Chem. Soc.* **1989**, *111*, 5141.

RECEIVED for review April 27, 1990. ACCEPTED revised manuscript August 21, 1990.

Theoretical Analysis of Energy-Gap Laws of Electron-Transfer Reactions

Distribution Effect of Donor–Acceptor Distance

Toshiaki Kakitani[1], Akira Yoshimori[1], and Noboru Mataga[2]

[1]Department of Physics, Faculty of Science, Nagoya University, Chikusaku, Nagoya 464–01, Japan
[2]Department of Chemistry, Faculty of Engineering Science, Osaka University, Toyonaka, Osaka 560, Japan

The electron-transfer rate as a function of the free energy gap (energy-gap law) was formulated by including the solvent nonlinear response effect and averaging over the distribution of donor–acceptor distances. Using the same parameter values, we fit the theoretical energy-gap laws to three independent experimental measurements: the photoinduced charge-separation (CS) rate as measured from fluorescence quenching in the stationary state, the actual photoinduced charge-separation rate as obtained from analysis of the transient effect in the fluorescence decay curve, and the charge-recombination (CR) rate of the geminate radical-ion pair. The different energy-gap laws among those reactions can be reproduced reasonably well by adopting different distributions of the donor–acceptor distance: that of the CS reaction covering those over various distances and more specified ones corresponding to the contact ion-pair (CIP) and solvent-separated ion-pair (SSIP) models for the CR reaction. The nonlinear effect in those homogeneous reactions is small when the CIP model applies and appreciable when the SSIP model applies.

ELECTRON-TRANSFER (ET) REACTIONS IN POLAR SOLUTIONS and electrode systems are among the most fundamental chemical processes. From a theoretical point of view, if the linear response of the solvent polarization is

0065–2393/91/0228–0045$07.25/0

assumed, the Gibbs energy-gap dependence of the ET rate (called the energy-gap law) becomes bell-shaped (1–5). There is a rapid rate increase in the small-energy-gap region (normal region), a maximum at an energy gap equivalent to the total reorganization energy, and a rapid rate decrease in the large-energy-gap region (inverted region). The linear response theory also predicts that the transfer coefficient at the standard electrode potential in electrochemical reactions should be ½ (6).

On the other hand, if dielectric saturation takes place around a charged reactant or product, a nonlinear solvent polarization response is predicted. In this case, the Gibbs energy curvature around the minimum in the initial state can be different from that in the final state (7–9). As a result, the energy-gap law of the charge-separation (CS) reaction (A* $\cdots$ D $\to$ A$^-$ $\cdots$ D$^+$) is broadened, in addition to a flattening around the maximum (7, 10). The inverted region exists so far as the Gibbs energy curve of the solvent is correctly calculated along the reaction coordinate (11–13). The energy-gap law of the charge-recombination (CR) reaction (A$^-$ $\cdots$ D$^+$ $\to$ A $\cdots$ D) is narrowed, with a shift of the peak to a smaller energy gap (11–16). The transfer coefficient of the ionization reaction [A^{z-} $\cdots$ M(e$^-$) $\to$ A$^{(z-)-1}$ $\cdots$ M, z being a valence ≥ 0] is smaller than ½ and that of the neutralization reaction [A^z $\cdots$ M(e$^-$) $\to$ A^{z-1} $\cdots$ M, $z \geq 1$] is larger than ½, where A^{z-} or A^z is the reactant and M the metal electrode (17). These results imply that the energy-gap law and the transfer coefficient are systematically modified by nonlinear response, depending on the type of the ET reaction (7–11, 14–18).

When a linear response applies, the energy-gap law does not depend on the type of the ET reaction. Therefore, the significance of the nonlinear response effect can be determined by comparing the energy-gap laws between the CS and CR reactions. Experimental data have demonstrated that bell-shaped energy-gap laws, which are obtained for the CR reaction of the geminate ion pair produced by fluorescence quenching (18), are qualitatively in agreement with the prediction of the linear response theory. For the CS reaction, the inverted region was not obtained up to an energy gap of 2.5 eV (i.e., the ET rate constant was so large as to be masked by the diffusion-limit rate constant of the reactants until the value of the energy gap was 2.5 eV) (19). Participation of the excited-state product was eliminated at an energy gap of 1.6 eV (20), but it may be possible in the 1.6–2.5-eV energy-gap region.

In the normal region, there seems to be a considerable difference between the CS and CR reactions. The normal region of the CS reaction is considerably shifted toward a small energy gap, as compared with that of the CR reaction (18). On the other hand, in electrochemical reactions, most of the experimental data on the transfer coefficient for the ionization reaction are smaller than ½ when the standard rate constant is less than 1 cm s^{-1}. Most data for the neutralization reaction are larger than ½ (17), consistent with the theoretical calculations based on the nonlinear response theory. All of these experimental data appear to indicate that the energy-gap law varies between the CS and CR reactions, and between the ionization and neutralization reactions.

One important factor was not taken into account in the foregoing theoretical treatments. So far, the distance between the donor and acceptor was implicitly assumed to be a constant. It may be possible for the neutral donor and acceptor to distribute over various distances. However, the ion-pair distribution might be localized (i.e., the manner of the distance distribution can be different between the CS and CR reactions). Therefore, it is important to investigate how such distributions, if they occur, can affect the observed energy-gap law. This distribution effect on the rate constant was first investigated by Marcus and Siders (*21*). Brunschwig et al. (*22*) made numerical calculations for the charge shift reaction of transition metal complexes by taking into account the distance dependence of the reorganization energy and the tunneling matrix element, as well as the distance-distribution function.

We made a detailed investigation as to whether the different energy-gap laws of the CS and CR reactions can be reproduced, by either the linear or nonlinear response theory by taking into account most possible effects attributable to the distance distribution between the donor and acceptor. In such analyses, we used three independent experimental measurements: the photoinduced CS rate constant as measured from fluorescence quenching in the stationary state (*19*), the actual photoinduced CS rate constant as obtained by analysis of the transient effect in fluorescence decay curves (*23, 24*), and the CR rate of the geminate radical-ion pair (*18*). The results demonstrate that the difference in the observed energy-gap laws arises mostly from the different distributions of donor–acceptor distances between the CS and CR reactions. The role of the nonlinear response is not as large in homogeneous reactions as expected (*7, 10, 14–16*), although it is still considerable.

Formulation of the ET Rate for a Fixed Distance

This section presents a theoretical formulation of the ET rate when the distance between the donor and acceptor is fixed.

Solvent Mode. First, the formula for the ET rate will be derived by considering only the solvent mode and assuming that thermal equilibrium is always attained for the solvent motion in the initial state. This assumption appears to be valid for such a solvent as acetonitrile, whose longitudinal relaxation time τ_L is as small as 2×10^{-13} s (*25*).

The solvent molecule in solution does not move on a definite potential surface; its potential is the sum of many long-range interactions that fluctuate randomly. Therefore, it is impossible to calculate its Franck–Condon factor directly as an overlap of vibrational wave functions. Instead, it is more appropriate to calculate the Gibbs energy curve along the reaction coordinate, which can be defined unambiguously.

Hamiltonians $H^I(r)$ and $H^F(r)$ for polar solutions in the initial and final states, respectively, are as follows:

$$H^I(r) = H^r(r) + E_{el}^r(r) \tag{1}$$

$$H^F(r) = H^p(r) + E_{el}^p(r) \tag{2}$$

where

$$H^r(r) = \sum_i \left(\frac{z_a^r e(\vec{\mu}_i \cdot \vec{r}_{ia})}{r_{ia}^3} + \frac{z_d^r e(\vec{\mu}_i \cdot \vec{r}_{id})}{r_{id}^3} \right)$$
$$+ \sum_{i>j} \sum H_{ij}^{d-d} + V_{ind}^r(r) + V^{c-c} \tag{3}$$

$$H^p(r) = \sum_i \left(\frac{z_a^p e(\vec{\mu}_i \cdot \vec{r}_{ia})}{r_{ia}^3} + \frac{z_d^p e(\vec{\mu}_i \cdot \vec{r}_{id})}{r_{id}^3} \right)$$
$$+ \sum_{i>j} \sum H_{ij}^{d-d} + V_{ind}^p(r) + V^{c-c} \tag{4}$$

$$E_{el}^r(r) = \frac{z_a^r z_d^r e^2}{r} + \epsilon^r \tag{5}$$

$$E_{el}^p(r) = \frac{z_a^p z_d^p e^2}{r} + \epsilon^p \tag{6}$$

and

$$H_{ij}^{d-d} = \frac{\vec{\mu}_i \cdot \vec{\mu}_j}{r_{ij}^3} - \frac{3(\vec{\mu}_i \cdot \vec{r}_{ij})(\vec{\mu}_j \cdot \vec{r}_{ij})}{r_{ij}^5} \tag{7}$$

where the superscripts r and p denote reactant and product, respectively; z_a^r, z_d^r, z_a^p, and z_d^p are valences of the acceptor and donor of reactants and products, respectively; e is a unit charge; $\vec{\mu}_i$ is a permanent dipole moment of solvent molecules; $\vec{r}_{ia}$ is the distance between the dipole and the acceptor; $\vec{r}_{id}$ is the distance between the dipole and the donor; $\vec{r}_{ij}$ is the distance between the two dipoles; the variable r is the distance between the donor and acceptor; ϵ^r and ϵ^p are electronic energies of reactants and products, respectively; $E_{el}^r(r)$ and $E_{el}^p(r)$ are electronic energies of reactants and products with Coulomb interaction in vacuum, respectively; $V_{ind}^r(r)$ and $V_{ind}^p(r)$

are the residual charge–dipole and dipole–dipole interactions arising from the electronic polarization of solute and solvent molecules; and V^{c-c} is the core–core interaction among solute and solvent molecules. The explicit forms of $V_{ind}^r(r)$, $V_{ind}^p(r)$, and V^{c-c} are not necessary in the present calculations. The reaction coordinate operator f is defined (11) as

$$f = H^r(r) - H^p(r) \tag{8}$$

Gibbs energy curves $G^r(x;r)$ and $G^p(x;r)$ as a function of the reaction coordinate x in the initial and final states are written as

$$e^{-\beta G^r(x;r)} = \beta^{-1} \int \delta(f - x)e^{-\beta H^r(r)} \, d\Gamma \tag{9}$$

$$e^{-\beta G^p(x;r)} = \beta^{-1} \int \delta(f - x)e^{-\beta H^p(r)} \, d\Gamma \tag{10}$$

where $\delta(f-x)$ is a delta function; β is $1/k_B T$ (k_B being the Boltzmann constant and T the temperature); and Γ is the normalized configuration space (orientation and position) of solvent molecules. The dimension of x is energy.

When the interaction between the donor and acceptor is weak, the nonadiabatic mechanism applies and the ET rate is expressed as

$$k(-\Delta G(r);r) = \overline{A}(r) \cdot e^{-\beta \Delta G^{\ddagger}(-\Delta G(r);r)} \tag{11}$$

with

$$\overline{A}(r) = \frac{2\pi}{\hbar} J^2(r)\beta \tag{12}$$

where $\overline{A}(r)$ is an effective frequency; $J(r)$ is the electronic tunneling matrix element; $\hbar$ is Planck's constant divided by 2π; and $-\Delta G(r)$ is the Gibbs energy gap of the reaction. An effect of mutual orientations of donor and acceptor molecules is neglected in eq 11. Such an effect might be important when the donor and acceptor come close. However, as discussed later, an adiabatic mechanism works for ion pairs in close proximity, and the amplitude of the electronic tunneling matrix element becomes insensitive to the ET rate.

The activation Gibbs energy $\Delta G^{\ddagger}(-\Delta G(r);r)$ is expressed (11) as

$$\Delta G^{\ddagger}(-\Delta G(r);r) = G^r(\Delta G(r) - \Delta G_s(r);r) - G^r(r) \tag{13}$$

where

$$-\Delta G_s(r) = G^r(r) - G^p(r) \tag{14}$$

with

$$e^{-\beta G^r(r)} = \int e^{-\beta H^r(r)} d\Gamma \equiv \beta \int_{-\infty}^{\infty} e^{-\beta G^r(x;r)} \, dx \qquad (15)$$

$$e^{-\beta G^P(r)} = \int e^{-\beta H^P(r)} d\Gamma \equiv \beta \int_{-\infty}^{\infty} e^{-\beta G^P(x;r)} \, dx \qquad (16)$$

In the case of the CS, $-\Delta G_s(r) > 0$ holds, and this can be regarded as the solvation energy of the products. This solvation energy is denoted as $-\Delta G_s^{CS}(r)$ hereafter.

As r becomes small, the interaction between the donor and acceptor becomes strong. In such a case, the adiabatic mechanism predominates. To cover both adiabatic and nonadiabatic regions, we tentatively adopt the following effective frequency

$$\nu(r) = \cfrac{1}{\cfrac{1}{\nu_{ad}} + \cfrac{1}{\overline{A}(r)}} \qquad (17)$$

where ν_{ad} is a frequency in the case of the adiabatic mechanism. The ET rate is then expressed as

$$k(-\Delta G(r);r) = \nu(r)e^{-\beta \Delta G^{\ddagger}(-\Delta G(r);r)} \qquad (18)$$

Intramolecular Vibrational Mode. The intramolecular vibration with high frequency can be adequately described by the quantum-mechanical harmonic-oscillator model (hereafter called the quantum mode). If we consider the quantum mode in addition to the solvent mode, the ET rate is written as a convolution of the two factors as follows.

$$k_e(-\Delta G(r);r) = \int_{-\infty}^{\infty} k(-\Delta G(r) - \epsilon;r)F_q(\epsilon) \, d\epsilon \qquad (19)$$

where $F_q(\epsilon)$ is the Franck–Condon factor of the quantum mode (5)

$$F_q(\epsilon) = \frac{1}{\hbar\omega} \exp\left[-S(2\overline{\nu} + 1)\right] \cdot I_p\left[2S\sqrt{\overline{\nu}(\overline{\nu} + 1)}\right] \cdot \left[\frac{\overline{\nu} + 1}{\overline{\nu}}\right]^{p/2} \qquad (20)$$

$$\overline{\nu} = \frac{1}{\exp(\beta\hbar\omega) - 1} \qquad (21)$$

$$p = \frac{\epsilon}{\hbar\omega} \qquad (22)$$

$$S = \frac{\delta^2}{2} \qquad (23)$$

where I_p is the modified Bessel function, ω is the effective angular frequency, $\bar{\nu}$ is the averaged vibrational number, S is the coupling strength, and δ is the shift of the dimensionless normal coordinate between the initial and final states of the quantum mode.

Mean Potentials.   The mean potentials $U^r(r)$ and $U^p(r)$, which work between the donor and acceptor in the initial and final states, respectively, are defined as

$$e^{-\beta U^r(r)} = \frac{\int e^{-\beta H^I(r)}\, d\Gamma}{\int e^{-\beta H^I(\infty)}\, d\Gamma} \tag{24}$$

$$e^{-\beta U^p(r)} = \frac{\int e^{-\beta H^F(r)}\, d\Gamma}{\int e^{-\beta H^F(\infty)}\, d\Gamma} \tag{25}$$

$U^r(r)$ and $U^p(r)$ are defined to be zero for $r = \infty$. Substituting eq 1 into eq 24 yields

$$U^r(r) = G^r(r) + E_{el}{}^r(r) - G^r(\infty) - E_{el}{}^r(\infty)$$

$$= G^r(r) - G^r(\infty) + \frac{z_a{}^r z_d{}^r e^2}{r} \tag{26}$$

Similarly,

$$U^p(r) = G^p(r) - G^p(\infty) + \frac{z_a{}^p z_d{}^p e^2}{r} \tag{27}$$

The Gibbs energy gap $-\Delta G(r)$ is given by

$$-\Delta G(r) = -\Delta G^0 + U^r(r) - U^p(r) \tag{28}$$

where $-\Delta G^0$ is the standard Gibbs energy gap, equivalent to $-\Delta G(\infty)$.

The mean potentials $U^r(r)$ and $U^p(r)$ differ greatly among the different types of the ET reaction. Therefore, it is useful to introduce mean potentials $U^{np}(r)$ and $U^{ip}(r)$, corresponding to the neutral-pair and ion-pair states of the donor and acceptor, respectively. That is, $U^r(r)$ and $U^p(r)$ are $U^{np}(r)$ and $U^{ip}(r)$, respectively, for the CS reaction. Similarly, $U^r(r)$ and $U^p(r)$ are $U^{ip}(r)$ and $U^{np}(r)$, respectively, for the CR reaction. Because the mean potential within the neutral pair is small, $U^{np}(r)$ is assumed to be 0.

Average over the Distance

Now the ET rate is averaged over the distance distribution. It is convenient to do this for the CS and CR reactions separately.

Charge-Separation Reaction.   With eq 28 and $U^{np}(r) = 0$, the Gibbs energy gap of the CS reaction is written explicitly as

$$-\Delta G^{CS}(r) = -\Delta G^0 - U^{ip}(r) \tag{29}$$

In general, $U^{ip}(r)$ is negatively large for small r. Then $-\Delta G^{CS}(r)$ is larger than $-\Delta G^0$ for small r. This fact indicates that the normal region of the curve of the CS rate constant drawn as a function of $-\Delta G^0$ shifts to the smaller energy gap.

Assuming that a thermal equilibrium is attained for the distance distribution in the initial state of the CS reaction, then the actual ET rate constant $k_{act}^{CS}(-\Delta G^0)$ is written as

$$k_{act}^{CS}(-\Delta G^0) = \frac{4\pi N_A}{1000} \int_{r_0}^{\infty} k_e(-\Delta G^{CS}(r);r)r^2 \, dr \tag{30}$$

where N_A is Avogadro's number. The lower limit r_0 is the closest distance between the donor and acceptor. Equation 30 is the same as used by Brunschwig et al. (22).

Soon after photoexcitation, the observed ET rate would be equivalent to $k_{act}^{CS}(-\Delta G^0)$. This value might be obtained by the measurement of the transient effect in the fluorescence decay process in the picosecond region.

On the contrary, later, when the stationary state is realized, the observed ET rate would be given (21, 26) by

$$k_{obs}^{CS}(-\Delta G^0) = \cfrac{1}{\cfrac{1}{k_{act}^{CS}(-\Delta G^0)} + \cfrac{1}{k_{diff}}} \tag{31}$$

where k_{diff} is the rate constant of the diffusional encounter of the donor and acceptor. This case would correspond to the fluorescence quenching experiment in the stationary state. Generally, $k_{obs}^{CS}(-\Delta G^0)$ is the rate constant for bimolecular reactions.

Charge-Recombination Reaction.   In the CR reaction, the ion pair of the initial state is prepared by the photoinduced CS reaction. Therefore, the donor and acceptor molecules are already close to each other. That is, the ion pair is temporarily trapped in the potential well after it is produced. If the CR rate is small, a certain fraction of the ion pairs may dissociate in competition with the CR reaction. The probability that those dissociated ion radicals can come together again is small, because the original ion-pair concentration is very low. Therefore, the CR reaction is possible only for the undissociated ion pair and can be treated essentially as a unimolecular reaction.

This ET rate, denoted as $k_{act}^{CR}(-\Delta G^0)$, is written as

$$k_{act}^{CR}(-\Delta G^0) = \int_{r_0}^{\infty} k_e(-\Delta G^{CR}(r);r)g(r)r^2 \, dr \tag{32}$$

where the function $g(r)$ represents the initial distribution of undissociated ion-pair distances for the CR reaction. Because this initial state is not in thermal equilibrium, the function is not proportional to $\exp[-\beta U^{ip}(r)]$. On the other hand, if $g(r)$ is a constant independent of r, eq 32 reduces to the same form as eq 30. In this case, when the linear response applies, the energy-gap law of the CR reaction becomes the same as that of the CS reaction. Substantially different energy-gap laws between the CS and CR reactions can be obtained only when we adopt considerably large nonlinear solvent response effects (7, 14–16). Under these situations, we phenomenologically consider the following two types of the distribution function: First, $g(r)$ has a maximum at $r = r_0$ and may be expressed as

$$g(r) = \frac{e^{-t(r-r_0)^2}}{\int_{r_0}^{\infty} e^{-t(r-r_0)^2} r^2 \, dr} \tag{33}$$

where t is a constant. This case may be called a contact-ion-pair model, which is abbreviated as CIP model. Second, $g(r)$ has a maximum at the solvent-separated ion-pair distance r_1 and may be expressed as

$$g(r) = \frac{e^{-u(r-r_1)^2}}{\int_{r_0}^{\infty} e^{-u(r-r_1)^2} r^2 \, dr} \tag{34}$$

where u is a constant. This case is called a solvent-separated ion-pair model, which is abbreviated as SSIP model.

With eq 28 and $U^{np}(r) = 0$, the energy gap of the CR reaction can be written as

$$-\Delta G^{CR}(r) = -\Delta G^0 - U^{ip}(r) \tag{35}$$

Contrary to $-\Delta G^{CS}(r)$, $-\Delta G^{CR}(r)$ is smaller than $-\Delta G^0$. This fact indicates that the normal region of the curve of the CR rate, drawn as a function of $-\Delta G^0$, shifts to the larger-energy-gap side.

Characterization of Gibbs Energy Curves and Reorganization Energy

For a concrete form of the Gibbs energy curve $G^r(x;r)$, if the linear response holds, $G^r(x;r)$ should be expressed in a quadratic form. The effect of the nonlinear response appears in the higher order terms. If the electronic polarizability is neglected [i.e., $V_{ind}{}^r(r) = V_{ind}{}^P(r) = 0$ in eqs 3 and 4], $G^r(x;r)$ is a symmetric function of x for the CS reaction (11). When the electronic polarizability effect is taken into account, $G^r(x;r)$ is no longer a symmetric function, and the reaction coordinate corresponding to the minimum of $G^r(x;r)$ is not zero (27). For the time being, in the CS reaction, we write $G^{np}(x;r)$ for $G^r(x;r)$ and $G^{ip}(x;r)$ for $G^P(x;r)$.

Within the linear response approximation, $G^{np}(x;r)$ can be expressed as $b(x - x_{0n}(r))^2$ without loss of generality, where b and $x_{0n}(r)$ are constants. Here we consider the simplest case, in which the nonlinear response effect is to add the nth-order term to this quadratic function as follows:

$$G^{np}(x;r) = a(x - x_{0n}(r))^n + b(x - x_{0n}(r))^2 \tag{36}$$

The exponent of the anharmonicity, n, is usually chosen as 6 in this chapter.

By using a general relation $G^P(x;r) = G^r(x;r) - x$, which was obtained from our previous work (11), $G^{ip}(x;r)$ can be written as follows:

$$G^{ip}(x;r) = a(x - x_{0n}(r))^n + b(x - x_{0n}(r))^2 - x \tag{37}$$

The coefficients a and b can be related to two parameters $\lambda_{out}{}^{CS}(r)$ and $\bar{\beta}$; $\lambda_{out}{}^{CS}(r)$ is the reorganization energy of the solvent for the CS reaction as defined (11) by

$$\lambda_{out}{}^{CS}(r) \equiv G^{ip}(x_{0n}(r);r) - G^{ip}(x_{0i}(r);r) \tag{38}$$

where $x_{0i}(r)$ is the reaction coordinate of the minimum of $G^{ip}(x;r)$. If the nonlinear response applies, the reorganization energy differs between the CS and CR reactions. The parameter $\bar{\beta}$ represents a scale of the curvature change of $G^{np}(x;r)$ along the reaction coordinate and it is defined by

$$\bar{\beta} \equiv \frac{C^{np}(x_{0n}(r);r)}{C^{ip}(x_{0i}(r);r) - C^{np}(x_{0n}(r);r)}$$

$$= \frac{C^{np}(x_{0n}(r);r)}{C^{np}(x_{0i}(r);r) - C^{np}(x_{0n}(r);r)} \tag{39}$$

with

$$C^{np}(x_{0n}(r);r) = \left. \frac{\partial^2 G^{np}(x;r)}{\partial x^2} \right|_{x = x_{0n}(r)} \tag{40a}$$

$$C^{ip}(x_{0n}(r);r) = \left.\frac{\partial^2 G^{ip}(x;r)}{\partial x^2}\right|_{x=x_{0n}(r)} \tag{40b}$$

$$C^{np}(x_{0i}(r);r) = \left.\frac{\partial^2 G^{np}(x;r)}{\partial x^2}\right|_{x=x_{0i}(r)} \tag{40c}$$

$$C^{ip}(x_{0i}(r);r) = \left.\frac{\partial^2 G^{ip}(x;r)}{\partial x^2}\right|_{x=x_{0i}(r)} \tag{40d}$$

where we have used the relationship $\partial^2 G^r(x;r)/\partial x^2 = \partial^2 G^p(x;r)/\partial x^2$, which holds generally (*11*). From the definition in eq 39, $\bar{\beta}$ is generally r-dependent, but its r-dependence is relatively small. Therefore, here $\bar{\beta}$ will be considered independent of r. $\bar{\beta}$ is infinite if the linear response applies, and $\bar{\beta}$ becomes small as the degree of nonlinear response increases.

Substituting eqs 36 and 37 into eqs 38 and 39 yields the following relationships

$$a = \frac{(n-1)^{n-1}(2+n\beta)^{n-1}}{2^{n-1}n^n(n\beta - \beta + 1)^n} \cdot \frac{1}{[\lambda_{out}{}^{CS}(r)]^{n-1}} \tag{41}$$

$$b = \frac{(n-1)^2\beta(2+n\beta)}{4n(n\beta - \beta + 1)^2} \cdot \frac{1}{\lambda_{out}{}^{CS}(r)} \tag{42}$$

In the case of the CR reaction, $G^r(x;r)$ and $G^p(x;r)$ should be read as $G^{ip}(-x;r)$ and $G^{np}(-x;r)$, respectively, and all the listed properties of $G^{np}(x;r)$ and $G^{ip}(x;r)$ can be used as they are. It is desirable to use the same parameter values of a and b or the same values of $\lambda_{out}{}^{CS}(r)$ and $\bar{\beta}$ for both the CS and CR reactions if the two reactions represent the forward and backward reactions of the same kind of species.

The reorganization energy $\lambda_{out}{}^{CS}(r)$ is related to the solvation energy $-\Delta G_s{}^{CS}(r)$ as follows (*11*)

$$\lambda_{out}{}^{CS}(r) \simeq -\Delta G_s{}^{CS}(r) - x_{0n}(r) \tag{43}$$

On the other hand, eqs 14, 26, and 27 yield

$$-\Delta G_s{}^{CS}(r) = U^{np}(r) - U^{ip}(r) - \frac{e^2}{r} + G^{np}(\infty) - G^{ip}(\infty) \tag{44}$$

Then, substituting eq 44 into eq 43 yields

$$\lambda_{out}{}^{CS}(r) \simeq -U^{ip}(r) - \frac{e^2}{r} - \Delta G_s{}^{CS}(\infty) - x_{0n}(r) \tag{45}$$

Equation 45 demonstrates that $\lambda_{out}^{CS}(r)$ can be calculated from the knowledge of $U^{ip}(r)$, $-\Delta G_s^{CS}(\infty)$, and $x_{0n}(r)$. Consider briefly the case in which we adopt the dielectric continuum model of the medium and the linear response approximation. Assuming that donor and acceptor are spheres with radii a_D and a_A, respectively, we can write (2)

$$U^{ip}(r) = \frac{e^2}{\epsilon_s r} \tag{46}$$

$$-\Delta G_s^{CS}(\infty) = \left(1 - \frac{1}{\epsilon_s}\right) \cdot \frac{e^2}{2} \cdot \left(\frac{1}{a_A} + \frac{1}{a_D}\right) \tag{47}$$

$$\lambda_{out}(r) = \left(\frac{1}{\epsilon_{op}} - \frac{1}{\epsilon_s}\right) \cdot \frac{e^2}{2} \cdot \left(\frac{1}{a_A} + \frac{1}{a_D} - \frac{2}{r}\right) \tag{48}$$

where ϵ_s and ϵ_{op} are the static and optical dielectric constants, respectively. We have dropped the superscript CS for the reorganization energy in eq 48 because the reorganization energy is independent of the kind of reactions in a linear response scheme. Substituting eqs 46–48 into eq 45 yields

$$x_{0n}(r) = \left(1 - \frac{1}{\epsilon_{op}}\right) \cdot \frac{e^2}{2} \cdot \left(\frac{1}{a_A} + \frac{1}{a_D} - \frac{2}{r}\right) \tag{49}$$

Equation 49 represents that $x_{0n}(r)$ is the solvation energy when only the electronic polarizability is considered.

With reference to eq 49, the following form is assumed for $x_{0n}(r)$ in general

$$x_{0n} = c - \frac{d}{r} \tag{50}$$

where c and d are constants.

When the donor and acceptor are organic molecules, the most probable, simple form of $U^{ip}(r)$ will be expressed as

$$U^{ip}(r) = -\frac{s}{r} \qquad \text{for } r > r_0 \tag{51}$$

where s is a constant. A result of Monte Carlo simulations (27) suggested a little more complex form with a deep minimum at $r \sim r_0$ and a shallow second minimum at the solvent-separated distance. A simpler form of eq 51 would be sufficient for the present theoretical analysis.

Substituting eqs 50 and 51 into eq 45 yields a typical form of $\lambda_{out}^{CS}(r)$, as follows

$$\lambda_{out}^{CS}(r) = c' - \frac{d'}{r} \tag{52}$$

where c' and d' are constants.

Finally, the r-dependence of $\overline{A}(r)$ is assumed to be

$$\overline{A}(r) = \overline{A}_0 e^{-\alpha(r-r_1)} \tag{53}$$

where $\overline{A}_0$ and α are constants and r_1 is the same as the solvent-separated ion-pair distance in eq 34.

Qualitative Features of Theoretical Energy-Gap Laws

The principle behind the present calculations is as follows. We try to fit the theoretical curves to the following three kinds of experimental data by simultaneously using the same parameter values for the CS and CR reactions:

1. The data of $k_{obs}^{CS}(-\Delta G^0)$ are used as obtained by Rehm and Weller (*19*) by measuring the fluorescence quenching rate due to the CS reaction when the lateral diffusion of reactants is in a stationary state. Those data are plotted in Figure 1 as crosses. A constant rate at the large energy gaps indicates that the reaction is diffusion-limited (i.e., $k_{diff} = 2 \times 10^{10}$ M^{-1} s^{-1}).

2. The data of $k_{act}^{CS}(-\Delta G^0)$ were recently obtained by Nishikawa et al. (*23*) by analyzing the transient effect in the fluorescence decay curves measured by the single-photon counting method with picosecond dye laser for excitation. The transient effect was analyzed by fitting the observed fluorescence decay curve to the theoretical curve as derived by Noyes (*26*). The results are listed in Table I. Those data are plotted as white circles in Figure 1. The rate constant does not change much over a wide region of the energy gap. The maximum of the rate constant does not exceed 10^{11} M^{-1} s^{-1} by much. A group of open circles of $k_{act}^{CS}(-\Delta G^0)$ smoothly connects a group of crosses in the normal region of $k_{obs}^{CS}(-\Delta G^0)$. Details of those experimental results, including more extensive fluorescence–quencher pairs, will be published elsewhere.

3. The data of $k_{act}^{CR}(-\Delta G^0)$, obtained by Mataga et al. (*18*) for the CR reaction of geminate radical-ion pairs produced by

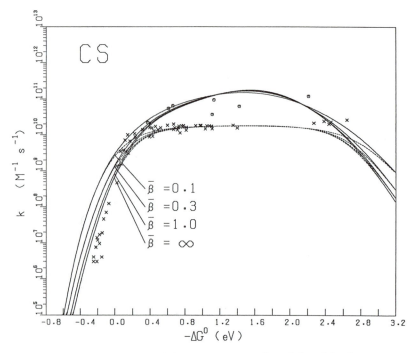

Figure 1. Energy-gap laws of the CS reaction. The solid and broken curves are the theoretical values of $k_{act}{}^{CS}(-\Delta G^0)$ *and* $k_{obs}{}^{CS}(-\Delta G^0)$, *respectively. The value of* $\bar{\beta}$ *represents the curvature change of the Gibbs energy curves in the initial and final states. The white circles are the experimental data for the photoinduced CS rate constant as obtained by analysis of the transient effect in the fluorescence decay curves (ref. 23). The crosses are the experimental data for the fluorescence quenching rate constant in the stationary state (ref. 19).*

Table I. Experimental Data for the Photoinduced
Charge-Separation Rate Constant with DPA
as Fluorescer

Quencher[a]	$-\Delta G^0$ (eV)	$k_{act}{}^{CS}$ $(-\Delta G^0)(M^{-1} s^{-1})$
DCNB	0.37	2.2×10^{10}
FN	0.61	5.5×10^{10}
PA	0.66	6.5×10^{10}
TCPA	1.11	3.7×10^{10}
MA	1.13	9.5×10^{10}
PMDA	1.42	6.3×10^{10}
TCNE	2.21	1.2×10^{11}

NOTE: DPA is 9,10-diphenylanthracene. Data were obtained by analysis of the transient effect in fluorescence decay curves (23).
[a]DCNB is dicyanobenzene; PA is phthalic anhydride; FN is fumaronitrile; TCPA is tetrachlorophthalic anhydride; MA is maleic anhydride; PMDA is pyromellitic dianhydride; and TCNE is tetracyanoethylene.

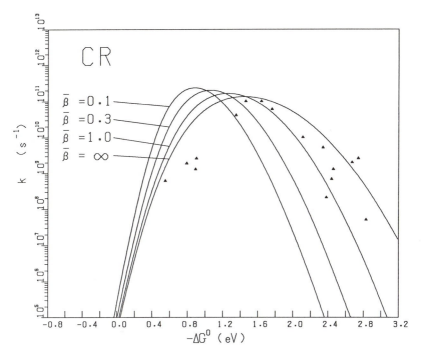

Figure 2. Energy-gap laws of the CR reaction. The solid curves are the values of $k_{act}{}^{CR}(-\Delta G^0)$ calculated by using the contact-ion-pair (CIP) model. The triangles are the experimental data for the CR reaction rate $k_{act}{}^{CR}(-\Delta G^0)$ for the geminate radical-ion pairs produced by the fluorescence quenching (ref. 18). The others are the same as shown in Figure 1.

fluorescence quenching, are plotted as triangles in Figure 2. The experimental data of $k_{act}{}^{CR}(-\Delta G^0)$ are quite different from those of $k_{act}{}^{CS}(-\Delta G^0)$. In particular, the normal region of $k_{act}{}^{CS}(-\Delta G^0)$ exists at the small energy gap with a very steep slope; that of $k_{act}{}^{CR}(-\Delta G^0)$ is found at a rather large energy gap with a dull slope.

When the distribution of donor–acceptor distances takes place in the initial state of ET reactions, the following three factors vary with the distance:

1. the reorganization energy $\lambda_{out}{}^{CS}(r)$, which is small for small r (such as for contact pairs) and large for large r (such as for solvent-separated pairs);

2. the effective frequency $\overline{A}(r)$ in the nonadiabatic mechanism in eq 17; and

3. the mean ion-pair potential $U^{ip}(r)$. This potential contributes to shift $-\Delta G^{CS}(r)$ and $-\Delta G^{CR}(r)$ in opposite ways (*see* eqs 29 and 35).

The energy-gap law would be modified by the variation of the stated factors with r. In the CS reaction, both reactants are neutral, and so the distance between the donor and acceptor in the initial state is distributed over a wide range. For the pair with small r, $\lambda_{out}^{CS}(r)$ is small, $\overline{A}(r)$ is large, and $U^{ip}(r)$ is negatively large. Those properties mainly affect the normal region, giving rise to a sharp increase in the rate constant and a considerable shift to the small-energy-gap side. When $\overline{A}(r)$ exceeds ν_{ad}, the adiabatic mechanism applies. For the pair with large r, $\lambda_{out}^{CS}(r)$ is large, $\overline{A}(r)$ is small, and $U^{ip}(r)$ is small. Those properties mainly affect the maximum and inverted region of the energy-gap law, leading to a very broad maximum and a gradual decrease of the rate constant at a large energy gap. Averaging those rate constants over the distance, we obtain the $-\Delta G^0$ dependence of the CS rate constant, whose normal region reflects solvent properties in the small r; the maximum and inverted region reflect solvent properties in the large r. Theoretically, this situation is represented by $k_{act}^{CS}(-\Delta G^0)$.

Under the condition that the lateral diffusion of the reactants is effective, $k_{obs}^{CS}(-\Delta G^0)$ is obtained by using eq 31. In the CR reaction, on the other hand, the ion pair of the initial state is produced by the photoinduced CS reaction. The distribution of ion-pair distances is not necessarily in thermal equilibrium in accordance with a mean ion-pair potential. If most of the ion pairs are solvent-separated in the initial state, $\overline{A}(r_1)$ is small, $\lambda_{out}^{CS}(r_1)$ is large, and $U^{ip}(r_1)$ is small. This case yields a broad bell-shaped energy-gap law with a peak at a rather large energy gap. If contact ion pairs (CIP) are mainly realized in the initial state, $\overline{A}(r_0)$ is large, $\lambda_{out}^{CS}(r_0)$ is small, and $U^{ip}(r_0)$ is negatively large. Then a rather narrow energy-gap law with a maximum located at a moderate energy gap results.

Numerical Calculations

Numerical calculations are performed in order to examine to what extent the experimental data can be reproduced quantitatively by the present theoretical scheme. For this purpose, the parameter values are fixed successively.

Some simple parameters are typically evaluated as

$$S = 3, \qquad \hbar\omega = 0.10 \text{ eV}, \qquad r_0 = 4.4 \text{ Å}$$

$$\alpha \text{ (defined in eq 53)} = 1.2 \text{ Å}^{-1}, \qquad T = 300 \text{ K} \qquad (54)$$

The parameter values for the intramolecular vibration S and $\hbar\omega$ are reasonable for aromatic solute molecules.

As discussed previously, the normal region and the onset of the plateau of $k_{act}^{CS}(-\Delta G^0)$ are mostly contributed from donor and acceptor pairs with small r where $\overline{A}(r)$ is very large. To reproduce the experimentally observed long plateau of $k_{act}^{CS}(-\Delta G^0)$ over an energy-gap range of 2 eV, it is necessary to assume that the adiabatic mechanism applies for small r. In this case, the heights of the ET rate constants in the normal region and the onset of the plateau of $k_{act}^{CS}(-\Delta G^0)$ are essentially determined by the value of the adiabatic frequency ν_{ad}. To fit the experimental data in those regions, ν_{ad} should not be larger than about 5×10^{12} s^{-1}. On the other hand, the heights of the maximum and inverted region of $k_{act}^{CR}(-\Delta G^0)$ are determined by the smaller of ν_{ad} and $\overline{A}_0$. To reproduce those experimental data, the value of the smaller one should be about 5×10^{12} s^{-1}. Judging from those results, ν_{ad} is most likely to be 5×10^{12} s^{-1}. The value of $\overline{A}_0$, which should be equal to or larger than 5×10^{12} s^{-1}, affects the inverted region. To reproduce the experimental data for the inverted region of $k_{act}^{CR}(-\Delta G^0)$, $\overline{A}_0$ should not be much larger than 5×10^{12} s^{-1}. Therefore, we tentatively choose it as $\overline{A}_0 \cong \nu_{ad}$. Summarizing these results yields

$$\nu_{ad} \cong 5 \times 10^{12} \text{ s}^{-1}, \qquad \overline{A}_0 \cong 5 \times 10^{12} \text{ s}^{-1} \qquad (55)$$

To fix the parameters in $U^{ip}(r)$ and $\lambda_{out}^{CS}(r)$ for organic solvent molecules such as acetonitrile, we make use of the onset of the plateau of $k_{obs}^{CS}(-\Delta G^0)$, which is found at a very small energy gap $-\Delta G^0 \sim 0.1$ eV. In this region, ion pairs with small r play a significant role. Reproduction of the experimental result requires that either $U^{ip}(r)$ be negatively large at $r = r_0$ or $\lambda_{out}^{CS}(r)$ be relatively small at $r = r_0$. On the other hand, because the solvent reorganization energy of the donor and acceptor linked by a rigid spacer molecule is more than 0.7 eV (28, 29), it looks unreasonable that $\lambda_{out}^{CS}(r)$ is substantially less than 0.5 eV at $r = r_0 \, (= 4.4 \text{ Å})$. By harmonizing these two conditions, we choose $\lambda_{out}^{CS} (4.4 \text{ Å}) = 0.5$ eV. Then $U^{ip} (4.4 \text{ Å})$ is determined to be -0.3 eV, which is substantially larger than predicted by the dielectric continuum model (eq 46). On the other hand, $\lambda_{out}^{CS}(r)$ for $r = \infty$ is theoretically estimated to be $2.5 \sim 3.0$ eV by the Monte Carlo simulation with a spherical hard-core model (30). We finally obtain

$$U^{ip}(r) = -\frac{1.3}{r} \quad \text{(eV)} \qquad (56)$$

$$\lambda_{out}^{CS}(r) = 2.7 - \frac{9.7}{r} \quad \text{(eV)} \qquad (57)$$

Calculated Results of $k_{act}^{CS}(-\Delta G^0)$ and $k_{obs}^{CS}(-\Delta G^0)$. The calculated curves of $k_{act}^{CS}(-\Delta G^0)$ and $k_{obs}^{CS}(-\Delta G^0)$ for $\beta = 0.1$, 0.3, 1.0, and ∞ are drawn in Figure 1 by solid curves and broken curves, respectively. The theoretical curves have very small curvatures around the maximum, nearly independent of the value of β. The experimental data (white circles) of Nishikawa et al. (23) are rather well fitted by those theoretical curves. At the edge of the diffusion limit of $k_{obs}^{CS}(-\Delta G^0)$ at about $-\Delta G^0 = 0.1$ eV, theoretical curves fit well to the experimental data (crosses) of Rehm and Weller (19). Theoretical curves in the normal region have a rather sharp slope, but the experimental data represent a much sharper slope than the calculated one.

The origin of this difference will be discussed in the last section by considering the possible participation of the exciplex formation. The normal region is found at a smaller energy gap as $\bar{\beta}$ decreases. The calculated curves show that a substantial decrease of $k_{obs}^{CS}(-\Delta G^0)$ starts at very large energy gaps. Thus, the experimental data by Rehm and Weller (19) at $-\Delta G^0 = 2.2$ ~ 2.6 eV can be nearly reproduced by the calculation described here, even if they are due to intrinsic electron-transfer mechanisms without contribution of the excited state of the products. Because the calculated curves do not differ substantially except when $\bar{\beta} = 0.1$, when $\bar{\beta}$ is decreased the nonlinear effect, even if it exists, is rather insensitive to the curves of $k_{act}^{CS}(-\Delta G^0)$ and $k_{obs}^{CS}(-\Delta G^0)$. It is unrealistic to evaluate $\bar{\beta}$ definitely by fitting the experimental data in the normal region of $k_{obs}^{CS}(-\Delta G^0)$, because theoretical curves in this region are easily shifted by modifying $U^{ip}(r)$ and $\lambda_{out}^{CS}(r)$.

Calculated Results of $k_{act}^{CR}(-\Delta G^0)$. Contact Ion-Pair (CIP) Model.

First, $k_{act}^{CR}(-\Delta G^0)$ was calculated by using the CIP model of the distribution function $g(r)$ of eq 33. The parameter in $g(r)$ was chosen as $t = 0.25$ Å^{-2} so that the theoretical curves may fit the experimental data of Mataga et al. (18). The calculated curves of $k_{act}^{CR}(-\Delta G^0)$ for $\bar{\beta} = 0.1$, 0.3, 1.0, and ∞ are shown in Figure 2 by solid curves. The maxima of the calculated curves are found at rather small values of the energy gap, except for $\bar{\beta} = \infty$. Owing to this, the experimental data in the normal region are considerably below the calculated curves. It is characteristic that the calculated curves differ greatly, depending on the value of $\bar{\beta}$. The experimental data in the inverted region are well fitted by theoretical curves for $\bar{\beta} = 1.0$ or ∞. These calculated results do not change appreciably even if the value of the parameter t, which represents the width of the r distribution, is changed in the range 0.1 Å$^{-2} < t < 0.5$ Å^{-2}. In summary, the experimental data of $k_{act}^{CR}(-\Delta G^0)$ are generally fitted well by a theoretical curve for $\bar{\beta} = \infty$ so long as we adopt the CIP model.

Solvent-Separated Ion-Pair (SSIP) Model.

Next, $k_{act}^{CR}(-\Delta G^0)$ was calculated by using the SSIP model and the distribution function of eq 34. The parameters in $g(r)$ were $u = 0.25$ Å^{-2} and $r_1 = 8.8$ Å. This value of u is the same as that of t. The calculated curves of $k_{act}^{CR}(-\Delta G^0)$ for $\bar{\beta} = 0.1$, 0.3, 1.0, and ∞ are shown in Figure 3. These curves are nearly bell-shaped, and the maxima are shifted to the larger energy gap, as compared with the case of the CIP model (Figure 2). The calculated curves differ greatly, depending on the value of $\bar{\beta}$. The experimental data are fitted very well by a theoretical curve for $\bar{\beta} = 1.0$. In contrast to the case of CIP model, the theoretical

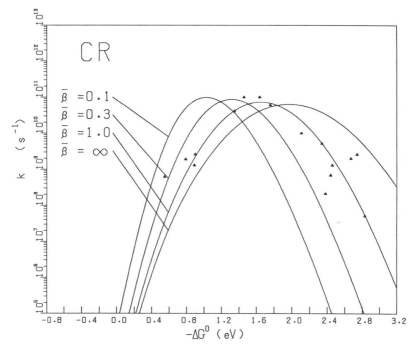

Figure 3. The energy-gap laws of the CR reaction. The solid curves are the values of $k_{obs}{}^{CS}(-\Delta G^0)$ calculated by using the solvent-separated ion-pair (SSIP) model. The others are the same as shown in Figure 2.

curve is slightly shifted to the large-energy-gap side with an increase of r_1. When $r_1 = 11$ Å, the theoretical curve for $\bar{\beta} = 0.5$ best fits the experimental data. More generally, when the reorganization energy at $r = r_1$ is made large, the maximum of $k_{act}{}^{CR}(-\Delta G^0)$ locates at a large energy gap and the experimental data can be fitted by theoretical curves with smaller values of $\bar{\beta}$. However, as long as we choose reasonable parameter values that seem to be consistent with molecular pictures, $\bar{\beta} = 1.0$ would be the best value in the SSIP model.

Discussion

This chapter has introduced different distribution functions of donor–acceptor distances for the CS and CR reactions, that of the CS reaction covering those over various distances and more specified functions corresponding to the CIP and SSIP models for the CR reaction. By averaging the ET rate over the distance distributions, we obtained quite different energy-gap laws between the CS and CR reactions, even in the case of the linear response ($\bar{\beta} = \infty$). In general, $k_{act}{}^{CS}(-\Delta G^0)$ and $k_{obs}{}^{CS}(-\Delta G^0)$ have broad shapes. Their rises in the normal region show a steep slope starting at the uphill energy gap, the curvature around the maximum is small, and the inverted region exists at a very large energy gap. The effect of the nonlinear response on $k_{act}^{CS}(-\Delta G^0)$ and $k_{obs}^{CS}(-\Delta G^0)$ is not so large.

In contrast to this, the energy-gap dependence of $k_{act}^{CR}(-\Delta G^0)$ is nearly bell-shaped, and the maximum is found at a fairly large energy gap. In this case, the breadth and the maximum position are very sensitive to the initial distribution function $g(r)$ and the $\bar{\beta}$ value. When the CIP model of $g(r)$ applies, the theoretical curve using a larger value of $\bar{\beta}$ (ca. $\bar{\beta} = \infty$) fits the experimental data rather well. When the SSIP model of $g(r)$ applies, the theoretical curve using a smaller value of $\bar{\beta}$ (ca. $\bar{\beta} = 1.0$) fits the experimental data very well. The experimental data are better fitted by the SSIP model than by the CIP model.

Recently Mataga and co-workers (31–33) obtained a quite different energy-gap law for the CR reaction of the contact ion pair produced by exciting the ground-state charge-transfer (CT) complexes in acetonitrile solution. The energy-gap law of these systems demonstrates that the CR rate is very large at a small energy gap and that it decreases exponentially with an increase in the energy gap. Ion pairs produced by the photoexcitation of these complexes, especially those with strong donor and acceptor corresponding to the small energy gap between the ion pair and the ground state, would probably be electronically delocalized and the donor and acceptor molecules would be in close contact (32).

Then the problem is how this close-contact ion pair can be differentiated from the CIP that might also be produced by the photoinduced charge separation at the diffusional encounter. In the latter case, as shown by the distribution of donor–acceptor distances, the CIP might not adopt a unique, specified donor–acceptor configuration in the pair, and the electronic delocalization appears to be small. Generally, in a diffusional encounter between the pair with a sufficiently positive energy gap of the CS reaction, electron transfer can take place without close contact and SSIP may be formed. When a relatively strong donor and acceptor are brought together in the ground state, they take a specific mutual configuration due to the weak CT interaction in close contact. Strong CT interaction takes place when they are excited, leading to the formation of close-contact ion pairs even in the acetonitrile solution.

However, when the energy gap for CS is nearly zero or a little positive, the donor–acceptor pair must be in close contact in order to realize the ET interaction. In this case a kind of CT complex, an exciplex, may be formed at the encounter. Namely, it is generally believed that when the energy level of the excited-state donor or acceptor is close to the energy level of the CT state, an exciplex is likely to be formed. Figure 1 shows a very sharp increase in the rate constant $k_{obs}^{CS}(-\Delta G^0)$ for the standard energy-gap region $-0.2 \text{ eV} \leq -\Delta G^0 \leq 0.1 \text{ eV}$. The actual energy gap $-\Delta G^{CS}(r)$ in this normal region is larger than $-\Delta G^0$ by about 0.3 eV. Therefore, this sharp slope occurs virtually in the energy-gap region 0.1 eV $< -\Delta G^{CS}(r) < 0.4$ eV.

Actually, a small amount of fluorescent exciplex formation in parallel with nonfluorescent ion-pair formation in acetonitrile solution was observed in the energy-gap region of $-\Delta G^{CS}(r) \sim 0.4$ eV (39). Although those reactions take place downhill of the Gibbs free energy, the energy gap is still rather small. Therefore, some mixing of the locally excited state with the CT state is possible, even though it may be minimal. A slope of the experimental data that is a little sharper than the theoretical curve will indicate this possibility.

As a result of the present theoretical analysis of the energy-gap laws, we found that the nonlinear effect due to the dielectric saturation could not be so large; the $\bar{\beta}$ value is 1.0 at most (corresponding to the SSIP model). Formerly, we expected (7, 10, 14–16) that a very large nonlinear effect would be required to reproduce quite different energy-gap laws between the CS and CR reactions. This supposition followed a model in which the donor–acceptor distance where the ET reaction takes place is uniquely fixed. By eliminating the restriction of a constant donor–acceptor distance and introducing different distributions of donor–acceptor distances for the CS and CR reactions, we need not relate the origin of the different energy-gap laws to the nonlinear effect.

On the other hand, our obtained value $\bar{\beta} \sim 1.0$ appears to be concomitant with the results of recent computer simulations by molecular dynamics (34–36) and Monte Carlo methods (37). Molecular dynamics studies (34, 35) for the charge-shift reaction with fixed reactant position in aqueous solution showed that the energy-gap law is nearly parabolic. This shape indicates a very small nonlinear response of the solvent polarization. On the other hand, the molecular dynamics study (36) using a spherical hard-core model for solute and solvent molecules represented a moderate nonlinearity corresponding to $\bar{\beta} = 1.6$. The spherical radii are chosen to be rather small in this calculation. Recently we did Monte Carlo simulations that involve a spherical hard-core model with and without the electronic polarizability effect of solvent molecules (37). In these calculations $\bar{\beta} = 1.4$ when the polarizability is not considered and $\bar{\beta} = 1.5–1.8$ when the various polarizability values are taken into account (37). This result is rather consistent with the one presented in this chapter.

It was suggested (34) that the quantum-mechanical character of intramolecular vibration will considerably modify the energy-gap law, with a dull decrease in the inverted region. Such an effect may be evident in the donor–acceptor-linked system. However, in unlinked systems the donor–acceptor distance distributes and the reorganization energy of the solvent mode is considerably larger than that of the quantum mode. When the energy-gap law for each donor–acceptor distance is averaged over its distribution, the quantum-mechanical nature of intramolecular vibration is nearly blotted out. The calculated (rather dull) decreases in the inverted region of $k_{act}^{CS}(-\Delta G^0)$ and $k_{act}^{CR}(-\Delta G^0)$ mostly reflect the distribution of the donor–acceptor distance.

We obtained nearly the same values for ν_{ad} and $\overline{A}_0$ as shown in eq 55. This result indicates that the adiabatic mechanism works for the distance $r < r_1$ (8.8 Å) and that the nonadiabatic mechanism works only for $r > r_1$. Therefore, the r dependence of $\overline{A}(r)$ plays no significant role in modifying the energy-gap law of $k_{act}^{CS}(-\Delta G^0)$ in the normal region and the plateau region. Similarly, a detail of the mutual orientation of the donor and acceptor molecules in this close distance is not important. But the r dependence of $\overline{A}(r)$ does play a role in the inverted region. The r dependence of λ_{out}^{CS} has been effective in the whole region. Our calculated value of ν_{ad} is very close to the prefactor $kT/\hbar$ ($= 6 \times 10^{12}$ s^{-1}) of the transition state theory.

In the theoretical treatment presented here, we have assumed that the thermal equilibrium for solvent motion is always attained. This assumption is guaranteed only when $\nu(r)$ is considerably smaller than $1/\tau_L$. However, the value of $1/\tau_L$ for acetonitrile is 5×10^{12} s^{-1} and coincides with the maximum of $\nu(r)$. Therefore, the assumption of thermal equilibrium is critical. On the other hand, if we take into account the dynamical effect of the solvent motion from the starting point, the effective frequency factor may be approximately written as

$$\nu^*(r) = \cfrac{1}{\cfrac{\tau_L}{\mu} + \cfrac{1}{\nu(r)}} = \cfrac{1}{\cfrac{\tau_L}{\mu} + \cfrac{1}{\nu_{ad}} + \cfrac{1}{\overline{A}(r)}} \tag{58}$$

where μ is the parameter of an order of unity (38). Our calculated value of 5×10^{12} s^{-1} for ν_{ad} would be $(\tau_L/\mu + 1/\nu_{ad})^{-1}$. To see whether τ_L/μ or $1/\nu_{ad}$ is dominant, experiments involving solvents with much smaller values of τ_L than that of acetonitrile would be useful.

We have examined how the theoretical curves are changed if we adopt the anharmonic exponent $n = 4$ instead of $n = 6$ in eq 36. As a result, we found that the theoretical curves obtained by using $n = 4$ do not appreciably change, except that the shoulder of $k_{act}^{CS}(-\Delta G^0)$ around $-\Delta G^0 = 0$ becomes a little flat.

The experimental data of $k_{act}^{CS}(-\Delta G^0)$, which can be obtained from the transient effect of the fluorescence decay, are important for obtaining direct information as to the magnitude and shape of the energy-gap law $[k_{act}^{CS}(-\Delta G^0)]$ of the CS reaction around its maximum. From those data, we could estimate the tunneling amplitude, the adiabatic frequency, and so on. It is desirable to accumulate much more data for quantitative discussions. Such experimental work will be reported in a forthcoming paper.

Next, the slope of the rate is defined by

$$\alpha^{CS} = \frac{1}{\beta} \left(\frac{\partial \ln k_{act}{}^{CS}(-\Delta G^0)}{\partial \ln (-\Delta G^0)} \right)_{\Delta G^0 = 0} \tag{59}$$

$$\alpha^{CR} = \frac{1}{\beta} \left(\frac{\partial \ln k_{act}{}^{CR}(-\Delta G^0)}{\partial \ln (-\Delta G^0)} \right)_{\Delta G^0 = 0} \tag{60}$$

We calculated α^{CS} and α^{CR} by using Figures 1–3; the result is listed in Table II. In all the $\bar{\beta}$ values α^{CS} is substantially smaller than 0.5 and α^{CR} is larger than 0.5. Because the initial distribution function $g(r)$ of the CR reaction is not in thermal equilibrium, a sum of α^{CS} and α^{CR} for the same value of $\bar{\beta}$ does not become unity. The values of α^{CS} and α^{CR} are not 0.5, even in the case of $\bar{\beta} = \infty$, where a linear response holds, because the normal region is shifted by $-U^{ip}(r)$ for the CS reaction and by $U^{ip}(r)$ for the CR reaction. When $\bar{\beta}$ becomes small, α^{CS} decreases a little and α^{CR} increases a little. α^{CR} in the CIP model is usually larger than α^{CR} in the SSIP model.

Table II. Calculated Slopes for Some Values of $\bar{\beta}$

$\bar{\beta}$	α^{CS}	$\alpha^{CR}(CIP)$	$\alpha^{CR}(SSIP)$
0.1	0.21	0.76	0.70
0.3	0.26	0.72	0.65
1.0	0.30	0.69	0.61
∞	0.31	0.67	0.60

NOTE: α^{CS} represents the slope of the CS reaction. $\alpha^{CR}(CIP)$ and $\alpha^{CR}(SSIP)$ denote the slopes of the CR reaction for the CIP and SSIP models.

It is of interest whether such a large deviation of the slope from ½ has been experimentally observed in the homogeneous reaction. Unfortunately, most available experimental data (*see* Table III in ref. 3) are for exchange reactions (or charge-shift reactions) for which the slopes are nearly 0.5. For the exchange reaction, we can theoretically prove that the slope is exactly 0.5 even if we take into account the distribution of the donor–acceptor distance. An accumulation of experimental data for the CS and CR reactions should confirm the role of $U^{ip}(r)$ in the slope.

The mechanism shifting the slope from 0.5 due to $U^{ip}(r)$ does not apply to the electrochemical reaction (in this case we call the slope a transfer coefficient), because $U^{ip}(r)$ as defined in the ion-pair system does not exist in the electrode systems. The transfer coefficient is obtained at a zero over-potential by adequate correction of the electric double-layer effect. In this case the transfer coefficient for $\bar{\beta} = \infty$ is ½, even if the distribution of the distance between the reactant and the metal electrode is taken into account. This theorem can be easily proved by using the fact that the transfer coefficient for $\overline{\beta} = \infty$ is always ½, independent of the distance r. The experimental data of transfer coefficients show substantially smaller values than 0.5 for the ionization reaction and larger values than 0.5 for the neutralization reaction (17). This variation indicates the existence of a considerable nonlinear response effect in electrochemical reactions. Why the nonlinear effect appears to be strong in the electrochemical reaction and moderate in the homogeneous reaction is not yet known. The mechanism responsible presents an interesting problem for the future.

Acknowledgment

T. Kakitani was supported by Grants-in-Aid (6330014, 01300009) from the Japanese Ministry of Education, Science, and Culture. N. Mataga acknowledges support by Grant-in-Aid (62065006) from the Japanese Ministry of Education, Science, and Culture. The computations were done by using a FACOM VP 200 in the Computation Center of Nagoya University.

References

1. Marcus, R. A. *J. Chem. Phys.* **1956**, *24*, 966.
2. Marcus, R. A. *Annu. Rev. Phys. Chem.* **1964**, *15*, 155.
3. Marcus, R. A.; Sutin, N. *Biochim. Biophys. Acta* **1985**, *811*, 265.
4. Levich, V. G. *Adv. Electrochem. Electrochem. Eng.* **1966**, *4*, 249.
5. Kestner, N. R.; Logan, J.; Jortner, J. *Phys. Chem.* **1974**, *78*, 2148.
6. Marcus, R. A. *J. Phys. Chem.* **1963**, *67*, 853.
7. Kakitani, T.; Mataga, N. *Chem. Phys.* **1985**, *93*, 381.
8. Kakitani, T.; Mataga, N. *Chem. Phys. Lett.* **1986**, *124*, 437.
9. Hatano, Y.; Saito, M.; Kakitani, T.; Mataga, N. *J. Phys. Chem.* **1988**, *92*, 1008.
10. Kakitani, T.; Mataga, N. *J. Phys. Chem.* **1985**, *89*, 8.
11. Yoshimori, A.; Kakitani, T.; Enomoto, Y.; Mataga, N. *J. Phys. Chem.* **1989**, *93*, 8316.
12. Tachiya, M. *Chem. Phys. Lett.* **1989**, *159*, 505.
13. Tachiya, M. *J. Phys. Chem.* **1989**, *93*, 7050.
14. Kakitani, T.; Mataga, N. *J. Phys. Chem.* **1985**, *89*, 4752.
15. Kakitani, T.; Mataga, N. *J. Phys. Chem.* **1986**, *90*, 993.
16. Kakitani, T.; Mataga, N. *J. Phys. Chem.* **1987**, *91*, 6277.
17. Yoshimori, A.; Kakitani, T.; Mataga, N. *J. Phys. Chem.* **1989**, *93*, 3694.
18. Mataga, N.; Asahi, T.; Kanda, Y.; Okada, T.; Kakitani, T. *Chem. Phys.* **1988**, *127*, 249.
19. Rehm, D.; Weller, A. *Isr. J. Chem.* **1970**, *8*, 259.
20. Mataga, N.; Kanda, Y.; Asahi, T.; Miyasaka, H.; Okada, T.; Kakitani, T. *Chem. Phys.* **1988**, *127*, 239.
21. Marcus, R. A.; Siders, P. *J. Phys. Chem.* **1982**, *86*, 622.
22. Brunschwig, B. S.; Ehrenson, S.; Sutin, N. *J. Am. Chem. Soc.* **1984**, *106*, 6858.
23. Nishikawa, S.; Asahi, T.; Hagihara, M.; Kanaji, K.; Okada, T.; Mataga, N.; Kakitani, T.; Yoshimori, A., unpublished results.

24. Kakitani, T.; Yoshimori, A.; Mataga, N., unpublished results.
25. Gennett, T.; Milner, D. F.; Weaver, M. J. *J. Phys. Chem.* **1985**, *89*, 2787.
26. Noyes, R. M. *Prog. React. Kinet.* **1961**, *1*, 129.
27. Patey, G. N.; Valleau, J. P. *J. Chem. Phys.* **1975**, *63*, 2334.
28. Wasielewski, M. R.; Niemczyk, M. P.; Svec, W. A.; Pewitt, E. B. *J. Am. Chem. Soc.* **1985**, *107*, 1080.
29. Miller, J. R.; Calcaterra, L. T.; Closs, G. L. *J. Am. Chem. Soc.* **1984**, *106*, 3047.
30. Enomoto, Y.; Kakitani, T.; Yoshimori, A.; Hatano, Y.; Saito, M. *Chem. Phys. Lett.*, in press.
31. Asahi, T.; Mataga, N. *J. Phys. Chem.* **1989**, *93*, 6575.
32. Asahi, T.; Mataga, N. *J. Phys. Chem.*, in press.
33. Ojima, S.; Miyasaka, H.; Mataga, N. *J. Phys. Chem.* **1990**, *94*, 7534.
34. Hwang, J. K.; Warshel, A. *J. Am. Chem. Soc.* **1987**, *109*, 715.
35. Carter, E. A.; Hynes, J. T. *J. Phys. Chem.* **1989**, *93*, 2184.
36. Kuharski, R. A.; Bader, J. S.; Chandler, D.; Sprik, M.; Klein, M. L.; Impey, R. W. *J. Chem. Phys.* **1989**, *89*, 3248.
37. Hatano, Y.; Enomoto, Y.; Kakitani, T.; Yoshimori, A. *Mol. Simul.* **1990**, in press.
38. Sumi, H.; Marcus, R. A. *J. Chem. Phys.* **1986**, *84*, 4894.
39. Hirata, Y.; Kanda, Y.; Mataga, N. *J. Phys. Chem.* **1983**, *87*, 1659.

RECEIVED for review April 27, 1990. ACCEPTED revised manuscript September 5, 1990.

Electron Transfer

From Model Compounds to Proteins

David N. Beratan[1] and José Nelson Onuchic[2,3]

[1]Jet Propulsion Laboratory, California Institute of Technology,
Pasadena, CA 91109
[2]Department of Physics, University of California at San Diego, La Jolla, CA
92093 and Instituto de Física e Química de São Carlos, Universidade de São
Paulo, São Carlos, SP, Brazil

We summarize the formulation of the protein-mediated electronic coupling calculation as a two-level system with weakly interacting bridge units. Using model compounds as a starting point from which to derive coupling parameters, we present a strategy for defining the pathways for electron tunneling in biological and biomimetic systems. The specific bonding and nonbonding interactions in cytochrome c and myoglobin that mediate the tunneling between the porphyrin and an attached transition metal probe are described. The method appears to succeed where traditional structureless tunneling barrier or periodic bridge models are not adequate. An algorithm to search for these tunneling pathways in proteins is described, and the nature of the paths is discussed.

T HE PROCESS OF ELECTRON TRANSPORT IS CENTRAL in chemistry, biology, and physics. This field is frequently subjected to detailed reanalysis and review (*1–4*). We begin the discussion here by presenting the Hamiltonian that has been used extensively to model the generic electron-transfer problem

$$H_{ET} = H_{rp}\sigma_x + \frac{1}{2}\delta\sigma_z + H_Q \tag{1}$$

[3]Current address: Department of Physics, University of California at San Diego, La Jolla, CA 92093

0065–2393/91/0228–0071$06.00/0

H_{rp} is the tunneling matrix element between donor and acceptor (reactants and products); σ_x and σ_z are the Pauli matrices, where the eigenvalue $\sigma_z = 1$ is associated with the reactant state and the eigenvalue $\sigma_z = -1$ is associated with the product state; H_Q supplies the dynamics for the nuclear motion (reaction coordinates and bath), and δ is the instantaneous energy difference between the reactants and products (5–8). This Hamiltonian leads to the ubiquitous rate equation for transfer between weakly coupled donors and acceptors (1–10).

$$k_{ET} = \frac{2\pi}{\hbar} |H_{rp}|^2 \, (\text{FC}) \qquad (2)$$

Assuming that this separation can be performed and that the process is not relaxation-controlled, the rate is proportional to the electronic coupling factor $|H_{rp}|^2$ times a nuclear Franck–Condon (FC) weighted density of states (activated) factor; $\hbar$ is Planck's constant divided by 2π.

Equation 2 gives the rate in the weak coupling limit, often called the nonadiabatic limit. Two important conditions must hold to write this equation. First, an energy separation is required to reduce the problem to the Hamiltonian given by eq 1 (i.e., a two-level system coupled to nuclear modes; renormalization procedure). A separation of electronic energies is also required so that the electronic problem can be reduced to a two-level (donor and acceptor) system. Second, even when the Hamiltonian of eq 1 is valid, the transfer must be nonadiabatic to write the rate in eq 2 (1–10). The nonadiabatic limit is valid for the model systems and proteins discussed in this chapter. Next we discuss why the simple Hamiltonian (eq 1) is appropriate for such complex problems.

Bridge-Mediated Electron Tunneling and Two-Level Systems

First, consider the electronic part of this problem. Because of the complexity of proteins, we hope to reduce it to smaller appropriate parts (if possible) that can be analyzed and understood. This is achieved by gradually eliminating higher energies. The first step in this procedure is to assume that the energies involved in chemical bonding are very small compared to core-electron excitations. This assumption allows the elimination of the core electrons. The core electrons provide a pseudopotential in which the valence electrons move. Next we make the further assumption that coupling energies associated with hopping between neighboring bonding orbitals are small compared to atomic excitation energies. This assumption leads to a tight-binding molecular orbital picture (11).

These assumptions are generally valid for the electron-transfer problem and are adopted throughout this chapter (4, 5). This modification justifies an effective one-electron Hamiltonian and permits computation of the tunneling matrix element H_{rp}. Finally, to reduce the one-electron Hamiltonian of the entire system to a two-level Hamiltonian, the electronic energy separation between donor (or acceptor) and bridge sites must be much larger than the coupling energy between donor and acceptor. If this is not the case, there are electronic excitations with energies of the same order as the donor–acceptor coupling, invalidating a two-level approach. [These energy comparisons are best made with bond orbitals rather than atomic orbitals (*12*).]

We now include the vibrational modes. If the energy scales associated with excitations of these modes are much smaller than the electronic excitation energies, we can use the Born–Oppenheimer approximation. This approximation allows us to solve the electronic problem for fixed nuclear coordinates, so the nuclear coordinates enter as parameters. A two-level system then results, with energies that are functions of nuclear coordinates. The tunneling matrix element is calculated by fixing the nuclear coordinates so that the reactant and product states have the same energy (Condon approximation) (*2, 4, 5*). If all of the energy separations discussed here are appropriate, the problem is reduced to eq 1. References 5 and 13 describe details of the electronic–nuclear energy separation. A tutorial showing the reduction of a three-level system to a two-level system is given in ref. 5. Electronic excitation energies are about 1–3 eV, so this separation is valid for most of the nuclear modes.

Next, we include the high-frequency nuclear modes (C＝O stretches, for example, with $\hbar\Omega$ in the range 0.1–0.25 eV; Ω is the frequency of the mode). In this case, $\hbar\Omega$ is much larger than other vibrational excitation energies and $k_B T$ (k_B is Boltzmann's constant and T is the temperature). These modes, which typically arise from local vibrations, have a nearly discrete spectrum [very low damping (*14*)], which should be treated in the quantum limit; they simply renormalize the tunneling matrix element and the driving force (*7*). For example, in a two-level system with thermodynamic driving force δ_0 coupled to one high-frequency mode, $|n_D\rangle$ ($|n_A\rangle$) represents the vibrational state of the high-frequency mode when the electron is on the donor (D) or acceptor (A). (The equilibrium position of this high-frequency mode shifts, depending on whether the electron is on the donor or acceptor.) Because $k_B T \ll \hbar\Omega$, the donor vibrational state is always $|0_D\rangle$. One of the acceptor states $|n_A\rangle$ will dominate the process, depending on δ_0. The renormalized parameters are

$$H_{rp}^{\text{eff}} = H_{rp} \langle 0_D | n_A \rangle \tag{3a}$$

$$\delta_0^{\text{eff}} = \delta_0 - n_A \hbar\Omega \tag{3b}$$

The effective donor state is $|D^{el}\rangle|0_D\rangle$, and the effective acceptor state is $|A^{el}\rangle|n_A\rangle$ (el signifies an electronic state). Finally, if the electronic excitations are of the same order as $\hbar\Omega$ and the reorganization energy is a few times $\hbar\Omega$, the renormalization procedure is a bit different and we must construct an energy cutoff that includes the electronic states and the high-frequency mode. We cannot separate this process into two stages as we did before (electronic part first, then the high-frequency mode). The final result is very similar to the present one (i.e., a single donor state and a set of discrete acceptor states), but these states will be mixtures of electronic and high-frequency nuclear states rather than simple products.

Energy separation is not the only requirement for the validity of the Born–Oppenheimer approximation. Although energy separation guarantees that we can neglect the donor (or acceptor) excited electronic states, care is required when computing the tunneling matrix element that depends on details of the electronic wave function tail. Formally, as the electron moves from donor to acceptor it spends an imaginary time (a traversal time) in the forbidden region (15). If the nuclear modes are slow compared to this time, the Born–Oppenheimer approximation works (i.e., the nuclei stay essentially fixed as the electron tunnels). The traversal time increases with the tunneling distance and decreases with the tunneling barrier height (5). For very-long-range transfer the Born–Oppenheimer approximation must break down, but this approximation is reasonable for the systems discussed here (4, 16).

To this point, we have described why the Hamiltonian in eq 1 is, in many cases, an appropriate starting point for the electron-transfer problem. We will now describe how to obtain the two-level representation of the electronic portion of the problem for bridged systems. The questions to be addressed are (1) What are these two states in a complex bridged system? (2) How is the coupling H_{rp} between the two states related to energy splittings that are obtained from electronic structure calculations?

The simplest example of bridge-mediated electron transfer in a tight-binding or molecular orbital model results in the Hamiltonian of eq 4. The donor and acceptor are only coupled by their mutual interactions with one bridge (B) orbital via the exchange interactions β_{DB} and β_{BA}. The Hamiltonian matrix in this case $(\alpha_B > \alpha_D, \alpha_A)$ is

$$H = \begin{pmatrix} \alpha_{D(\vec{y})} & \beta_{DB} & 0 \\ \beta_{DB} & \alpha_B & \beta_{BA} \\ 0 & \beta_{BA} & \alpha_{A(\vec{y})} \end{pmatrix} \tag{4}$$

The nuclear coordinates are represented by $\vec{y}$, and eq 4 is written in the Born–Oppenheimer approximation. The $\vec{y}$ dependence of the site energies reflects the separation between electronic and nuclear motion and the assumption that only the donor and acceptor orbitals are coupled to nuclear distortions. Because the donor and acceptor (unmixed) orbitals are degenerate at the crossing of the nuclear surfaces, $\alpha_D(\vec{y}) = \alpha_A(\vec{y}) = \alpha$ (Condon approximation). The symmetric–antisymmetric splitting (ΔE) between the two lowest states localized dominantly on donor and acceptor (eq 4) is

$$\Delta E = \sqrt{\frac{(\alpha_B - \alpha)^2}{4} + \beta_{DB}{}^2 + \beta_{BA}{}^2}$$

$$- \frac{(\alpha_B - \alpha)}{2} \simeq \frac{(\beta_{DB}{}^2 + \beta_{BA}{}^2)}{(\alpha_B - \alpha)} \tag{5}$$

This splitting is nonzero even when there is no donor–bridge or bridge–acceptor coupling! Contrary to the common claim, this splitting is not proportional to H_{rp}. The resolution of this issue arises from the fact that we have calculated the splitting between the wrong states. The splitting in eq 3 is nonzero because it includes contributions of pure donor–bridge and bridge–acceptor mixing. The net bridge-mediated donor–acceptor interaction, H_{rp}, is not the splitting between states in the overall Hamiltonian with $\alpha_D = \alpha_A$. Also, $2|H_{rp}|$ is the energy associated with mixing the donor *plus* bridge state with the acceptor *plus* bridge state (i.e., the splitting of the states in the corresponding two-level system). The splittings between eigenvalues of the full Hamiltonian of eq 4 are not directly related to H_{rp}. From the standpoint of perturbation theory, the donor–acceptor degeneracy at the crossing point of the nuclear surfaces is broken only in second order by the bridge, so that the coupling between the states in this order is $-\beta_{DB}\beta_{AB}/(\alpha_B - \alpha)$ (*17*). Only in a true two-site model is there direct equivalence between ΔE and $2|H_{rp}|$. Also, strictly speaking, the orbital energies that were made equivalent in the Condon approximation should be the energies of the two-level system, not the individual site energies of the donor and acceptor.

A general technique to reduce a bridged donor–acceptor system to the corresponding two-level system is Löwdin diagonalization (*18, 19*). Working in a basis diagonal in the bridge orbitals, the total Hamiltonian is

$$H = \begin{pmatrix}
\alpha_D & \beta_{DA} & \beta'_{D1} & \cdots & \cdots & \beta'_{DN} \\
\beta_{AD} & \alpha_A & \beta'_{A1} & \cdots & \cdots & \beta'_{AB} \\
\hline
\beta'_{1D} & \beta'_{1A} & \alpha_{B1} & 0 & \cdots & 0 \\
\beta'_{2D} & \beta'_{2A} & 0 & \alpha_{B2} & \cdots & 0 \\
\vdots & \vdots & \vdots & \vdots & \ddots & \vdots \\
\beta'_{ND} & \beta'_{NA} & 0 & 0 & \cdots & \alpha_{BN}
\end{pmatrix} \tag{6}$$

Primes denote interactions between a single atomic orbital and a molecular orbital and N is the number of bridge orbitals. Unprimed interactions are between single atomic orbitals. The exact corresponding two-level Hamiltonian is

$$
H = \begin{pmatrix} \alpha_D - \sum_i \left[\dfrac{\beta_{Di}'^2}{\alpha_{Bi} - E} \right] & \beta_{DA} - \sum_i \left[\dfrac{\beta_{Di}' \beta_{Ai}'}{\alpha_{Bi} - E} \right] \\ \beta_{DA} - \sum_i \left[\dfrac{\beta_{Di}' \beta_{Ai}'}{\alpha_{Bi} - E} \right] & \alpha_A - \sum_i \left[\dfrac{\beta_{Ai}'^2}{\alpha_{Bi} - E} \right] \end{pmatrix} \tag{7}
$$

The off-diagonal elements in eq 7 are the electron tunneling matrix elements of the corresponding two-level system. The tunneling energy E is determined by the diagonal energies (these are donor plus bridge and acceptor plus bridge energies) and the vibronic coupling in the molecule (a simple average is appropriate, for example, if the vibronic coupling on the two sites is identical) (4).

There are other methods of calculating tunneling matrix elements in bridged systems. An elegant method that is experiencing growing interest is the Green's function technique. The matrix elements of the bridge Green's function contain the effective coupling between sites in the bridge (20–22). Numerical techniques applicable to Green's functions are somewhat different from those usually applied in a Schrödinger equation approach, and some powerful theorems allow both exact and perturbation evaluation of the couplings for tight-binding Hamiltonians. The Green's function for a system, G, is defined by Dyson's equation:

$$
(E - H)G = 1 \tag{8}
$$

If the Green's function of the isolated bridge is given by $\widetilde{G}$, the donor is coupled to bridge orbitals i with strength β_{Di}, and the acceptor is coupled to sites n with strength β_{iA}.

$$
H_{rp} = \beta_{DA} + \sum_i \sum_n \beta_{Di} \widetilde{G}_{in} \beta_{nA} \tag{9}
$$

$\widetilde{G}_{in}$ describes the propagation of amplitude within the bridge from site i to site n; β_{DA} is the direct "through-space" donor–acceptor coupling and can generally be neglected relative to the bridge-mediated terms for distant electron transfer.

Information Learned from Model Compounds

Donor plus bridge and acceptor plus bridge states are needed for a two-level calculation of H_{rp}. As such, techniques that calculate this mixing reliably

were the first targets of study. Ab initio techniques are now being successfully applied to relatively small bridged electron-transfer model compounds (23–25) and idealized systems (26). Our approach has relied on one-electron and effective potential methods because these methods are adequate for addressing issues of tunneling energy dependence and bridge topology effects and because it is possible to perform these calculations in very weakly coupled systems without serious concern about basis set artifacts. Qualitative issues related to through-bond and through-space coupling are addressed conveniently with carefully parameterized exactly soluble square barrier models (27).

The generic results of the bridge studies are summarized in Figure 1. Most bridges can be "reduced" to chains of interacting pairs of orbitals with two characteristic interactions. The details of the reduced orbitals are determined by the topology of the chain and energetics of the bonds in the bridge (28). Tunneling through a bridge of such repeating units where the mixing into the bridge is weak and decay is rapid enough (decay per bridge unit squared is small compared to 1, not a very stringent condition) allows H_{rp} to be written as in eq 10. Writing the decay of H_{rp} per bond as ϵ (12, 28–32)

$$H_{rp} = \frac{\beta_A \beta_D \beta_1}{(E - \alpha_L^1)(E - \alpha_R^1) - \beta_1^2} \prod_{i=1}^{N} \epsilon_i \qquad (10a)$$

Neglecting backscattering between bonds,

$$\epsilon_i \simeq \left[\frac{\beta_i \gamma_i}{(E - \alpha_{iL})(E - \alpha_{iR}) - \beta_i^2} \right] \qquad (10b)$$

For $|\epsilon| > 0.4$, corrections for backscattering must be incorporated in the calculation of ϵ itself (12, 31). Here, E is the tunneling energy, L and R refer to the left and right hybrid atomic orbitals in the bonds, $(N + 1)$ is the total number of bonds in the bridge, β is the interaction within bonds, γ is the interaction between bonds, α is the energy of the (hybrid) orbitals in the bonds, and β_D and β_A are the coupling matrix elements between the donor and acceptor and the first and last bridge units, respectively. As an example, in a linear extended hydrocarbon chain $\gamma/\beta \simeq 0.25$ and $\beta \simeq -9$ eV. Equations 10a and 10b are generalized in the next section for the case in which the bond types in the bridge may be chemically different.

Most of the electron-transfer model compounds aimed at testing the distance dependence of the transfer rate are of the form DB_nA, where n is variable. The potential in such linkers is, to a good approximation, periodic (12, 28–30). The boundary conditions on the periodic potential contain the details of the donor and acceptor structure, but the periodic nature of the bridge allows relatively simple calculations to make predictions about the energy and symmetry dependence of the coupling within broad classes of linkers. These predictions, which typically include more details than were used to calculate eqs 10a and 10b, are reliable as long as the decay within the bridge is sufficiently rapid and the net mixing onto the bridge is weak. Predictions for σ-bond-coupled electron transfer included pointing out the enhanced mediation properties of bridges with convergent pathways of equal length, such as exist in corner-fused rings (vs. edge-fused rings) and other effects (28–30). Although the theoretical calculations seem to be in fair agreement with experiment, there are several questions begging to be addressed synthetically.

1. For fixed reaction free energy, ΔG, but donor and acceptor energies varied in an absolute sense, will the decay length of H_{rp} change the parameter $\beta/2$ in $H_{rp} \propto \exp[-R\beta/2]$ (where R is the donor–acceptor separation distance)? Does a hole or electron-transfer mechanism dominate in chemical systems?

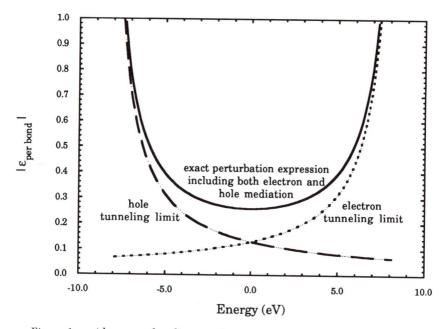

Figure 1a. ϵ *(decay per bond) vs. E plots are shown for a C–C chain with* β *= –8.5,* $\alpha_C = 0$, *and* $\gamma_C = -2.2$ *eV. The infinite chain result (U-shaped curve) is shown (28, 29), as well as the hole- and the electron-mediation limits (eq 13). The approximate curves are adequate in energetic regimes expected for typical model compounds (*$|\epsilon| \sim 0.4$–0.6*).*

2. For fixed donor and acceptor but varied bridge, will the net coupling show the predicted topological effects (28–30)?

3. In saturated systems coupling π-donors, does σ or π symmetry coupling into the bridge dominate the net interaction, H_{rp}?

4. How important are hydrogen bonds for mediating electron transfer? Surely there is a role for model building here. Is the picture of hydrogen bonds as preferentially assisting hole mediation (12) accurate?

5. How costly are the symmetry demands of σ/π interactions in proteins? Do π groups assist transfer or not? Our current thinking is that the π systems must be aligned in special ways for significant enhancements.

6. The distance dependence of ΔG and λ (reorganization energy) complicate the interpretation of bridge and tunneling energy dependence studies because these parameters cannot be held fixed with transfer distance. Can ΔG and λ studies be per-

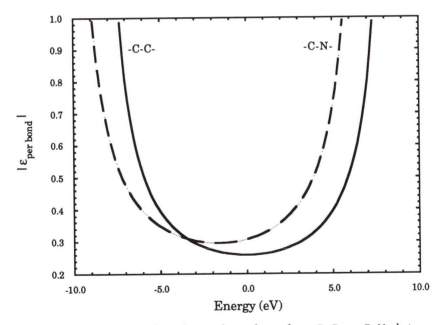

Figure 1b. The energy dependence of ε is shown for a C–C vs. C–N chain with β = –8.5, α_N = –3.3, γ_N = –3.1, and γ_C = –2.2 eV. The C–N plot is centered at lower energy because of the greater electron affinity of nitrogen. The "U" shape of the curves is characteristic, where ε shows distinct electron- and hole-mediation regimes.

formed as a function of distance to unambiguously deconvolute
the bridge structure dependence of the rates?

Answers to these questions are within synthetic and spectroscopic reach,
but obtaining them will require a coordinated effort.

The model compounds and the theoretical studies have taught us about
the typical length scales for decay of tunneling interactions in saturated and
unsaturated organic bridges. Counting bonds along the shortest path from
donor to acceptor in well-characterized model compounds suggests that the
decay of H_{rp} is about a factor of 0.4–0.6 per bond. Reference 3 summarizes
experimentally measured values of these parameters and their dependence
on structural details of the bridge. We have learned from the theory that
"decay per bond" strategies work rather well for these typical decays, with
some qualifications (12, 28–30). The propagation of the donor and acceptor
states can be built up by sequentially introducing single bonds (or groups
of bonds in strongly delocalized systems) to the chain of orbitals. Following
the addition of each bond, the amplitude leaking onto it is calculated as a
2×2 problem. Interference effects can be treated within this strategy (12,
28–32) if intersecting pathways bearing similar amplitudes are handled carefully.

Protein-Mediated Electron Transfer Interpreted with Decay-per-Bond Methods

Although intriguing questions remain in the model compound area, our aim
in pursuing that work was to learn how to piece together and parameterize
a model for protein-mediated electron transfer in photosynthetic and res-
piratory reactions. A *physical tunneling pathway* is defined as a collection
of interacting bonds in a protein around and between the donor and acceptor
that make some contribution to the donor–acceptor interaction. A few specific physical pathways may or may not dominate the electronic coupling
between donor and acceptor. Whether a relatively small number of pathways
is adequate to describe the coupling in proteins is actually a deep theoretical
issue. We argue (on the basis of rapid decay of through-space interactions
for typical tunneling energies, the relatively low density of residues, and
the anisotropic packing of bonds) that a relatively small number of pathways
is likely to be important.

The decay-per-bond approach leads to eq 11 for the contribution to the
tunneling matrix element arising from a single physical pathway with N_B
covalent couplings between bridge bonds, N_S through-space contacts, and
N_H hydrogen bonds (12, 31, 32)

$$t_{da} = \frac{\beta_A \beta_D \beta_1}{(E - \alpha_L^1)(E - \alpha_R^1) - \beta_1^2} \prod_{i=2}^{N_B} \epsilon_i^B \prod_{j=1}^{N_S} \epsilon_j^S \prod_{k=1}^{N_H} \epsilon_k^H \qquad (11)$$

β_D or β_A couples the donor or acceptor into the first or last bond of the bridge, respectively; t_{da} is the contribution to H_{rp} arising from a single pathway; and β_1 is the coupling between orbitals in the first bond. Values for ϵ can often be approximated by using perturbation theory. As an example, the lowest order contribution to ϵ is given by eq 10b. This limit totally neglects backscattering between bonds, and corrections to it need to be included for large ϵ. For a particular interaction, ϵ can be dissected into contributions from electron and hole mediation across a bond (*12*) as

$$\epsilon = \epsilon^e + \epsilon^h \tag{12}$$

In the limit where hole mediation through the bond dominates, for example, and the two coupled covalent bonds are the same

$$\epsilon \simeq \epsilon^h = \frac{\gamma}{E - \alpha_{bond}} \tag{13}$$

One can also write the propagator G_{1M}, which is proportional to H_{rp}, for a donor coupled to site 1 and acceptor coupled to site M, as

$$G_{1M} \propto \prod_{i=1}^{M} \epsilon_i \tag{14a}$$

The exact expressions for ϵ_i can be written (*20, 21*)

$$\epsilon_i = \frac{\gamma_i}{E - \alpha_i - \Delta_i} \tag{14b}$$

where Δ_i is a site-energy correction that takes into account the influence of all residues off (as well as on) the physical pathway between sites 1 and M (*20, 21*). Strategies that include the influence of all of the higher order corrections to the coupling neglected in our decay-per-bond (eq 11) formulation exist for the calculation of the Δ values. The exciting aspect of these methods is that they provide a way of interpreting the impact of specific residues anywhere in the protein on the coupling between two sites. The challenge now is to implement calculations of Δ values and related quantities for realistic but tractable protein Hamiltonians such as that of ref. 32. This approach still neglects interference between physical pathways. A new approach that includes multiple interacting pathways has been developed and will be used to test the present assumption that a few pathways dominate the coupling in many proteins.

Software that will include the calculation of interaction parameters combined with a search algorithm using these strategies is under development in our group and in other groups that are using somewhat different approaches (33). We recently wrote software (34) in an effort to understand the dependence of electron-tunneling mediation in proteins on details of the primary, secondary, and tertiary structure. The software makes the following assumptions:

1. All covalent bonds in the path cause equivalent decay, $\bar{\epsilon}_B$. A typical value of this factor is 0.6. Model compounds would suggest typical per-bond decay factors of 0.4–0.6 (this is the decay of H_{rp} per bond; square it to see the effect on the transfer rate).

2. All through-space interactions have the same orientation prefactor (σ) and decay length (β'), $\epsilon_S = \sigma\bar{\epsilon}_B \exp[-\beta'(R - R_C^{eq})]$ (where R is the through-space distance and R_C^{eq} is the reference covalent separation distance). Typically, σ is fixed at 0.5 for pathway surveying and β' is fixed between 1.7 Å^{-1} (10-eV binding energy for transferring electron) and 1.0 Å^{-1} (5-eV binding energy).

3. Hydrogen bonds couple as strongly as two covalent bonds, when scaled to reference covalent bond lengths, $\epsilon_{HB} = \bar{\epsilon}_B^2 \exp[-\beta'(R - R_{HB}^{eq})]$.

4. Interactions between pathways are neglected during the search. Interference effects due to the addition of amplitude arriving at the acceptor from multiple pathways can be included by summing the contributions independently.

We choose to neglect orientation factors in hydrogen bonds. Discussion of these parameters is found in ref. 12. In an extended-Hückel calculation, the β values in eqs 10a, 10b, and 11 depend on the orbital binding energies. References 12 and 27 show that it is actually more appropriate to use the electron tunneling energy to calculate β. In any case, the pathways are not strongly dependent on the particular chosen tunneling energy, as long as a realistic value is selected (34).

This strategy for pathway mapping intentionally neglects differences among bond types and orientations. The method for including bond differences and angular effects is described in ref. 12. Although angular effects require greater attention, the differences among decay factors for different bond types should not cause gross changes in the pathways. The strategy presented here would be meaningless if the qualitative aspects of the predictions were dependent on fine details of the decay parameters.

The σ value can be purposely varied to find pathways that exclude through-space segments. The decay factors include the qualitative aspects of the coupling, such as the similarity between covalent and hydrogen-bonded coupling (12), as opposed to through-space coupling. The rough choice of parameters is sufficient for a qualitative understanding of dominant pathways.

Realistic values of $\bar{\epsilon}_B$ are defined by the resonance integral for the bond and tunneling electron energy relative to the bond energy. As discussed in refs. 27 and 34, typical values of $\bar{\epsilon}_B$ are 0.4–0.6 for the bonds of interest. The value of β, the decay length of the through-space interactions, is determined by the binding energy of the tunneling electron.

The density of physical pathways found is sufficiently low that identification of individual paths is sensible (i.e., key residues can have an impact on the net coupling). The limitations of these admittedly simplistic assumptions are discussed in refs. 28–32. Now we justify, or at least explain, the approach. Square barrier models of protein-mediated electron tunneling (9, 35) are cruder than the calculation described here because they neglect the atomic graininess and the inhomogeneity of the bridging medium (36, 37). The present model includes these features. However, all of these simple models contain the essential physics of the electron-mediation problem and have provided excellent guidance for designing and interpreting experiments. We are confident that the models presented here will be supplanted by less naive ones in the future. In the meantime, we hope that they will allow the rational design of target proteins for site-directed mutagenesis and semi-synthesis-based electron-transfer studies, along with the interpretation of experimental results not anticipated by existing structureless barrier tunneling models.

Using these assumptions, we searched for the physical pathways with a graph search algorithm (34) in well-characterized proteins with known donor–acceptor couplings and transfer rates. We focus here on ruthenated myoglobin and cytochrome c (38–40) because these systems are so well-defined and well-characterized, and the coupling is clearly polypeptide-mediated (41–57). Detailed discussion of this work is presented in ref. 34. Here we will review some of the qualitative conclusions of the ruthenated protein studies and present the strongest evidence in support of the pathway search method.

The pathway search algorithm is not sensitive to special orientation or aromatic residue effects. For this reason, it is useful to look at the family of best pathways to draw qualitative conclusions and to compare relative path lengths between isomers for the best paths found. Myoglobin is a highly helical protein. The best family of pathways (34) between His 81 and the porphyrin are shown in Figure 2. Pathways follow the α-helix from the His to the porphyrin. In the His 12 derivative, the physical paths are roughly orthogonal to two portions of α-helix between His and porphyrin. Important paths follow prominent secondary structures only to the extent that they provide rather direct connections between donor and acceptor. In myoglobin, there seem to be abundant "good" pathways differing from one another in only minor ways. Hence, induction of a rate change in myoglobin by

Ru(His 81) Myoglobin

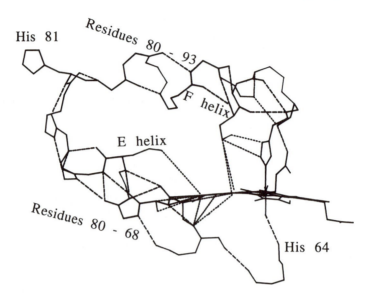

Ru(His 12) Myoglobin

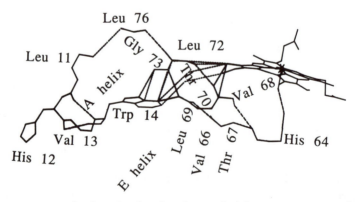

Figure 2. Top: The best family of pathways (34) between His 81 and the porphyrin are shown (myoglobin). The paths follow the α-helix from the His to the porphyrin and the through-space connections onto the ring. Bottom: In the His 12 derivative, the Trp 14 ring bridges most of the pathways between two α helices that are hard to identify because the paths move orthogonally to the helix axes.

changing specific protein atoms or bonds would probably require exquisite planning.

Another interesting aspect of the myoglobin pathway analysis is that the His 116 and His 12 derivatives have through-space contacts in all of their best pathways, as opposed to the His 48 and His 81 derivatives that have purely bond-mediated paths available. This observation led to reanalysis of the experimental data and questioning of whether the quenching in these isomers results from intramolecular electron transfer. The experiments are now being carefully reexamined (58).

The relatively large number of paths in myoglobin (hundreds within a factor of 10 coupling of the best) is not found in other proteins. In cytochrome c (Figure 3) only about 10–30 strongly coupled pathways are found, most without any through-space connections. This is probably a result of the less helical and less compact nature of cytochrome c. The measured rates in cytochrome c are known with greater certainty because they are sufficiently fast [the ^{3}Zn–porphyrin experiments in particular (38–40)] and provide an interesting study of the utility of the pathway model. Table I reports $\Pi_i \epsilon_i$ for the pathways and the equivalent calculated effective number of sequential covalent bonds.

These effective transfer distances track quite well with the measured rates. However, the measured rates do not track well with structureless-medium models that predict simple exponentially decaying rates proportional to $\exp[-\beta R_{DA}]$. For typical choices of β, the simple exponential scaling is off by at least an order of magnitude for these isomers. This result is the best experimental evidence so far for the importance of the pathways. Not only does the pathway analysis predict the proper ordering of the rates in the three isomers, it also predicts the relative couplings rather well. The factor of 0.6 decay per bond was chosen to give the ratio of His 48/His 81 myoglobin rates of $\sim 10^3$ for paths that differ by approximately five bonds (41).

To summarize the results of the pathway analysis, we are finding that (1) there are qualitative differences in the kinds of coupling pathways in different proteins, (2) transfer rates in cytochrome c seem to correlate well with the effective number of steps in the pathway but not with the through-space distance, and (3) hydrogen bonds appear to be crucial for linking covalent legs of the paths. Although the method does not differentiate between saturated and aromatic residues, there is no evidence from this family of experiments that aromatic residues provide any special rate enhancement. This finding may reflect the fact that the orbital overlap cost of mixing onto and off such groups can be rather large. Tests of these calculations can be carried out by site-directed mutagenesis of the protein pathways and careful temperature-dependence studies in ranges where fluctuations that facilitate through-space coupling (28) interactions (not gating) are shut down.

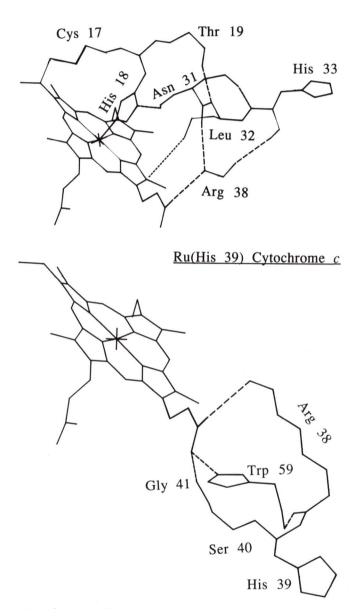

Figure 3. Pathways in His 33 (top), His 39 (bottom), and His 62 modified cytochrome c (next page). The His 33 derivative is from horse heart, His 39 is from Candida krusei, and His 62 is from Saccharomyces cerevisiae. Table I correlates the experimental values of the electronic couplings for these isomers with the pathway predictions.

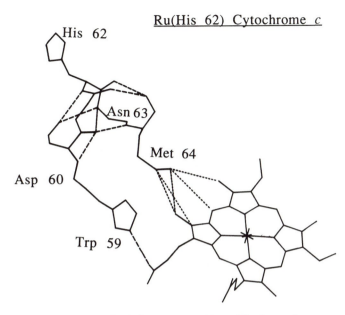

Figure 3. Continued. Pathway in His 62 modified cytochrome c.

Table I. Dominant Paths in Cytochrome c

Isomer	Through-Bond Links[a]	Effective Through-Bond Links	$\Pi_i\, \epsilon_i^{2}$ [b] (Relative)	Relative $|H_{rp}|^2$ Fit for $Ru^{2+}{\to}FeP^+$ and $Zn\ P^*{\to}Ru^{3+}$ Reactions	R_{DA} (Å)
His 39[c]	15 (1 H bond)	16.3	664	441	13.0
His 33[c]	18 (1 H bond)	17.6	180	144	13.2
His 62[d]	20 (3 H bond)	22.1	1	1	15.5

NOTE: All results were calculated with $\sigma = 0.5$, $\beta' = 1.7$ Å^{-1}, and counting hydrogen bonds as two through-bond connections from heteroatom to heteroatom. See ref. 38 for description of the experiments. None of the isomers had through-space links.
[a] Bonds were counted from Ru to the porphyrin ring edge or to the porphyrin metal atom for paths involving a ligand of the porphyrin metal (His 33 cytochrome c only).
[b] Relative coupling squared gives predicted relative transfer rates, assuming equal activation parameters, donor–bridge couplings, and acceptor–bridge couplings.
[c] *C. krusei.*
[d] *S. cerevisiae.*

Conclusions

We conclude with a summary of where the bridge-mediated problem now stands for complex bridges. Typical binding energies of π-electron systems are roughly 6–10 eV. Optical excitation of 2 eV or less decreases this effective binding energy somewhat. However, binding energies like this would result in values of β, the distance decay of the rate (proportional to the square root of the tunneling energy), of roughly 2–3 Å^{-1}, roughly twice that observed in any well-characterized model compound or protein. Apparently, coupling mediated by the bridge dominates the coupling that would result at equal distance in the absence of the bridge. This order-of-magnitude argument has been confirmed by several groups in more detailed calculations. In the weak coupling, relatively rapid decay regime, perturbation-theory approaches are adequate and expressions for the decay per bond in saturated linkers can be estimated. For relevant tunneling energies, these predictions are in very good agreement with calculations that consider the entire bridge. However, predictions of energetic and topological effects on the transfer rates remain to be unambiguously tested.

Experimental and theoretical studies have made rapid progress over the last 5 years. Of particular theoretical interest has been the synthesis and physical study of donor–bridge–acceptor systems with fixed separation distances. The theoretical framework for understanding bridge-mediated coupling in model compounds seems to be well in place; yet many key experimental tests of generic predictions remain to be performed as described in the foregoing discussion. As experimental tests emerge, more detailed theoretical analysis will undoubtedly be warranted.

The situation with proteins is more complex. Site-directed mutagenesis techniques, redox-active protein labeling, semisynthesis techniques, and the solution of the photosynthetic reaction center structure have introduced unanticipated theoretical challenges. Simple questions like, "How well does the ubiquitous hydrogen bond mediate electron transfer?" remain nearly unaddressed. Meanwhile, strategies are still needed to reliably treat coupling mediated by large and complex bridges, as are more systematic experimental tests of the theory. Although the protein tunneling pathway model proposed here seems compelling with existing data, the model makes real predictions that can be directly tested experimentally. Such experimental work might involve the design of systems with particularly weak or strong coupling at fixed transfer distance, analysis of pathway-induced temperature dependences, and tunneling energy effects on the coupling.

Acknowledgments

This work was performed, in part, at the Jet Propulsion Laboratory, California Institute of Technology, and was sponsored in part by the Department of Energy's Energy Conversion and Utilization Technologies Division–ECUT, through an agreement with the National Aeronautics and Space Administration. We thank the National Science Foundation and Conselho Nacional de Desenvolvimento Científico e Tecnológico (CNPq) (Brazil) for a binational research grant that allowed international visits during which much of this work was performed and the Brazilian agencies Fianciadora de Estudos e Projetos (FINEP) and CNPq for additional support. We thank our collaborators at Caltech, J. J. Hopfield and H. B. Gray, for many years of exciting collaborative work in this area. We are also grateful to M. J. Therien and H. B. Gray for discussion of the cytochrome *c* and other protein electron-transfer experiments prior to their publication. We also thank E. Canel of Rockefeller University for enjoyable discussions of this problem.

References

1. Marcus, R. A.; Sutin, N. *Biochim. Biophys. Acta* **1985**, *811*, 265.
2. Newton, M. D.; Sutin, N. *Annu. Rev. Phys. Chem.* **1984**, *35*, 437.
3. Mikkelson, K. V.; Ratner, M. A. *Chem. Rev.* **1988**, *87*, 113.
4. Onuchic, J. N.; Beratan, D. N.; Hopfield, J. J. *J. Phys. Chem.* **1986**, *90*, 3707.
5. Bialek, W.; Bruno, W. J.; Joseph, J.; Onuchic, J. N. *Photosynth. Res.* **1989**, *22*, 15.
6. Garg, A.; Onuchic, J. N.; Ambegaokar, V. *J. Chem. Phys.* **1985**, *83*, 4491.
7. Onuchic, J. N. *J. Chem. Phys.* **1987**, *86*, 3925.
8. Onuchic, J. N.; Wolynes, P. G. *J. Phys. Chem.* **1988**, *92*, 6495.
9. Hopfield, J. J. *Proc. Natl. Acad. Sci. U.S.A.* **1974**, *71*, 3640.
10. Jortner, J. *Biochim. Biophys. Acta* **1980**, *594*, 139.
11. Ballhausen, C. J.; Gray, H. B. *Molecular Orbital Theory*; Benjamin/Cummings: Reading, MA, 1964.
12. Onuchic, J. N.; Beratan, D. N. *J. Chem. Phys.* **1990**, *92*, 722.
13. Joseph, J.; Bialek, W., private communication, 1990.
14. Bialek, W.; Onuchic, J. N. *Proc. Natl. Acad. Sci. U.S.A.* **1988**, *85*, 5908.
15. Caldeira, A. O.; Leggett, A. J. *Ann. Phys. (N.Y.)* **1983**, *149*, 374.
16. Beratan, D. N.; Hopfield, J. J. *J. Chem. Phys.* **1984**, *81*, 5753.
17. Schiff, L. I. *Quantum Mechanics*, 3rd ed.; McGraw Hill: New York, 1984; Chapter 8.
18. Riemers, J. R.; Hush, N. S. *Chem. Phys.* **1989**, *134*, 323.
19. Larsson, S. *J. Am. Chem. Soc.* **1981**, *103*, 4034.
20. da Gama, A. A. S. *J. Theor. Biol.* **1990**, *142*, 251.
21. de Andrade, P. C. P.; Onuchic, J. N.; Beratan, D. N., unpublished results.
22. Ratner, M. A. *J. Phys. Chem.* **1990**, *94*, 4877.
23. Balaji, V.; Ng, L.; Jordan, K. D.; Paddon-Row, M. N.; Patney, H. K. *J. Am. Chem. Soc.* **1987**, *109*, 6957.
24. Falcetta, M. F.; Jordan, K. D.; McMurry, J. E.; Paddon-Row, M. N. *J. Am. Chem. Soc.* **1990**, *112*, 579.
25. Farazdel, A.; Dupuis, M.; Clementi, E.; Aviram, A. *J. Am. Chem. Soc.* **1990**, *112*, 4206.
26. Cave, R. J.; Baxter, D. V.; Goddard, W. A., III; Baldeschwieler, J. D. *J. Chem. Phys.* **1987**, *87*, 926.
27. Beratan, D. N.; Onuchic, J. N.; Hopfield, J. J. *J. Chem. Phys.* **1985**, *83*, 5325.
28. Onuchic, J. N.; Beratan, D. N. *J. Am. Chem. Soc.* **1987**, *109*, 6771.
29. Beratan, D. N.; Hopfield, J. J. *J. Am. Chem. Soc.* **1984**, *106*, 1584.
30. Beratan, D. N. *J. Am. Chem. Soc.* **1986**, *108*, 4321.
31. Beratan, D. N.; Onuchic, J. N.; Hopfield, J. J. *J. Chem. Phys.* **1987**, *86*, 4488.
32. Beratan, D. N.; Onuchic, J. N. *Photosynth. Res.* **1989**, *22*, 173.
33. Kuki, A.; Wolynes, P. G. *Science (Washington, D.C.)* **1987**, *236*, 1647.

34. Beratan, D. N.; Onuchic, J. N.; Betts, J.; Bowler, B. E.; Gray, H. B. *J. Am. Chem. Soc.*, **1990**, *112*, 7915.
35. Siders, P.; Cave, R. J.; Marcus, R. A. *J. Chem. Phys.* **1984**, *81*, 5613.
36. Davydov, A. S. *Phys. Status Solidi B* **1987**, *90*, 457.
37. McConnell, H. *J. Chem. Phys.* **1961**, *35*, 508.
38. Therien, M. J.; Bowler, B. E.; Selman, M. A.; Gray, H. B.; Chang, I-J.; Winkler, J. R. In *Electron Transfer in Inorganic, Organic, and Biological Systems*; Bolton, J. R.; Mataga, N.; McLendon, G. L., Eds.; Advances in Chemistry Series 228; American Chemical Society: Washington, DC, 1991; Chapter 12.
39. Bowler, B. E.; Meade, T. J.; Mayo, S. L.; Richards, J. H.; Gray, H. B. *J. Am. Chem. Soc.* **1989**, *111*, 8757.
40. Therien, M. J.; Selman, M. A.; Gray, H. B.; Chang, I-J.; Winkler, J. R. *J. Am. Chem. Soc.* **1990**, *112*, 2420.
41. Cowan, J. A.; Upmacis, R. K.; Beratan, D. N.; Onuchic, J. N.; Gray, H. B. *Ann. N.Y. Acad. Sci.* **1988**, *550*, 68.
42. Moore, J. M.; Case, D. A.; Chazin, W. J.; Gippert, G. P.; Havel, T. F.; Powls, R.; Wright, P. E. *Science (Washington, D.C.)* **1988**, *240*, 314.
43. Bowler, B. E.; Raphael, A. L.; Gray, H. B. *Prog. Inorg. Chem.*, in press.
44. Gray, H. B.; Malmström, B. G. *Biochemistry* **1989**, *28*, 7499.
45. Liang, N.; Pielak, G. J.; Mauk, A. G.; Smith, M.; Hoffman, B. M. *Proc. Natl. Acad. Sci. U.S.A.* **1987**, *84*, 1249.
46. Liang, N.; Mauk, A. G.; Pielak, G. J.; Johnson, J. A.; Smith, M.; Hoffman, B. M. *Science (Washington, D.C.)* **1988**, *240*, 311.
47. Elias, H.; Chou, M. H.; Winkler, J. R. *J. Am. Chem. Soc.* **1988**, *110*, 429.
48. Conrad, D. W.; Scott, R. A. *J. Am. Chem. Soc.* **1989**, *111*, 3461.
49. Pan, L. P.; Durham, B.; Wolinska, J.; Millett, F. *Biochemistry* **1988**, *27*, 7180.
50. Durham, B.; Pan, L. P.; Long, J. E.; Millett, F. *Biochemistry* **1989**, *28*, 8659.
51. Farver, O.; Pecht, I. *FEBS Lett.* **1989**, *244*, 379.
52. Jackman, M. P.; McGinnis, J.; Powls, R.; Salmon, G. A.; Sykes, A. G. *J. Am. Chem. Soc.* **1988**, *110*, 5880.
53. Osvath, P.; Salmon, G. A.; Sykes, A. G. *J. Am. Chem. Soc.* **1988**, *110*, 7114.
54. Faraggi, M.; Klapper, M. H. *J. Am. Chem. Soc.* **1988**, *110*, 5753.
55. Hazzard, J. T.; McLendon, G.; Cusanovich, M. A.; Das, G.; Sherman, F.; Tollin, G. *Biochemistry* **1988**, *27*, 4445.
56. Cusanovich, M. A.; Meyer, T. E.; Tollin, G. *Adv. Inorg. Biochem.* **1987**, *7*, 37.
57. Raphael, A. L.; Gray, H. B.; Raphael, A. L.; Gray, H. B. *Proteins* **1989**, *6*, 338.
58. Upmacis, R.K., unpublished results.

RECEIVED for review April 27, 1990. ACCEPTED revised manuscript August 16, 1990.

6

Photoinduced Charge Separation and Charge Recombination of Transient Ion-Pair States

Ultrafast Laser Photolysis

Noboru Mataga

Department of Chemistry, Faculty of Engineering Science, Osaka University, Toyonaka, Osaka 560, Japan

The mechanisms and dynamics underlying photoinduced charge separation (CS) and charge recombination (CR) of the produced charge-transfer (CT) or ion-pair (IP) state are discussed on the basis of the results obtained by femtosecond–picosecond laser photolysis and time-resolved spectral studies on various donor–acceptor (D–A) systems combined by spacers or directly, on the uncombined fluorescer–quencher pairs, and on CT complexes. By comparing the results for these various systems concerning the effects of electronic interaction between D and A, energy gap of electron transfer, and solvent dynamics on the photoinduced CS and CR of the produced CT or IP state, a much deeper insight into the nature of the electron-transfer mechanism prevailing among those different kinds of systems has been obtained.

THE MECHANISMS AND DYNAMICS regulating the photoinduced charge separation (CS) and charge recombination (CR) of the produced charge-transfer (CT) or ion-pair (IP) states, as well as related phenomena in liquid solutions, rigid matrices, molecular assemblies, and biological systems, are the most fundamental and important problems in the photophysical and photochemical primary processes in the condensed phase (*1–5*).

0065–2393/91/0228–0091$07.25/0

In general, the rate of the photoinduced CS and that of the CR of the CT or IP state are regulated by the magnitude of the electronic interaction responsible for the electron transfer (ET) between D (electron donor) and A (electron acceptor) groups, the Franck–Condon (FC) factor (which is related to the energy gap for the ET reaction), the reorganization energies originating from the vibrational freedoms within D and A as well as the polarization motions of the solvent molecules surrounding D and A, and also the solvent-orientation dynamics in the course of ET in polar solvent.

Depending upon the strength of the electronic interaction between D and A responsible for the ET, the reaction will be nonadiabatic (a) or adiabatic (b). If the electronic interaction is sufficiently strong in case b and the energy gap relation is also favorable, the ET process will become barrierless (c) and will be governed by both the solvent-orientation dynamics and intramolecular vibrations of D and A groups (6, 7). When the effect of the solvent dynamics on the ET process is dominant, it is believed that the longitudinal dielectric relaxation time, τ_L, or solvation time, τ_S, will be important as a factor controlling the ET rate. In a limit of strong electronic interaction between D and A groups combined rigidly, its excited singlet state can be regarded as a very polar single molecule, and a large fluorescence Stokes shift due to solvation in polar solvents can be observed (d). The first theoretical formula for the Stokes shift in case d, given by Mataga et al. (8) and Lippert (9), has been extended recently by Bagchi et al. (10) and others to take into consideration its dynamical aspects.

However, in many actual systems of strongly interacting D and A, the electronic structure or the extent of the CT from D to A can change gradually, accompanied by some geometrical change in the course of extensive solvation. This is the case between c and d; that is, when the electronic interaction between D and A groups is increased beyond case c, the simple electron-transfer mechanism based on the two-state model (locally excited initial state and final IP state due to one electron transfer) will become invalid. Because this case does not seem to be well-recognized in general, we discuss this problem mainly in relation to the photoinduced CS process of a strongly interacting D–A system, considering the results of our femtosecond–picosecond laser photolysis studies of D–A systems combined by spacers or directly by single bonds (11–20) and similar studies, as well as some previous investigations of CT complexes between aromatic hydrocarbons and various electron acceptors (21–30).

The photoinduced CS between fluorescer and quencher groups and excitation of the CT complex in such strongly polar solvents as acetonitrile leads to the formation of the IP or CT state. Those IPs generally undergo CR and dissociation into free ions or some chemical reactions. The behaviors of those CT or IP states are of crucial importance from various viewpoints; in many cases the fate of successive reactions is determined by intermediate CT or IP states. For example, the very rapid photoinduced CS among the special dimer of bacteriochlorophyll, bacteriochlorophyll monomer, and bacteriopheophytin, and the much slower CR rate of produced IP in the bacterial photosynthetic reaction center (*31*) are directly responsible for its ultrafast and extremely efficient redox reaction.

One of the most important factors regulating the CR rate of the IP leading to the formation of the ground state seems to be the energy gap between the relevant states. For the CR rate constant k_{CR} of IP produced by the fluorescence-quenching reaction in polar solution, only the results for the inverted region were available previously (*32–35*); no results had been published for the normal region. We made systematic studies of this problem by directly observing the CR deactivation of the geminate IP in acetonitrile solution with ultrafast laser spectroscopy. Our results cover the inverted region, the normal region, and the top region, establishing a bell-shaped relationship (*17, 36, 37*).

As in the case of the photoinduced CS reaction, we examined the k_{CR} of the IP that is produced by excitation of CT complexes where D and A interact more strongly than in the fluorescence-quenching reaction in acetonitrile solution. We found a k_{CR} energy-gap dependence quite different from the bell-shaped result (*29, 30*). We have confirmed that the relationship, $\log k_{CR} \propto -|\Delta G_{ip}^{0}|$, is observed in acetonitrile solution for a wide range of $-\Delta G_{ip}^{0}$ values ($-\Delta G_{ip}^{0}$ is the free-energy gap between the IP and the ground states). This remarkable difference of the IP energy-gap dependence of k_{CR} seems to be related to the difference in its structure, depending on the mode of its formation. The IP formed by exciting the CT complex may have a tighter structure with stronger interaction between D^{+} and A^{-} ions in the pair than in the IP produced by CS at the encounter between fluorescer and quencher. We have compared the behaviors of these different kinds of IPs and discuss the relevant mechanisms in this chapter.

In addition, the existence of different kinds of IPs is crucially important in some organic photochemical reactions that proceed via the IP states produced by photoinduced CS. The chemical reactivity of the IP formed by CT complex excitation seems to be quite different from that of the IP produced by the CS through an encounter between an excited molecule and an electron-donating or -accepting quencher molecule.

Photoinduced CS and CR of the Produced IP State of Combined D–A Systems

We investigated the Pn [p-$(CH_3)_2$N-C_6H_4-$(CH_2)_n$-(1-pyrenyl), $n = 1, 2, 3$], An [p-$(CH_3)_2$N-C_6H_4-$(CH_2)_n$-(9-anthryl), $n = 0, 1, 2, 3$], 9,9'-bianthryl, and their derivatives, 1,2-dianthrylethanes, with femtosecond–picosecond laser photolysis and time-resolved transient absorption spectral measurements in alkanenitrile and viscous alcohol solvents (*11–20*). The time-resolved ultrafast absorption spectral measurements give direct information on the electronic structures of the system undergoing photoinduced ET. Such structural information is extremely important for discriminating various cases of ET processes, as described in the introductory section. This discussion deals mainly with results obtained in alkanenitrile solutions of Pn and An.

As an example, time-resolved absorption spectra of Pn in acetonitrile (ACN) are shown in Figure 1 (13). The rapid rise of the characteristic sharp absorption band at 500 nm indicates the intramolecular IP state, and the rapid decay of the absorption around 470 nm indicates the $S_n–S_1$ (from the lowest excited singlet to the higher excited singlet state) transition localized in the pyrene part. Very similar time-resolved spectra with slightly slower rise and decay processes have been observed also in n-butyronitrile (BuCN) and n-hexanenitrile (HexCN) solutions. In the case of An ($n = 1, 2, 3$) in alkanenitrile solutions, we can observe the rapid rise of the characteristic absorption band at 480 nm due to the DMA (N,N-dimethylaniline) cation of the intramolecular IP state. In these systems, the rise curves of the IP state converge to constant values. These flexible-chain compounds may have some distribution of ground-state conformations that might affect the CS process in the excited state. Nevertheless, the rise curves of the IP-state absorbance can be reproduced approximately by an exponential function with rise times of ca. 1–10 ps, as shown in Table I (13).

The τ_{CS} (rise time of the IP state) values of Pn and An ($n = 1, 2, 3$) in Table I are much longer than the solvent τ_L value. This comparison suggests that the photoinduced CS in these systems are not directly controlled by the solvent-reorientation dynamics. Even if we use the solvation time τ_S (38) estimated from the dynamic fluorescence Stokes shift of the polar probe molecule, this conclusion is not altered except in the case of A_1, where τ_{CS} is close to τ_S, a result suggesting the possibility of control by solvation dynamics. However, at ~1 ps delay time after excitation of aromatic molecules, the intramolecular vibrational relaxation (cooling) is not yet completed (39). Therefore, possibly the CS is taking place from the state with excess vibrational energy in A_1.

These observations on Pn and An ($n = 1, 2, 3$) show that the photoexcitation is initially localized in the pyrene or anthracene portion; then ET takes place to produce the IP state. In the case of model compounds used previously for investigating the effects of solvent dynamics on the photoinduced intramolecular ET reaction, this criterion was not necessarily clear (40).

As shown in Table I, the τ_{CR} values are almost 3 orders of magnitude longer than τ_{CS} values. Contrary to the case of τ_{CS}, they generally become shorter with an increase of the intervening chain number n.

It seems possible to give a satisfactory account of these results in terms of the usual nonadiabatic ET mechanism (12, 13). In general, the electron-transfer rate constant k can be written in terms of the factor $\overline{A}$, which

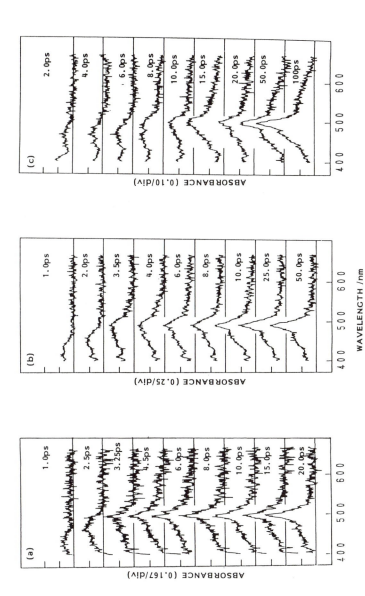

Figure 1. Time-resolved transient absorption spectra of P_1 (a), P_2 (b), and P_3 (c) in ACN, measured with the femtosecond laser photolysis method. (Reproduced from ref. 13. Copyright 1990 American Chemical Society.)

Table I. Rise and Decay Times of the Intramolecular IP State of Pn and An in Alkanenitrile Solutions

	τ_{CS} (ps)			τ_{CR} (ns)		
Compound	ACN	BuCN	HexCN	ACN	BuCN	HexCN
P_1	1.7	2.5	4.5	7.0	11	11
P_2	6.1	7.7	—	3.3	10	—
P_3	11	25	—	1.1	4.1	—
A_1	0.65	1.0	1.4	4.0	7.1	8.0
A_2	2.1	4–5	—	1.0	3.2	—
A_3	2.7	10	—	0.7	5.2	—
A_0	—	2.7	5.0	48[a]	—	—
τ_L	0.2	0.5	($\gtrsim$0.7)			

[a]Inverse of nonradiative rate constant from the (CT) fluorescent state, of which the observed lifetime was 31 ns. Data from ref. 53.
SOURCE: Reproduced from ref. 13. Copyright 1990 American Chemical Society.

incorporates the tunneling matrix element H_{rp} and the FC factor, including contributions from both intramolecular vibrations and solvent orientation as follows.

$$k = \overline{A} \cdot (FC) \tag{1}$$

$$\overline{A} = \frac{2\pi H_{rp}^2}{\hbar^2 \langle \omega \rangle} \tag{2}$$

where (FC) is related to the free-energy gap $-\Delta G^0$, $\langle \omega \rangle$ is the average angular frequency of the intramolecular vibrational mode, and $\hbar$ is Planck's constant divided by 2π.

The free-energy gap for CS ($-\Delta G_{CS}^0$) of the D–A pairs of these systems in ACN, for example, is ~0.5 eV, where the k_{CS} (τ_{CS}^{-1}) vs. $-\Delta G_{CS}^0$ relationship is in the normal region and rather close to the top region. This placement agrees with both theoretical predictions (33, 41–45) and the experimental estimation of k_{CS} values by means of the transient effect in the bimolecular fluorescence-quenching reaction (46). Therefore, the difference of k_{CS} values among compounds with varying chain numbers may be ascribed not only to the small difference in $-\Delta G_{CS}^0$ but also to the factor $\overline{A}$ containing the tunneling matrix element H_{rp}. By taking $\overline{A} = 10^{12}–10^{13}$ s^{-1}, k_{CS} can be estimated to be $10^{11}–10^{12}$ s^{-1} (33). On the other hand, $-\Delta G_{CS}^0$ will decrease slightly in the order of ACN > BuCN > HexCN because of the decrease of solvation energy of the IP state. Because the k_{CS} vs. $-\Delta G_{CS}^0$ relationship is in the normal region at $-\Delta G_{CS}^0 \sim 0.5$ eV, this decrease of $-\Delta G_{CS}^0$ results in a slight decrease of k_{CS}. Nevertheless, as the solvent reorganization energy also decreases in the same order, the effect will be rather small.

The much smaller rate of CR in the intramolecular IP state can be well understood as a result of the overwhelming influence of the FC factor. As already discussed, we have established by ultrafast laser photolysis measurements the bell-shaped k_{CR} (τ_{CR}^{-1}) vs. $-\Delta G_{ip}^0$ (free-energy gap between the IP and the neutral ground state) relationship for the IP produced by fluorescence-quenching reaction in acetonitrile solution (*17, 36, 37*). For both Pn and An, the k_{CR} vs. $-\Delta G_{ip}^0$ relationship is in the inverted region at the large energy gap around $-\Delta G_{ip}^0 \sim 2.8$ eV and the energy gap for An is slightly smaller than that for Pn. Taking $\overline{A} = 10^{13}$ s^{-1} for the $n = 1$ compounds yields $k_{CR} = 10^8$ s^{-1}, in agreement with observation. For ($n = 3$) compounds, configurational change to sandwich form can take place; this change increases $\overline{A}$ and decreases $-\Delta G_{ip}^0$, leading to the faster CR with $k_{CR} \sim 10^9$ s^{-1}.

Because the τ_{CS} of A_1 is rather close to the solvation time τ_S in the alkanenitrile solvent, the photoinduced CS of this system may be controlled approximately by solvent dynamics. In A_0, the D and A groups are directly combined by a single bond, and the electronic interaction between them will be much stronger than in A_1. The photoinduced CS in A_0 will be truly controlled by solvent orientation dynamics or, in strongly interacting D and A groups, its CS process cannot be described by the simple two-state model, $A^*-D \rightarrow A^--D^+$, but will be explained by assuming a gradual change of electronic structure toward increasing polarity, accompanied by some geometrical change and extensive solvation. The spectra in Figure 2 show a gradual change from an absorption band that is somewhat analogous to but broader than that of the S_1 state of anthracene to one that indicates the CT state, with its characteristic DMA cation band around 450 nm. The approximate decay time of the initial state or rise time of the CT state can be estimated as shown in Table I. These values are much longer than the corresponding τ_{CS} values of A_1.

These results on A_0 imply that A_0 is the case between c and d, as discussed in the opening section. For such systems with very strong electronic interaction between D and A groups, the photoinduced CS is not readily determined by the solvent dynamics. Similarly, these results cannot be interpreted simply by the two-state model based on the usual ET theories that assume weak interaction. A possible interpretation for this fact may be that the CS in such a strongly interacting system becomes a little slower because of intramolecular geometrical rearrangements and that extensive solvation is necessary to prevent the electronic delocalization interaction in the IP state.

Photoinduced CS can take place not only in systems with definite D and A groups combined by spacer or directly [as in the intramolecular exciplex compounds already discussed (*11–13, 17, 18*)], but also in some systems with identical halves (*14–16, 19, 20*). The most well-known example is the solvation-induced broken symmetry (*19*) of excited 9,9′-bianthryl (BA) in polar solvent (*14, 19, 47*). The picosecond time-resolved transient absorption spectra of BA in 1-pentanol, for example, can be reproduced ap-

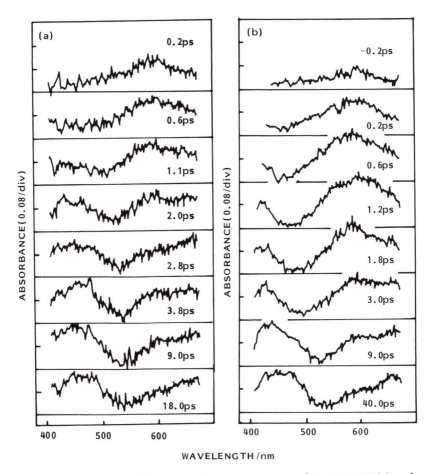

Figure 2. Time-resolved transient absorption spectra of A_0 in BuCN (a) and HexCN (b), measured with the femtosecond laser photolysis method. (Reproduced from ref. 13. Copyright 1990 American Chemical Society.)

proximately as a linear combination of the spectra in hexane (nonpolar S_1 state a little delocalized over two anthracene rings) and acetonitrile (intramolecular CT state), and the time-dependent change of the spectra converges to an equilibrium (*14*). From such analysis, the photoinduced CS of BA in 1-pentanol has been confirmed to take place with $\tau_{CS} \sim 170–180$ ps at 23 °C (*14*). A similar result, very close to $\tau_L = 174$ ps, has been obtained by picosecond time-resolved fluorescence measurement (*14*).

Although the time-resolved transient absorption spectra of BA in 1-pentanol can be reproduced approximately by linear combination of the spectra in hexane and acetonitrile, the spectrum in acetonitrile is much broader and peak positions are shifted compared with the superposition of the absorption bands of anthracene cation and anion radicals in acetonitrile solution. This means that complete CS seems to be difficult, even in acetonitrile solution, because of the strong interaction between two anthracene moieties. Moreover, the time-dependent change of the spectra converges to an equilibrium, and a considerable proportion of the nonpolar state seems to be populated in this equilibrium state.

Therefore, even though the two anthracene planes in BA are twisted, which decreases the delocalization interaction between the two moieties, a considerable amount of interaction still exists. Owing to this interaction, the photoinduced CT process of BA in polar solvents is deemed a gradual change of electronic structure from nonpolar to polar, probably accompanied by a small change of twisting angle in the course of solvation. An approximately quantitative description of such change of electronic structure along the solvation coordinate was given previously (*47*) without consideration of the change of twisting angle by means of the generalized Langevin equation.

We have examined also the photoinduced CS of 10-chloro-9,9'-bianthryl (BACl), in which the symmetry of BA is slightly perturbed by substitution at the 10 position. In this case, τ_{CS} has been confirmed to be 140 ps, which is considerably shorter than that of BA. This result means that the intramolecular CS of this slightly perturbed BA is not determined easily by the solvent reorientation, but the slightly presolvated state for this symmetry-disturbed compound will facilitate the CS process (*14*).

If two anthracene moieties are separated by methylene chains, an almost completely charge-separated state may be realized in strongly polar solvents, as we observed in the Pn and An (*n* = 1, 2, 3) systems. Actually, we have observed by means of picosecond laser photolysis such a solvation-induced CS in the excited state of 1,2-di-1-anthrylethane [D(1-A)E] and 1,2-di-9-anthrylethane [D(9-A)E], both of which seem to have partially and weakly overlapped configurations between two anthracene rings, as indicated in Figure 3 (*15*, *16*, *20*). In the case of D(1-A)E in acetonitrile, for example,

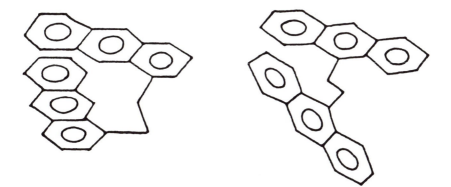

Figure 3. Schematic diagram of partially and weakly overlapped configurations of 1,2-di-1-anthrylethane (left) and 1,2-di-9-anthrylethane (right).

the formation of the completely charge-separated state within ~30 ps has been observed by picosecond time-resolved absorption spectral measurements ([15, 16]). This rise time of the CS state may not be unreasonable in view of the relevant energy gap $-\Delta G_{CS}^0 \sim 0.33$ eV ([15, 16]).

The photoinduced CS in the bacterial photosynthetic reaction center seems to start at the special pair (SP) of bacteriochlorophylls and lead to a series of redox processes. In the SP, two bacteriochlorophyll chromophores interact weakly and overlap only partially with each other. The photoinduced CS in the SP might be induced by some environmental effect in proteins, on which the chromophores are held. This circumstance is different from the photoinduced CS in composite systems with two aromatic groups in perpendicular configuration, as in BA and TICT (twisted intramolecular charge transfer) compounds. Rather, it is very similar to the example presented of the photoinduced CS of 1,2-dianthrylethanes in acetonitrile.

On the other hand, $-\Delta G_{ip}^0 = 2.85$ eV for the CR of the intramolecular IP state of D(1-A)E in acetonitrile, which leads to $k_{CR} \sim 10^7$–10^8 s^{-1} on the basis of the observed bell-shaped energy-gap dependence of the CR reaction of the geminate IP ([37]). However, the observed lifetime of the IP state becomes much shorter than 10 ns as a result of conversion to the excimer state ([15, 16]). The excimer formation may be facilitated by a slight mutual approach of two chromophores in the IP state. If the mutual approach of the chromophores in the present system is prevented by fixing them with a rigid spacer, the lifetime of the IP state may become much longer.

CS Processes in the Excited CT Complexes

Another extreme case of photoinduced CS in the strongly interacting D–A system is provided by the excited CT complexes. A brief discussion follows of the results of femtosecond–picosecond laser photolysis and time-resolved absorption spectral studies on aromatic hydrocarbon–TCNB (1,2,4,5-tetracyanobenzene), –acid anhydrides, –TCNQ (tetracyanoquinodinomethane), and –TCNE (tetracyanoethylene) complexes. We compare the photoinduced CS processes of these various CT complexes with D and A of different strengths. In addition, we compare the results on CT complexes with those of the D–A systems combined directly by single bond or by spacer.

Previous luminescence measurements, nanosecond laser photolysis studies on the TCNB–toluene system, and some MO (molecular orbital) theoretical investigations on its electronic structures in the ground and excited states indicated a large change of geometrical structure within the complex and the surrounding solvents in the course of relaxation from the excited FC to the equilibrium CS state ([21–24, 48]). Nevertheless, the details of this change were still unclear. We observed it directly with femtosecond laser spectroscopy ([12, 25–27]).

In general, the rate of the CS in the excited state of CT complexes will depend on the configurations of D and A in the complex in the ground state, excited FC state, and relaxed IP state; the strength of interactions between D and A; and the nature of the environment. The dynamics and mechanisms of the changes from the excited FC to the relaxed IP state can be demonstrated by time-resolved absorption spectral measurements, as shown in Figures 4 and 5 for TCNB in toluene solution.

Immediately after excitation, a slight change of absorption intensity accompanied by a slight sharpening of the band shape toward the free TCNB anion band takes place with a time constant of 1.5 ps. This spectral change can be ascribed to the configuration change of D and A within the 1:1 complex from the FC excited state with asymmetrical configuration toward a more symmetrical overlapped sandwich-type configuration as indicated in Figure 6, which increases the extent of CS according to the previous MO predictions (22–24). This structural change within the 1:1 complex does not lead to the complete CS, but further interaction with donor and formation of the 1:2 complex $^1(A^-{\cdot}D_2{}^+)^*$ is of crucial importance for it. As shown in Figure 5b, the spectrum at 170 K is very similar (even at 100 ps delay time) to that at 4.5 ps at room temperature. The positive hole created by removing an

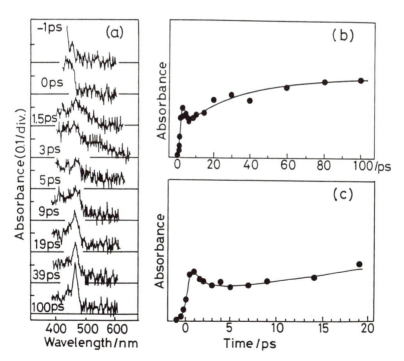

Figure 4. (a) Time-resolved transient absorption spectra of TCNB in toluene solution measured with the femtosecond laser photolysis method. (b,c) Time profiles of absorbance at 465 nm; c is an enlargement of the first section of b. (Reproduced from ref. 25. Copyright 1989 American Chemical Society.)

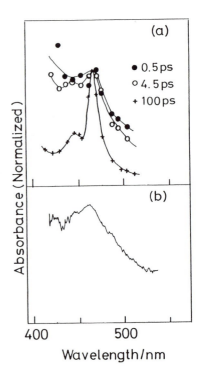

Figure 5. (a) Transient absorption spectra at several delay times corrected for the chirping of the monitoring white pulse. The absorption intensity is normalized at 465 nm. (b) Transient absorption spectra at 100 ps delay time observed at 170 K. (Reproduced from ref. 25. Copyright 1989 American Chemical Society.)

electron from an aromatic hydrocarbon is stabilized by dimer cation formation. Therefore, the formation of the unsymmetrical 1:2 complex $^1(A^-{\cdot}D_2{}^+)^*$ in the excited state facilitates the CS.

In the case of TCNB in toluene solution, the rise time of this 1:2 complex formation was determined to be 30 ps (25, 26). Similar results were obtained

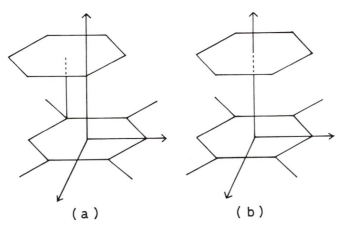

Figure 6. Possible geometrical structures of 1:1 TCNB complex. (a) FC excited state; (b) relaxed excited state with symmetrical overlapped configuration.(Reproduced from ref. 26. Copyright 1990 American Chemical Society.)

for other TCNB solutions in benzene and mesitylene. In summary, the photoinduced CS in these solutions is given by

$$^1(A^{-\delta} \cdot D^{+\delta}) \xrightarrow{h\nu} {}^1(A^{-\delta'} \cdot D^{+\delta'})^* \xrightarrow[\substack{\text{structural change} \\ \text{within 1:1 complex}}]{\tau_d}$$

$$^1(A^{-\delta''} \cdot D^{+\delta''})^* \xrightarrow[\text{D}]{\tau_r} {}^1(A^- \cdot D_2^+)^* \quad (3)$$

where δ, δ', and δ'' represent the degree of partial charge transfer, $\tau_d = 2$ ps, 1.5 ps, and 550 fs, and τ_r (time constant of the 1:2 complex formation) = 20 ps, 30 ps, and 40 ps, respectively, for benzene, toluene, and mesitylene solutions (25, 26).

We also examined the photoinduced CS of the TCNB complexes in polar solvents (25, 27). The results shown in Figure 7 for the TCNB–toluene complex in ACN indicate that the solvent reorientation can induce CS with a time constant shorter than 1 ps to a considerable extent, but not completely.

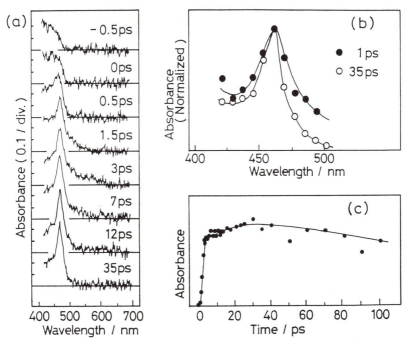

Figure 7. (a) Time-resolved transient absorption spectra of the TCNB–toluene complex in acetonitrile measured with the femtosecond laser photolysis method. (b) Transient absorption bands at 1- and 35-ps delay times, corrected for the chirping of the monitoring white pulse. The intensity is normalized at 462 nm (peak wavelength in acetonitrile). (c) Time profile of absorbance of corrected spectra at 462 nm. (Reproduced from ref. 25. Copyright 1989 American Chemical Society.)

For the complete CS leading to IP formation, further intracomplex structural change and solvation, which take place with a time constant (τ_{CS}) of $\sim$20 ps, seem to be necessary.

$$^1(A^{-\delta} \cdot D^{+\delta})_S \xrightarrow{h\nu} {}^1(A^{-\delta'} \cdot D^{+\delta'})_S^{FC} \xrightarrow[\text{solvation}]{<1 \text{ ps}}$$

$$^1(A^{-\delta''} \cdot D^{+\delta''})_{S'}^* \xrightarrow[\substack{\text{structural change} \\ \text{and further solvation}}]{\tau_{CS}} {}^1(A^- \cdots D^+)_{S''} \quad (4)$$

The τ_{CS} value in ACN has been confirmed as becoming shorter with lowering of the oxidation potential of the donor (that is, τ_{CS} = 41, 20, 13, 12, $\sim$7–8, and $\sim$5–6 ps, respectively, for the TCNB complexes with benzene, toluene, mesitylene, p-xylene, durene, and hexamethylbenzene donors, with oxidation potential decreasing in this order). These τ_{CS} values are much longer than τ_L of ACN. Accordingly, this CS process seems to involve a considerable intracomplex structural change within the 1:1 complex. Presumably, this change includes a slight increase in the distance between the charged D and A, assisted by strong solvation. The extent of this structural change will be smaller in the case of a donor with lower oxidation potential. We have also confirmed that the τ_{CS} value becomes shorter with an increase in the solvent dielectric constant.

The IP formed by a reorientation of the surrounding solvent and an intracomplex structural change in the course of relaxation from the FC excited state of the complex, as already discussed, seems to be a CIP (contact IP) without intervening solvent between D^+ and A^- in the pair. The formation of the CIP from the FC excited state becomes faster with an increase of the strength of the donor (i.e., with a decrease of the oxidation potential of the donor, in the case of the TCNB complexes). We have observed a similar effect when the reduction potential of the acceptor becomes higher, as follows (29, 30).

Compared with TCNB, PMDA (pyromellitic dianhydride) is a little stronger electron acceptor. By direct observation of CIP formation from the excited FC state of the PMDA–toluene complex in acetonitrile with femtosecond laser photolysis, we found that τ_{CS} = 7 ps (30), compared with τ_{CS} = 20 ps for the TCNB–toluene complex. Similar measurements on the PMDA–hexamethylbenzene complex in acetonitrile solution and analysis of the results have indicated a few picoseconds as the formation time of CIP (30). When we use stronger electron acceptors, such as TCNE and TCNQ, it has been confirmed that the formation of the CIP state becomes much faster. For example, in the case of the pyrene–TCNE and perylene–TCNE complexes in acetonitrile solution, the observed time profiles of the transient absorbance of the CIP state can be reproduced by convolution of the exciting femtosecond laser pulse and the decay curve of CIP with a short decay time of a few hundred femtoseconds, without taking into account the finite formation time of CIP (30).

By summarizing these results on the CS processes of relatively weak CT complexes like TCNB–benzene and TCNB–toluene systems, one may conclude that the photoinduced CS process is much slower than the solvent reorientation dynamics because of the intracomplex configurational change. This change seems to be necessary to cut the strong delocalization interaction between ions in the IP state, analogous to the case of A_0 discussed in the combined D and A systems. The extent of the intracomplex configurational change in the course of the relaxation from the FC excited state of the complex to the CIP becomes smaller in the case of the stronger CT complex. Specifically, the character of the CIP state changes, depending on the nature of D and A. The electronic and geometrical structures of CIP will become closer to those of the excited CT state of the complex itself with an increase in the strengths of D and A (*30*).

CR Deactivation of Geminate IP

As discussed, the CR process of the intramolecular IP state produced by photoinduced CS of Pn and An ($n = 1, 2, 3$) is almost 3 orders of magnitude slower than the photoinduced CS itself in alkanenitrile solutions. We interpreted this large difference between the CS and CR rate constants as due to their energy-gap dependence. The CRs in these systems are in the inverted region and are quite slow because of the large energy gap between the IP and ground state (*13*).

As mentioned, we have made systematic studies of the energy-gap dependence of the CR rate of geminate IP produced by CS at the encounter between fluorescer and quencher in a strongly polar solvent. Our studies involved directly observing the CR deactivation process competing with the dissociation by means of ultrafast laser spectroscopy and monitoring the time dependence of the absorbance of geminate IP, $^1(A_S^- \cdots D_S^+)$.

$$^1A^* + D \text{ or } A + {}^1D^* \rightarrow {}^1(A_S^- \cdots D_S^+) \xrightarrow{k_{diss}} A_S^- + D_S^+$$

$$\searrow h\nu \qquad \Big\downarrow k_{CR}$$

$$A + D \leftarrow A \cdots D \qquad (5)$$

This investigation established the bell-shaped energy-gap dependence of this type of CR process of IP (*17, 36, 37, 49*), as indicated in Figure 8.

On the other hand, very few systematic studies such as this have been made on the CR processes of the IPs formed by excitation of the CT complexes in strongly polar solutions. We discussed the CS processes in the excited state of various aromatic hydrocarbon–electron acceptor CT complexes in the section "CS Processes in the Excited CT Complexes". We concluded that the intracomplex configuration change in the excited state is more or less necessary for the CS of these strongly interacting electron

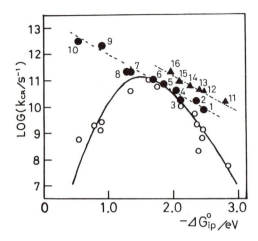

Figure 8. Energy-gap dependence of the CR rate constant of IP produced by CT complex excitation (●, ▲) compared with that formed by fluorescence-quenching reaction (○, data taken from ref. 8c) in acetonitrile solution. (1) Py^+-PA^-, (2) An^+-PA^-, (3) Per^+-PA^-, (4) $Naph^+-PMDA^-$, (5) Chr^+-PMDA^-, (6) Py^+-PMDA^-, (7) Per^+-PMDA^-, (8) $Naph^+-TCNQ^-$, (9) Py^+-TCNE^-, (10) Per^+-TCNE^-, (11) Bz^+-PMDA^-, (12) Tol^+-PMDA^-, (13) m-Xyl^+-PMDA^-, (14) p-Xyl^+-PMDA^-, (15) Du^+-PMDA^-, (16) HMB^+-PMDA^-. Py: pyrene; An: anthracene; Per: perylene; Naph: naphthalene; Chr: chrysene; Bz: benzene; Tol: toluene; m-Xyl: m-xylene; p-Xyl: p-xylene; Du: durene; HMB: hexamethylbenzene; PA: phthalic anhydride; PMDA: pyromellitic dianhydride; TCNE: tetracyanoethylene; TCNQ: tetracyanoquinodimethane. (Reproduced from ref. 29. Copyright 1989 American Chemical Society.)

donor–acceptor systems (25–27, 30). The extent of the configuration change depends on the strengths of D and A. CS accompanied by configuration change has been shown to be a rather slow process in the relatively weak CT complexes such as TCNB–benzene and TCNB–toluene systems. However, it is an ultrafast process in such strong CT complexes as TCNE–pyrene and TCNE–perylene systems (30).

Such CS processes taking place in the strongly interacting D–A systems are quite different from the simpler one that usually occurs in weakly interacting D–A systems. The distinction suggests a difference in the structure of geminate IP between the two cases of CT complex excitation and fluorescence-quenching reaction by diffusional encounter. The structural difference in the IP in these two cases will profoundly affect the behaviors of the IP, such as CR deactivation and dissociation into free ions. Actually, we previously observed that the k_{CR} vs. $-\Delta G_{ip}^{0}$ relationship of pyrene–TCNE IP produced by the fluorescence-quenching reaction in acetonitrile was in the normal region and $k_{CR} = 2.6 \times 10^9$ s^{-1}. In contrast, CR of the same system but with IP produced by CT complex excitation was much faster; $k_{CR} = 2 \times 10^{12}$ s^{-1} (17, 29, 30, 36, 37, 49).

As we discussed in the previous section, the IP produced by CT complex excitation will be the CIP without intervening solvent molecules between D^+ and A^- in the pair. The IP formed by the fluorescence-quenching reaction by diffusional encounter will be the so-called SSIP (solvent-separated IP) with solvent molecules intervening between D^+ and A^-. The much stronger electronic interaction between D^+ and A^- in the CIP seems to result in remarkably different k_{CR} values in the pyrene–TCNE system.

In view of these results, we have made a systematic study of the CR decay of the IP formed by excitation of various CT complexes and obtained the results indicated in Figure 8 (29, 30), together with the results for the IP formed by the fluorescence-quenching reaction. As an example, results of the PMDA (pyromellitic dianhydride) complex in acetonitrile, measured with the femtosecond laser photolysis method, are shown in Figure 9 (29, 30).

In Figure 8, the energy-gap dependence of the CR rate constant of the IP formed by the excitation of CT complexes with D and A systems of various strengths is demonstrated, together with that of the IP produced by the CS at encounter in the fluorescence-quenching reactions between similar D and A systems in the same solvent, acetonitrile. The $-\Delta G_{ip}{}^0$ values are evaluated in both cases by the same conventional method. The $-\Delta G_{ip}{}^0$ values for CIP

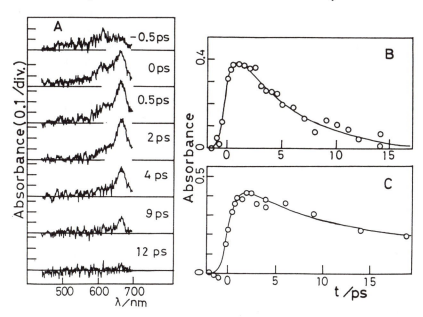

Figure 9. Femtosecond time-resolved absorption spectra of PMDA–HMB complex excited at 355 nm (A) in acetonitrile solution, and time profiles of absorbance at 665 nm observed for the PMDA–HMB (B) and PMDA–Naph (C) systems. (Reproduced from ref. 29. Copyright 1989 American Chemical Society.)

can be estimated empirically by using experimental results of the fluorescent CT complex of tetrachlorophthalic anhydride (TCPA) and hexamethylbenzene (50). By extrapolating the solvent-polarity effect on the fluorescence Stokes shift (50), the wave number of the CT fluorescence band peak $(\bar{\nu}_{max}{}^f)$ of the TCPA–hexamethylbenzene complex in acetonitrile, where this complex is practically nonfluorescent, can be obtained. From the wave number of the CT absorption band peak $(\bar{\nu}_{max}{}^a)$ and $\bar{\nu}_{max}{}^f$ value, we can estimate the sum $(\Delta\bar{\nu})$ of the FC destabilization energies in the excited and ground state by

$$\bar{\nu}_{max}{}^a - \bar{\nu}_{max}{}^f = \Delta\bar{\nu} \tag{6}$$

Assuming this $\Delta\bar{\nu}$ value to be typical of the CIP produced by exciting a relatively strong CT complex in acetonitrile, a rough approximation of the value of $-\Delta G_{ip}{}^0$ for the CIP state formed by exciting the nonfluorescent complex with the lowest CT absorption peak at $\bar{\nu}_{max}{}^a$ may be given by

$$-\Delta G_{ip}{}^0 \sim hc \left[\bar{\nu}_{max}{}^a - \frac{1}{2} \Delta\bar{\nu} \right] \tag{7}$$

The $-\Delta G_{ip}{}^0$ values estimated by eq 7 are shifted about 0.2 eV to the higher energy side as a whole, compared with the values obtained by the conventional method of using the oxidation potential of D and reduction potential of A in acetonitrile. Because the $-\Delta G_{ip}{}^0$ values shift as a whole, the functional form of the energy-gap dependence of k_{CR} may not be seriously affected by the method for evaluation of $-\Delta G_{ip}{}^0$.

The reaction scheme analogous to eq 5 in the case of the excitation of the CT complex may be given by eq 8 or 9.

$$^1(A^{-\delta'} \cdot D^{+\delta'})_S{}^{FC} \rightarrow {}^1(A^- \cdots D^+)_S \xrightarrow{k_{diss}} A_S{}^- + D_S{}^+$$

$$hv \qquad \qquad k_{CR}{}^{CIP}$$

$$^1(A^{-\delta} \cdot D^{+\delta})_S \tag{8}$$

$$^1(A^{-\delta'} \cdot D^{+\delta'})_S{}^{FC} \rightarrow {}^1(A^- \cdots D^+)_S \xrightarrow{k_{solv}} (A_S{}^- \cdots D_S{}^+) \xrightarrow{k_{diss}} A_S{}^- + D_S{}^+$$

$$hv \qquad \qquad k_{CR}{}^{CIP} \qquad k_{CR}{}^{SSIP}$$

$$^1(A^{-\delta} \cdot D^{+\delta})_S \tag{9}$$

As already discussed, the CIP formed by excitation of the CT complex in acetonitrile solution generally shows a quite different CR deactivation rate, compared with that of the SSIP of the same D,A pair produced by CS at diffusional encounter in the fluorescence-quenching reaction in the same solvent. This difference can be ascribed to the structural difference between these IPs.

In the course of the relaxation from the FC excited state of the complex to the IP and finally to dissociated ions competing with the CR deactivation from the IP state, an SSIP state similar to that formed by diffusional encounter in the fluorescence-quenching reaction may be produced from CIP as the precursor state of the ionic dissociation. However, there was practically no direct observation of the reaction scheme of eq 9, except our recent results of picosecond laser spectroscopic studies on TCNB–toluene, –benzene, and –xylenes in acetonitrile (28). Many other systems do not show such behavior, but the observed result can be reproduced well by the reaction scheme of eq 8 (28–30).

Moreover, in almost all cases examined in Figure 8, k_{CR} of CIP is much larger than its dissociation rate constant (k_{diss}), which leads to the negligible dissociation yield. In the PMDA complexes in Figure 9, $k_{CR} = 3.8 \times 10^{10}$ s^{-1} and $k_{diss} = 2 \times 10^{9}$ s^{-1} for the PMDA–naphthalene CIP in acetonitrile and $k_{CR} = 1.9 \times 10^{11}$ s^{-1} and $k_{diss} = 2 \times 10^{9}$ s^{-1} for the PMDA–HMB CIP in acetonitrile. The dissociation yield in the latter system is practically zero.

Although we assume a loose structure with intervening solvent molecules between A^{-} and D^{+} ions for the geminate IP formed by the fluorescence-quenching reaction in acetonitrile solution, this geminate IP will be quite different from the cation–electron pairs in nonpolar solvents that are frequently investigated in radiation chemistry. In such loose geminate pairs, they can undergo a wide range of thermal motions before CR and dissociation. The geminate SSIP formed by fluorescence-quenching reaction in acetonitrile solution will have a rather definite structure. The fact that the k_{CR} vs. $-\Delta G_{ip}^{0}$ relationship for those SSIPs shows a rather typical bell shape indicates strongly that the SSIPs of various D–A systems have a definite structure with similar interionic distance and similar solvent reorganization energies. This similarity suggests that the interaction between each ion in the IP and the surrounding polar solvents is strong and rather specific.

Energy-Gap Dependence of $k_{CR}{}^{CIP}$

Contrary to the bell-shaped energy-gap dependence of the k_{CR} of SSIP, the k_{CR} vs. $-\Delta G_{ip}^{0}$ relationship for CIP of similar D–A systems is quite different and can be given by

$$k_{CR}{}^{CIP} = \alpha \exp\left[-\gamma \left|\Delta G_{ip}^{0}\right|\right] \qquad (10)$$

where α and γ are constants independent of ΔG_{ip}^{0}. The energy-gap dependence of eq 10 is qualitatively analogous to that of the radiationless transition probability in the so-called "weak coupling" limit (51). In this respect, we examined the effect of deuteration on the k_{CR} by using perdeuterated toluene and benzene. We detected no effect of deuteration. It also seems difficult to give a reasonable interpretation for the observed γ value of eq 10, the slope in the plot of log $k_{CR}{}^{CIP}$ against $|\Delta G_{ip}{}^0|$, on the basis of the usual theory of radiationless transition. A new theoretical interpretation of this CR process is needed.

We give here a tentative interpretation of this $k_{CR}{}^{CIP}$ vs. $-\Delta G_{ip}{}^0$ relationship on the basis of the idea derived from our previous studies on exciplexes and excited CT complexes (1, 21, 52), as well as the recent studies on the ultrafast dynamics of the excited CT complexes and CIPs (12–15, 25–30). The electronic and geometrical structures of these strongly interacting D–A systems vary with the strengths of D and A, as well as the solvent polarity.

In the previous sections, we discussed the facts that, as the strength of D and A increases, the formation of CIP from the excited FC state of the CT complex becomes faster and the absorption spectrum of the transient CIP state becomes broader compared with that of dissociated ions and of the SSIP in acetonitrile solutions. This result indicates that the extent of the electronic and geometrical structure change, including surrounding solvent in the CIP formation process, is smaller in the stronger CT complexes. In other words, the stronger the complex, the closer is the position of the potential minimum of the CIP state on the reaction coordinate including the geometrical configuration of D and A, as well as the extent of solvation to that of the CT complex itself, as shown in Figure 10.

Figure 10 shows that observation of the normal region is difficult in the energy-gap dependence of k_{CR} for such a small horizontal shift of the potential minimum of the CIP state relative to the ground state. With a much larger horizontal shift of the potential minimum of the IP state along the reaction coordinate, observation of the normal region will become possible at small $-\Delta G_{ip}{}^0$ values, as indicated in Figure 11. The large horizontal shift represents a large structural change of the CIP, including the solvation state. The shift corresponds to the formation of the SSIP state for such D–A systems with small $-\Delta G_{ip}{}^0$ values as pyrene–TCNE and perylene–TCNE in acetonitrile solutions. Actually, we have observed the relatively small k_{CR} values of the SSIP of these systems formed by diffusional encounter in the fluorescence-quenching reaction in the normal region; the k_{CR} values of CIP of the same system are extremely large.

This interpretation of the CR mechanisms from the CIP and SSIP states is somewhat analogous to the radiationless transition mechanisms in the weak coupling and strong coupling limit (51), respectively. However, it seems to be difficult to interpret quantitatively the energy-gap dependence of k_{CR} covering a wide energy-gap range and with a very mild slope in the plot against $-\Delta G_{ip}{}^0$, in terms of the mechanism of a "weak coupling" limit. In spite of this it seems possible, at least qualitatively, to interpret the energy-gap dependence of k_{CR} by taking into consideration the change of the structure of CIP as it depends on the strength of D and A (i.e., depending on the $-\Delta G_{ip}{}^0$ value, as demonstrated in Figure 10). The analogy between the

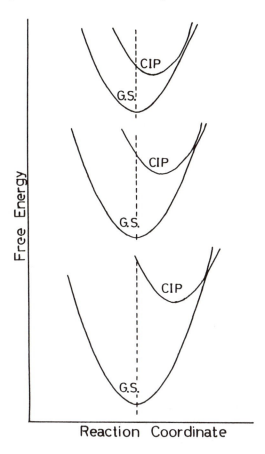

Figure 10. Free-energy curves for the IP states and the ground state (GS) of D–A systems vs. reaction coordinate. Change of the position of the potential minimum of the CIP depends on the change of the $-\Delta G_{ip}^0$ value, a result illustrating that the CR reaction of CIP is in the inverted region for all $-\Delta G_{ip}^0$ values. (Reproduced from ref. 30. Copyright 1991 American Chemical Society.)

theories of radiationless transition and of electron-transfer reaction has been recognized for a long time. The result shown in Figure 8 may be an experimental counterpart for this analogy, illustrating the two cases in electron transfer qualitatively corresponding to the "weak coupling" and "strong coupling" cases in radiationless transition.

Concluding Remarks

We have discussed the mechanisms of photoinduced CS and CR of the produced IP or CT state on the basis of our results obtained by femtosecond–picosecond laser photolysis studies, mainly on the strongly inter-

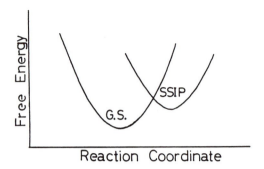

Figure 11. Relationship between the free-energy curves between the ground state and the SSIP state corresponding to the small $-\Delta G_{ip}^{0}$ value, where the large horizontal shift of the potential minimum of SSIP against ground state brings the CR rate constant of SSIP to the normal region. (Reproduced from ref. 30. Copyright 1991 American Chemical Society.)

acting D–A systems combined by spacers or directly by single-bond and CT complexes. The most important conclusions are as follows:

1. Photoinduced CS processes in such strongly interacting D–A systems cannot be described by the simple two-state model, $(D^* \cdots A$ or $D \cdots A^*) \rightarrow D^+ \cdots A^-$. This model assumes that weak interaction is responsible for the electron transfer, as is the case in the conventional electron-transfer theories. In such systems, the CS proceeds by gradual change of electronic structure due to the extensive solvation of the D–A system, accompanied by some change of its geometrical structure, which decreases the electronic delocalization interaction between D and A that facilitates the CS. Such behavior in the CS process of the intramolecular exciplexes and CT complexes with D and A interacting strongly is in accordance with the idea I proposed many years ago that the electronic and geometrical structures of exciplexes and excited CT complexes vary with the strengths of D and A and with the solvent polarity (21, 52). That is, as the oxidation potential of D decreases and the reduction potential of A and the solvent polarity increase, the electronic delocalization interaction between D and A decreases and the structure of the exciplex comes close to that of the ion pair.

2. Direct observation of the CIP formation process from the excited FC state of the CT complex in acetonitrile solution has revealed that the extent of the electronic and geometrical structural changes, including the surrounding solvent configurations in the course of CIP formation, is smaller for D with

the smaller oxidation potential and A with the higher reduction potential. This relationship also means that the extent of the geometrical structural change necessary to decrease the delocalization interaction in the course of CIP formation is smaller in the case of stronger D and A, in agreement with the reasoning given in paragraph 1.

3. The CIP formed by excitation of the CT complex in acetonitrile solution undergoes CR deactivation. The CR rate constant shows a peculiar energy-gap dependence [i.e., a monotonous (exponential) increase of the CR rate with decrease of the free-energy gap $-\Delta G_{ip}^{0}$ between the CIP and the ground state]. This dependence contrasts with the CR rate of SSIP formed by CS at diffusional encounter between fluorescer and quencher, which shows a bell-shaped energy-gap dependence. The peculiar energy-gap dependence of the CR rate of CIP has been interpreted on the basis of the reasoning given in paragraph 2. In this view, the structural changes following the excitation of the CT complex affect the position of the potential minimum of the CIP state on the reaction coordinate, which involves the D,A configuration and solvation. Thus the CR rate of SSIP comes closer to that of the CT complex itself with increase of the strength of D and A, which makes it difficult to observe the normal region in the energy-gap dependence of the CR rate.

4. The similarity between the energy-gap dependence of the CR rate of the CIP and that of the radiationless transition probability in the so-called "weak coupling" limit has been pointed out.

Acknowledgments

The studies discussed in this chapter were supported by a Grant-in-Aid (No. 6265006) from the Japanese Ministry of Education, Science, and Culture and were made in collaboration with following individuals, whose contribution I appreciate: S. Nishikawa, S. Ojima, T. Asahi, H. Miyasaka, and T. Okada.

References

1. Mataga, N.; Ottolenghi, M. In *Molecular Association*; Foster, R., Ed.; Academic: New York, 1979; Vol. 2, p 1.
2. Marcus, R. A.; Sutin, N. *Biochim. Biophys. Acta* **1985**, *811*, 265.
3. Rips, I.; Klafter, J.; Jortner, J. In *Photochemical Energy Conversion*; Norris, J. R.; Meisel, D., Eds.; Elsevier: New York, 1988; p 1.

4. Mataga, N. In *Photochemical Energy Conversion*; Norris, J. R.; Meisel, D., Eds.; Elsevier: New York, 1988; p 32.
5. Maroncelli, M.; MacInnis, J.; Fleming, G. R. *Science (Washington, D.C.)* **1989**, *243*, 1674.
6. Sumi, H.; Marcus, R. A. *J. Chem. Phys.* **1986**, *84*, 4894.
7. Nadler, W.; Marcus, R. A. *J. Chem. Phys.* **1987**, *86*, 3906.
8. Mataga, N.; Kaifu, Y.; Koizumi, M. *Bull. Chem. Soc. Jpn.* **1955**, *28*, 690; *Bull. Chem. Soc. Jpn.* **1956**, *29*, 465.
9. Lippert, E. Z. *Naturforsch.* **1955**, *10a*, 541; *Ber. Bunsen-Ges. Phys. Chem.* **1957**, *61*, 962.
10. Bagchi, B.; Oxtoby, D. W.; Fleming, G. *Chem. Phys.* **1984**, *86*, 257.
11. Mataga, N. *Pure Appl. Chem.* **1984**, *56*, 1255.
12. Mataga, N.; Miyasaka, H.; Asahi, T.; Ojima, S.; Okada, T. *Ultrafast Phenomena VI*; Springer Verlag: Berlin, 1988; p 511.
13. Mataga, N.; Nishikawa, S.; Asahi, T.; Okada, T. *J. Phys. Chem.* **1990**, *94*, 1443.
14. Mataga, N.; Yao, H.; Okada, T.; Rettig, W. *J. Phys. Chem.* **1989**, *93*, 3383.
15. Yao, H.; Okada, T.; Mataga, N. *J. Phys. Chem.* **1989**, *93*, 7388.
16. Mataga, N.; Yao, H.; Okada, T. *Tetrahedron* **1989**, *45*, 4683.
17. Mataga, N. *Acta Phys. Pol. A* **1987**, *71*, 767.
18. Okada, T.; Mataga, N.; Baumann, W.; Siemiarczuk, A. *J. Phys. Chem.* **1987**, *71*, 4490.
19. Nakashima, N.; Murakawa, M.; Mataga, N. *Bull. Chem. Soc. Jpn.* **1976**, *49*, 854.
20. Hayashi, T.; Suzuki, T.; Mataga, N.; Sakata, Y.; Misumi, S. *J. Phys. Chem.* **1977**, *81*, 420.
21. Mataga, N.; Murata, Y. *J. Am. Chem. Soc.* **1969**, *91*, 3144.
22. Kobayashi, T.; Yoshihara, K.; Nagakura, S. *Bull. Chem. Soc. Jpn.* **1971**, *44*, 2603.
23. Egawa, K.; Nakashima, N.; Mataga, N.; Yamanaka, C. *Bull. Chem. Soc. Jpn.* **1971**, *44*, 3287.
24. Masuhara, H.; Mataga, N. Z. *Phys. Chem. (Munich)* **1972**, *80*, 113.
25. Miyasaka, H.; Ojima, S.; Mataga, N. *J. Phys. Chem.* **1989**, *93*, 3380.
26. Ojima, S.; Miyasaka, H.; Mataga, N. *J. Phys. Chem.* **1990**, *94*, 4147.
27. Ojima, S.; Miyasaka, H.; Mataga, N. *J. Phys. Chem.* **1990**, *94*, 5834.
28. Ojima, S.; Miyasaka, H.; Mataga, N. *J. Phys. Chem.* **1990**, *94*, 7534.
29. Asahi, T.; Mataga, N. *J. Phys. Chem.* **1989**, *93*, 6575.
30. Asahi, T.; Mataga, N. *J. Phys. Chem.*, in press.
31. Deisenhofer, J.; Michel, H. *Science (Washington, D.C.)* **1989**, *245*, 1463 and references therein.
32. Wasielewski, M. R.; Niemczyk, M. P.; Svec, W. A.; Pewitt, E. B. *J. Am. Chem. Soc.* **1985**, *107*, 1080.
33. Kakitani, T.; Mataga, N. *J. Phys. Chem.* **1986**, *90*, 993.
34. Harrison, R. J.; Pearce, B.; Beddard, G. S.; Cowan, J. A.; Sanders, J. K. M. *Chem. Phys.* **1987**, *116*, 429.
35. Gould, I. R.; Ege, D.; Mattes, S. L.; Farid, S. *J. Am. Chem. Soc.* **1987**, *109*, 3794.
36. Mataga, N.; Kanda, Y.; Okada, T. *J. Phys. Chem.* **1986**, *90*, 3880.
37. Mataga, N.; Asahi, T.; Kanda, Y.; Okada, T.; Kakitani, T. *Chem. Phys.* **1988**, *127*, 249.
38. Kahlow, M. A.; Kang, T. J.; Barbara, P. F. *J. Phys. Chem.* **1987**, *91*, 6452.
39. Laermer, F.; Elsaesser, T.; Kaiser, W. *Chem. Phys. Lett.* **1989**, *156*, 381.
40. Kosower, E. M.; Huppert, D. *Annu. Rev. Phys. Chem.* **1986**, *37*, 127.
41. Kakitani, T.; Mataga, N. *Chem. Phys.* **1985**, *93*, 381.
42. Kakitani, T.; Mataga, N. *J. Phys. Chem.* **1985**, *85*, 8.

43. Kakitani, T.; Mataga, N. *J. Phys. Chem.* **1987**, *91*, 6277.
44. Yoshimori, A.; Kakitani, T.; Enomoto, Y.; Mataga, N. *J. Phys. Chem.* **1989**, *93*, 8316.
45. Kakitani, T.; Yoshimori, A.; Mataga, N. In *Electron Transfer in Inorganic, Organic, and Biological Systems*; Bolton, J. R.; Mataga, N.; McLendon, G., Eds.; American Chemical Society: Washington, DC, 1991; Chapter 4.
46. Nishikawa, S.; Asahi, T.; Hagihara, M.; Kanaji, K.; Okada, T.; Mataga, N.; Kakitani, T., unpublished results.
47. Barbara, P. F.; Kang, T. J.; Jarzeba, W.; Fonseca, T. In *Perspectives in Photosynthesis*; Jortner, J.; Pullman, B., Eds.; Kluwer-Academic: Dordrecht, Netherlands, 1990; pp 273–292.
48. Nagakura, S. In *Excited State*; Lim, E. C., Ed.; Academic: New York, 1975; Vol. 2, p 322.
49. Mataga, N.; Kanda, Y.; Asahi, T.; Miyasaka, H.; Okada, T.; Kakitani, T. *Chem. Phys.* **1988**, *127*, 239.
50. Czekalla, J.; Meyer, K.-O. *Z. Phys. Chem. (Munich)* **1961**, *27*, 184.
51. Englman, R.; Jortner, J. *Mol. Phys.* **1970**, *18*, 145.
52. Mataga, N.; Okada, T.; Yamamoto, N. *Chem. Phys. Lett.* **1967**, *1*, 119.

RECEIVED for review April 27, 1990. ACCEPTED revised manuscript August 17, 1990.

Solvent, Temperature, and Bridge Dependence of Photoinduced Intramolecular Electron Transfer

James R. Bolton[1], John A. Schmidt[1], Te-Fu Ho[1], Jing-yao Liu[1], Kenneth J. Roach[1], Alan C. Weedon[1], Mary D. Archer[2], Jacquin H. Wilford[2], and Victor P. Y. Gadzekpo[2]

[1]Photochemistry Unit, Department of Chemistry, University of Western Ontario, London, Ontario N6A 5B7, Canada
[2]Department of Chemistry, University of Cambridge, Lensfield Road, Cambridge CB2 1EP, United Kingdom

Photoinduced intramolecular electron-transfer rate constants were determined for several PLQ (tetraarylporphine linked to p-benzoquinone) molecules. Rate constants for PAQ (porphyrin–amide–quinone) vary significantly with solvent and temperature. Most results can be explained within the context of the high-temperature-limit semiclassical Marcus equation. Analysis of the temperature-dependent data reveals that the electronic coupling energy H_{rp} varies significantly with solvent. This variability explains the considerable scatter found in the solvent-dependent studies. Electron-transfer rate constants, determined for five other PLQ molecules, exhibit the following characteristics: (1) solvent dependence is broadly similar to that of PAQ; (2) peptide linkages are much more effective than a saturated linkage, such as bicyclobutane; (3) a phenyl linkage is the most effective, generating rate constants 100–1000 times that of a bicyclooctane linkage; and (4) the strained cyclobutane bridge is more effective than a corresponding unstrained saturated linkage.

THE DESIGN OF COVALENTLY LINKED donor–acceptor molecules to mimic the primary electron-transfer process in photosynthesis has received increasing interest over the past decade; ref. 1 is a comprehensive review of

0065–2393/91/0228–0117$06.00/0

this field up to 1988. The primary objective has been to design donor–acceptor (D–A) molecules in which the forward photoinduced electron-
transfer (PET) process is very rapid while the reverse electron-transfer (ET)
rate back to the original state is very slow. To the extent that this reverse
reaction is important, the overall efficiency and quantum yield of any energy-
storing process will be reduced.

These model compounds have provided an excellent "laboratory" to
study the PET mechanism and the dependence of ET rate constants on
various factors in the molecular structure and environment. The Marcus
theory of electron transfer (see Chapter 2 for an exposition) has proven to
be an excellent framework for the interpretation of the ET data. The dependence factors include

1. Excited-state energy. Generally, the higher the excited-state
   energy, the faster the rate constant will be.

2. Exergonicity ($-\Delta G^0$). As predicted from Marcus theory, ET
   rate constants increase with increasing exergonicity up to a
   maximum where $-\Delta G^0 = \lambda$ (the reorganization energy). The
   ET rate constants then decrease for larger exergonicities in
   the so-called "Marcus inverted region".

3. Distance between donor D and acceptor A. ET rate constants
   have been found to fit well to an exponential dependence on
   the edge-to-edge distance between D and A, with the rates
   decreasing about $1/e$ for every ~1-Å increase in distance.

4. Orientation of D with respect to A. An orientation effect has
   been found in some rigid model compounds, but not enough
   is known yet to provide a full understanding of this effect.

5. Nature of the linkage. The molecular structure of the linkage
   between D and A has been found to play a very important
   role in mediating ET from D to A. In most cases ET occurs
   through the bonds of the bridge and not through the surrounding medium. Often ET appears to be mediated by a
   superexchange mechanism involving the antibonding orbitals
   of the bridge. Aromatic and unsaturated bridges are therefore
   expected to be more effective than saturated bridges, although
   there are some surprises.

6. Solvent. The surrounding solvent changes two parameters that
   can markedly alter ET rates. First, the exergonicity can be
   altered through different solvation of the product ion pair
   D^+-A^-. Second, the external contribution to the reorgani-
   zation energy λ_{out} is altered through changes in ϵ_{op} and ϵ_s, the
   optical and static dielectric constants (see eq 10, Chapter 2).

7. Temperature. As with any rate constant, ET rate constants usually exhibit an Arrhenius temperature dependence, but the activation energies are usually small.

During the past several years our groups have been studying the intramolecular PET rates in a series of PLQ (tetraarylporphine linked to *p*-benzoquinone) molecules. Chart 1 illustrates the structures of the various PLQ

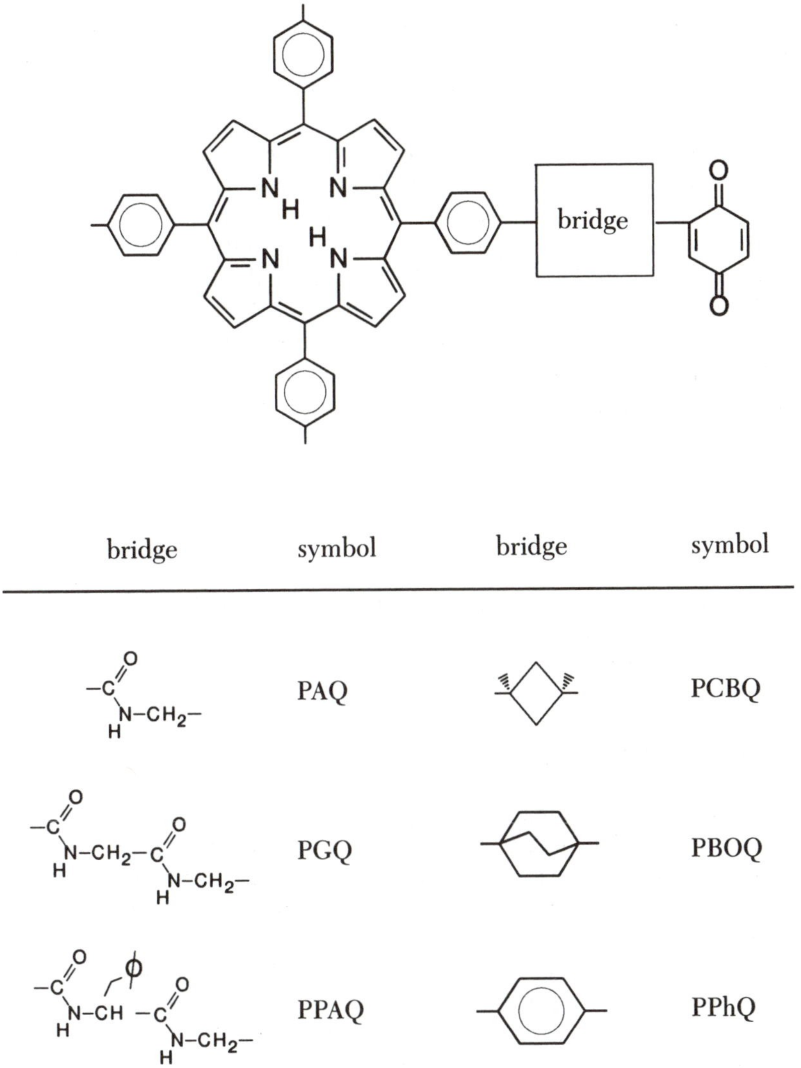

bridge	symbol	bridge	symbol
	PAQ		PCBQ
	PGQ		PBOQ
	PPAQ		PPhQ

Chart 1. Structures of porphyrin–quinone molecules containing various bridges.

molecules. This chapter is intended to review this work, and in so doing we shall attempt to come to some conclusions regarding the influence of the factors listed, particularly factors 5–7. The data for this review come from several published papers (2–8) and unpublished work (9–11).

Fluorescence Lifetimes

The compound that we have studied most is the porphyrin–amide–quinone molecule PAQ (*see* Chart 1 for structure). Although the bridge in this molecule has some flexibility, molecular modeling (6) has shown that the center-to-center distance varies only from 12.7 Å in the most compact structure to 14.3 Å in the most extended structure. The fluorescence decay, as determined from the time-correlated single photon counting technique (2, 7, 8), is biexponential with a short and a long component. The long component is of low amplitude (<5%) and almost certainly arises from a residual fraction of the corresponding hydroquinone compound PAQH$_2$. The short component is assumed to be shortened by the intramolecular ET process. Provided that the internal photophysical rate constants of the excited porphyrin (k_f for fluorescence, k_{ic} for internal conversion, and k_{isc} for intersystem crossing) are the same for PAQ and PAQH$_2$, the rate constant k_{ET}^S for PET from the porphyrin excited singlet state to the quinone is given by (12)

$$k_{ET}^S = \frac{1}{\tau_1} - \frac{1}{\tau_2} \tag{1}$$

where τ_1 and τ_2 are the fluorescence lifetimes of PAQ and PAQH$_2$. The error in τ_1 and τ_2 is about 0.1 ns, which produces an error in the k_{ET}^S values that varies from 3 to 10%.

Energy-Level Diagram

In addition to determining the PET rate constant in PAQ, we have determined (6) most of the energies and rate constants in the energy-level diagram of PAQ in benzonitrile. Figure 1 illustrates the various pathways and corresponding rate constants in that solvent. The energy of the ^{1}P*AQ excited state is 1.90 eV, as determined from the overlap of the normalized absorption and fluorescence spectra. The energy of the ^{3}P*AQ triplet state was determined to be 1.43 eV at 77 K (13). Finally, the energy of the charge-separated radical-ion pair (P$^+$AQ$^-$) was estimated as 1.41 ± 0.05 eV by using redox potentials obtained from differential pulse voltammetry (5). This estimation includes a small correction (–0.04 eV) for coulombic stabilization of the radical-ion pair relative to the separated ions.

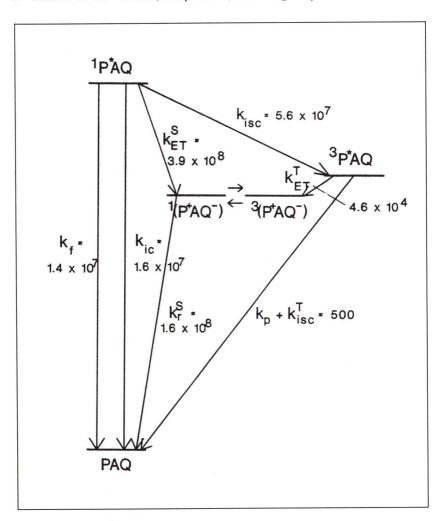

Figure 1. *Energy-level diagram and rate constants (s^{-1}) for PAQ in benzonitrile. The ET rate constants are $k_{ET}{}^S$ from the porphyrin excited singlet state to the singlet radical ion pair $^1(P^+AQ^-)$; $k_{ET}{}^T$ from the porphyrin excited triplet state to the triplet radical ion pair $^3(P^+AQ^-)$; and $k_r{}^s$, the reverse ET rate constant from the $^1(P^+AQ^-)$ state to the ground state. The other rate constants are defined in the text.*

Solvent Dependence

PET rate constants for PAQ were measured in 17 solvents, in which the redox potentials were also determined by differential pulse voltammetry (*4*, *5*, *7*). The data can be analyzed in terms of the high-temperature limit form of the semiclassical Marcus equation. [This equation is valid only in the normal Marcus region; a more complex equation must be used in the inverted region (14).]

$$k_{ET} = \frac{2\pi}{\hbar} \frac{H_{rp}^{2}}{(4\pi\lambda k_B T)^{1/2}} \exp\left[-\frac{(\Delta G^0 + \lambda)^2}{4\lambda k_B T}\right] \tag{2}$$

In this equation $\hbar$ is the Planck constant divided by 2π; H_{rp} is the electronic coupling energy (see eq 12, Chapter 2) between reactant $^1P^*AQ$ and product P^+AQ^-; λ is the reorganization energy; k_B is the Boltzmann constant; and ΔG^0 is the Gibbs energy of reaction. For our analysis of $k_{ET}{}^S$ for PAQ, we reorganized eq 2 into the following linearized form

$$\ln[k_{ET}{}^S\lambda^{1/2}] = C_1 - \frac{\Delta G^*}{k_B T} \tag{3a}$$

where

$$C_1 = \ln\left[\frac{2\pi}{\hbar} \frac{H_{rp}^{2}}{(4\pi k_B T)^{1/2}}\right] \tag{3b}$$

and the Marcus Gibbs energy of activation is

$$\Delta G^* = \frac{(\Delta G^0 + \lambda)^2}{4\lambda} \tag{3c}$$

The reorganization energy λ has two contributions

$$\lambda = \lambda_{in} + \lambda_{out} \tag{4a}$$

where λ_{in} is the contribution to λ from structural changes in the PAQ molecule itself and λ_{out} is the contribution from the surrounding solvent. The solvent dependence of λ_{out} can best be expressed by the relation

$$\lambda_{out} = B\left[\frac{1}{\epsilon_{op}} - \frac{1}{\epsilon_s}\right] \tag{4b}$$

where B is a parameter whose value depends on the model chosen and the molecular dimensions (see eqs 10 and 11, Chapter 2). Although B can be calculated from a specific model (see eq 10, Chapter 2), these values are not very realistic. Thus, to express the solvent dependence of λ_{out}, we treat B as an empirical parameter. From our previous PAQ analysis (7), we found that a value of $B = 1.8$ eV gave reasonable results. The exergonicity $(-\Delta G^0)$ varies with solvent from 0.44 eV in acetone to 0.77 eV in chloroform. (These values were corrected by using the coulombic attraction term $-e^2/(4\pi\epsilon_0\epsilon_s r_{DA})$, where the center-to-center distance $r_{DA} \sim 13.5$ Å; *see* ref. 7 for details.) The data for the analysis are given in Table I.

Figure 2 shows a plot of $\ln[k_{ET}{}^S\lambda^{1/2}]$ vs. $\Delta G^*/k_BT$ for the 17 solvents studied. The solid line is constrained to have a slope of -1 to indicate the expected Marcus behavior. There is considerable scatter in the data; however, they conform reasonably well to the Marcus predicted behavior. In drawing Figure 2, H_{rp} was assumed to be solvent-independent. We shall see later in this chapter that this assumption will have to be modified.

The y intercept of a least-squares fit of the data in Figure 2 occurs at ca. -0.5. From this value of C_1 we can calculate that $H_{rp} \sim 0.00030$ eV. This result clearly places the PET rate constants for PAQ in the nonadiabatic region where eq 2 is expected to hold.

Temperature Dependence

The temperature dependence of PET rate constants reveals further information concerning the effect of solvent on the ET rates. The PET rate

Table I. Rate Constants, Reorganization Energies, and Gibbs Energies for Electron Transfer in PAQ

Solvent	Label in Fig. 2	$k_{ET}{}^S$ $(10^7\,s^{-1})$	ϵ_{op}	ϵ_s	λ^a (eV)	$-\Delta G^{0\,b}$ (eV)
Acetonitrile	1	4.8	1.800	35.94	1.15	0.52
Propionitrile	2	5.6	1.910	24.83	1.07	0.54
Benzonitrile	3	39	2.328	25.20	0.90	0.49
Acetone	4	2.9	1.839	20.56	1.09	0.44
1-Butanol	5	24	1.953	17.51	1.02	0.54
1,2-Dichloroethane	6	56	2.080	10.37	0.89	0.67
Methylene chloride	7	80	2.020	8.93	0.89	0.61
2-Methyltetrahydrofuran	8	2.3	1.974	7.60	0.87	0.62
1,1,1-Trichloroethane	9	49	2.062	7.25	0.83	0.57
1,2-Dimethoxyethane	10	2.0	1.899	7.20	0.90	0.47
Ethyl acetate	11	2.2	1.876	6.02	0.86	0.53
Ethyl ether	12	1.39	1.842	4.38	0.77	0.53[c]
Chlorobenzene	13	82	2.316	5.62	0.66	0.57
α-Chloronaphthalene	14	74	2.667	5.04	0.52	0.63
Chloroform	15	227	2.082	4.81	0.69	0.77
1,2-Dibromoethane	16	126	2.369	4.78	0.58	0.70
Anisole	17	33	2.293	4.33	0.57	0.57

NOTE: Data measured at 295 K were taken from ref. 7.
[a]Calculated from eq 4 with $B = 1.8$ eV and $\lambda_{in} = 0.2$ eV.
[b]Corrected for the work of bringing the product ions together by using the coulombic term $-e^2/(4\pi\epsilon_0\epsilon_s r_{DA})$; *see* ref. 7. This correction varies from -0.03 eV in acetonitrile to -0.25 eV in anisole.
[c]Not measured; ΔG^0 was taken to be the same as in ethyl acetate because the two solvents have similar k_{ET}, ϵ_s, and ϵ_{op}.

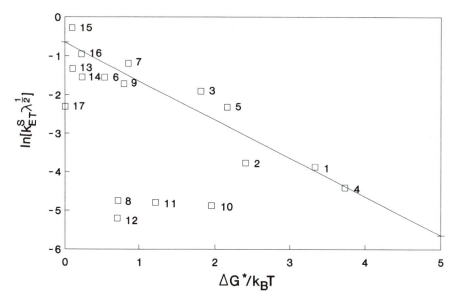

Figure 2. ln[k$_{ET}$^{S}λ$^{1/2}$] vs. ΔG/k$_B$T for PAQ in 17 different solvents at 295 K. The line is a least-squares fit to all the data, with the slope constrained to be –1.00. The solvent key is given in Table I.*

constants of PAQ were measured (10) as a function of temperature in five of the solvents included in the preceding section. These data can be analyzed by using equations similar to eq 3, namely

$$\ln [k_{ET}{}^S(\lambda T)^{1/2}] = C_2 - \frac{\Delta G^*}{k_B T} \tag{5a}$$

where

$$C_2 = \ln \left[\frac{2\pi}{\hbar} \frac{H_{rp}{}^2}{(4\pi k_B)^{1/2}} \right] \tag{5b}$$

In our initial analysis we assumed that ΔG^0 is independent of temperature. (This assumption is arbitrary; however, Gunner et al. (15) found that ΔG^0 is independent of temperature over a wide range in photosynthetic reaction centers.) The temperature dependence of λ_{out} may be accounted for through the use of literature values of ϵ_{op} and ϵ_s at each temperature.

If B in eq 4b is taken to be a constant at 1.8 eV [the value used in our previous analysis (7)] in all solvents, plots of ln $[k_{ET}{}^S(\lambda T)^{1/2}]$ vs. $\Delta G^*/k_B T$ are linear, but the slopes differ significantly from –1. We must conclude either that eqs 5a–5b are not valid or that ΔG^0 varies with temperature. We chose the latter assumption and assumed a corrective linear temperature coefficient for ΔG^0 [i.e., $\Delta G^0(T) = \Delta G^0(298) + K(T - 298)$; K is then varied until the slope is exactly -1.00; K is an arbitrary coefficient in units of reciprocal kelvins]. Fortunately, the ΔG^0 values required by this approach do not have to vary a great deal; over the temperature range studied, the required change is ≤ 0.10 eV. ΔG^0 is required to be less negative as the temperature decreases. This direction is expected if the only temperature effect on ΔG^0 is the change in the coulombic attraction term $[-e^2/(4\pi\epsilon_o\epsilon_s r_{DA})]$ arising from the change in ϵ_s with temperature.

Figure 3 shows plots of $\ln[k_{ET}{}^S(\lambda T)^{1/2}]$ vs. $\Delta G^*/k_B T$ for ethyl acetate and for acetonitrile, adjusted as described so that the slopes are -1. Table II summarizes the parameters obtained from such plots for all five solvents.

The electronic coupling energy H_{rp} can be obtained from the intercept C_2 (eq 5b); the results are given in Table II. ΔG^0 might appear to be adjusted somewhat arbitrarily to obtain the required slope of -1; however, simulation indicates that these adjustments decrease all H_{rp} by 40% at most. Thus, the more substantial variations in H_{rp} shown in Table II must be real.

Figure 2 shows that those solvents (ethyl ether, ethyl acetate, and 2-methyltetrahydrofuran) in which small values of H_{rp} are found also exhibit low points. This explanation for the scatter in Figure 2 can be tested quantitatively by a further rearrangement of the logarithmic form of eq 2, namely

$$\ln [k_{ET}{}^S\lambda^{1/2}H_{rp}{}^{-2}] = C_3 - \frac{\Delta G^*}{k_B T} \tag{6a}$$

where

$$C_3 = \ln \left[\frac{2\pi}{\hbar} \frac{1}{(4\pi k_B T)^{1/2}} \right] \tag{6b}$$

The plot of eq 6a for the five solvents in Table II is given in Figure 4. The slope (-1.10) is very close to the expected -1.00, and the intercept (102.7) is also very close to C_3 computed from the fundamental constants (102.3). Thus we conclude that the scatter in Figure 2 arises largely from solvent dependence of H_{rp}. This result is rather remarkable because it has been assumed in all other studies to date, including our own (7), that H_{rp} is a property only of the PLQ molecule itself. However, there are at least two possible explanations for why H_{rp} for a given molecule might vary with solvent.

1. Because the definition of H_{rp} (*see* eq 12, Chapter 2) involves an overlap of the wave functions of the reactant and product states, solvent sensitivity of H_{rp} could arise from a change in

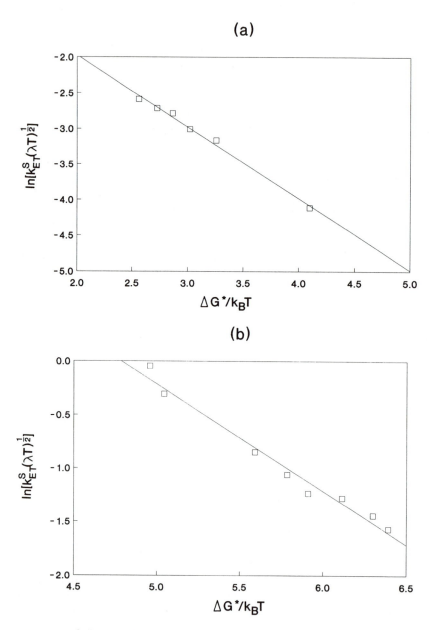

Figure 3. $ln[k_{ET}{}^S(\lambda T)^{1/2}]$ vs. $\Delta G^*/k_BT$ for PAQ in (a) ethyl acetate and (b) acetonitrile. The lines are least-squares fits to the data with the slope constrained to be -1.00 by assuming a small temperature dependence in ΔG^0.

Table II. Marcus Analysis of the Temperature Dependence of the PET Rate Constants of PAQ in Five Solvents

Solvent	Temperature Range (K)	$d\Delta G^0/dT^a$ $(10^{-3}\ eV\ K^{-1})$	Intercept, C_2	r^b	$H_{rp}^{\ c}$ $(10^{-3}\ eV)$
Ethyl ether	185–293	−1.060	−2.00	0.996	0.034
Ethyl acetate	200–293	−0.624	−1.40	0.997	0.046
Acetonitrile	245–342	−0.555	2.55	0.988	0.332
2-Methyltetrahydrofuran	230–293	−0.934	−1.48	0.914	0.044
Acetone	250–290	−1.307	1.96	0.942	0.248

[a]Chosen so that the slopes of plots such as Figure 3 are exactly −1.00.
[b]Absolute value of the correlation coefficient.
[c]From eq 5b.

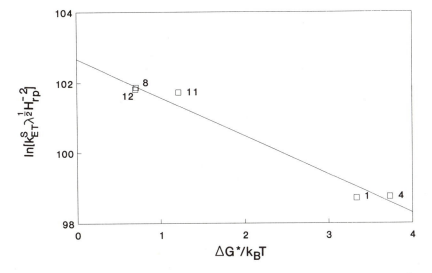

Figure 4. ln[$k_{ET}{}^S\lambda^{1/2}H_{rp}{}^{-2}$] vs. $\Delta G^*/k_BT$ for PAQ. The line is an unconstrained least-squares fit to the data. The solvent labels are given in Table I.

the average conformation of the PAQ molecule from one solvent to another. For example, in solvents with high dielectric constants such as acetone and acetonitrile, the PAQ molecule might adopt a more folded average conformation, which would bring the porphyrin and quinone units closer together and thus enhance H_{rp}. In solvents with low dielectric constants, such as ethyl acetate and 2-methyltetrahydrofuran, PAQ might adopt a more extended average conformation. Other solvent properties, such as the size and shape of the solvent molecules, may also play important roles in determining the value of H_{rp}.

2. There may be a "through-solvent" pathway for ET in some conformations that would enhance H_{rp} in certain solvents.

The analysis of the temperature dependence of the PAQ rate constants given here is preliminary (a full treatment will be given in ref. 10). However, future work should take account of the possible solvent dependence of H_{rp}. It would also be highly desirable to carry out a similar solvent and temperature study on a molecule with a rigid bridge, for which the complication of varying conformation is eliminated.

Bridge Dependence

We measured the PET rate constants of a number of D–A molecules in which the donor (tetraarylporphine) and acceptor (p-benzoquinone) are kept constant while the linkage is varied. The results are summarized in Table III. All compounds, except PPhQ, exhibit broadly similar solvent dependence. In terms of eq 2, this result is to be expected if, for the various compounds, ΔG^0 and λ_{out} have solvent dependence broadly similar to that of PAQ and there are no significant specific solvent effects. For PPhQ, the rate constants are so large that the system is approaching the adiabatic region ($H_{rp} \geq 0.025$ eV), where eq 2 is no longer valid.

A comparison of PAQ with PGQ demonstrates that the ET rates are not strongly attenuated by increasing the length of the chain in this peptide linkage. Even though PGQ has three more atoms in its peptide chain than

Table III. Bridge Dependence of Porphyrin–Quinone Rate Constants in Various Solvents

Solvent	PAQ[a]	PGQ[b]	PPAQ[c]	PCBQ[d]	PBOQ[e]	PPhQ[f]
1,2-Dimethoxyethane	0.20	0.15	0.29			38
Ethyl acetate	0.22	0.16	0.68			37
2-Methyltetrahydrofuran	0.23	0.11	0.33		0.016	33
Acetone	0.29	0.08	0.89		0.024	48
Acetonitrile	0.48	0.20	0.99	1.0	0.000	57
Propionitrile	0.56	0.18			0.330	
Anisole	3.30	0.71	1.30			81
1-Butanol	2.39	1.20		8.4		
Benzonitrile	3.90	1.2	1.2	8.4		226
Methylene chloride	8.00	2.10	3.00	8.4	0.150	27
Chloroform	24.0	3.90	3.90	27.0		12

NOTE: All values are k_{ET}^S (10^8 s^{-1}) measured at 295 K.
[a] From ref. 2.
[b] From ref. 8.
[c] From ref. 10.
[d] From ref. 11.
[e] From ref. 3.
[f] From ref. 9.

PAQ, the rate constants decrease by only a factor of ca. 2–5. The replacement of a hydrogen atom in the central methylene group of the peptide linkage in PGQ by the phenyl group in PPAQ enhances the rate constant by a factor of ca. 1–4. This enhancement could be due to the interposition of an aromatic group in the bridge, but it could also be due to a change in the conformation of the bridge.

Isied and co-workers (*16–18*) studied ground-state ET in a series of Ru–Co binuclear complexes, in which the two metal centers are linked via amino acids, involving three atoms in the linkage, or by peptide groups with two or more linked amino acids. They found that k_{ET} drops by a factor of $\sim 10^2$ for each additional proline in the peptide bridge. For a three-atom increase in the linkage length, this factor is considerably larger than in our case.

In contrast, Schanze and Sauer (*19*), in a study of PET in a series of polypyridyl Ru(II) complexes linked to *p*-benzoquinone by proline peptide bridges ($n = 0$–4), found that each additional proline decreased the rate constant by ~ 10. Cabana and Schanze (*20*) found similar results in a study of PET in a series of polypyridyl Re(I) complexes linked to 4-(*N,N*-dimethylamino)benzoate by proline peptide bridges ($n = 0$–2). The results of these latter two studies are closer to our findings.

The series PCBQ, PBOQ, and PPhQ reveals some important insights into the role of bridge orbitals in mediating intramolecular ET. The rate constants in PPhQ are ca. 10^2–10^3 times greater than in PBOQ. This difference almost certainly arises from the availability of low-lying empty antibonding levels in the phenyl group, which are not present in the bicyclooctane bridge. The results for PCBQ are noteworthy. The cyclobutane bridge in PCBQ has one fewer bond than the bicyclooctane bridge in PBOQ, and the 56-fold increase observed in $k_{ET}{}^S$ in methylene chloride is larger than might be expected for such a modest bond decrease.

H_{rp} is expected to depend on the center-to-center distance r according to

$$H_{rp}(r) = H_{rp}(r_0) \exp \left[\frac{-\beta(r - r_0)}{2} \right] \tag{7}$$

where r_0 is the center-to-center distance when donor and acceptor are at the van der Waals contact distance. β has been found to be ~ 1 Å^{-1} from a number of studies (*1*). The edge-to-edge distance of the bicyclooctane bridge is ca. 1 Å greater than that of the cyclobutane bridge. Thus if distance were the only factor, the rate constant should increase by a factor of only ca. $e = 2.7$ on changing the bridge from bicyclooctane to cyclobutane. It may be that the strained aliphatic orbitals in the cyclobutane bridge mediate ET more effectively, or that this bridge orients P and Q more favorably.

Support for the strained-orbital concept is provided from a recent study by Sakata et al. *(21)* of two rigid porphyrin–quinone molecules. They found that the ET rate constant for a strained spiro[4.4]nonane bridge is about 5 times faster than a *trans*-decalin bridge, although each bridge has the same number of saturated bonds between the donor and acceptor. Moreover, Onuchic and Beratan *(22)* predicted theoretically that a spiro cyclobutane linkage should exhibit higher rate constants than those with other types of aliphatic hydrocarbon linkages with the same number of bridge bonds.

Conclusions

The data and analyses in this review allow the following conclusions to be drawn:

1. The ET rate-constant data as a function of solvent for PAQ fit tolerably well (with a considerable scatter in the points) to the high-temperature limit of the Marcus equation (eq 2), provided that λ_{out} and ΔG^0 are determined in each solvent.

2. When the PAQ rate constants are examined as a function of temperature, a reasonable fit with the high-temperature limit of the Marcus equation is again obtained; however, the electronic coupling energy H_{rp} varies significantly with solvent. This variability explains most of the scatter found in the solvent-dependence analysis.

3. A peptide bridge is a very effective linkage between the porphyrin and quinone, the rate constant dropping only slowly with increasing chain length. The introduction of an aromatic side group increases the rate slightly, but not dramatically.

4. An unsaturated bridge (e.g., phenyl) allows ET at rates 100–1000 times faster than a saturated bridge (e.g., bicyclooctane); however, a cyclobutane bridge allows ET faster than other saturated bridges of similar dimensions.

This chapter has provided strong evidence that the solvent and temperature and the nature of the bridge are very important factors in mediating intramolecular ET in model compounds. This and other studies should provide considerable insight into understanding the mechanisms of intramolecular electron transfer in natural and artificial systems.

References

1. Connolly, J. S.; Bolton, J. R. In *Photoinduced Electron Transfer;* Fox, M. A.; Chanon, M., Eds.; Elsevier: New York, 1989; pp 303–393.
2. Schmidt, J. A.; Siemiarczuk, A.; Weedon, A. C.; Bolton, J. R. *J. Am. Chem. Soc.* **1985,** *107,* 6112.

3. Bolton, J. R.; Ho, T.-F.; Liauw, S.; Siemiarczuk, A.; Wan, C. S. K.; Weedon, A. C. J. *Chem. Soc., Chem. Commun.* **1985**, 559.
4. Wilford, J. H.; Archer, M. D.; Bolton, J. R.; Ho, T.-F.; Schmidt, J. A.; Weedon, A. C. J. *Phys. Chem.* **1985**, *89*, 5395.
5. Archer, M. D.; Gadzekpo, V. P. Y.; Bolton, J. R.; Schmidt, J. A.; Weedon, A. C. J. *Chem. Soc., Faraday Trans. 2* **1986**, *82*, 2305.
6. Schmidt, J. A.; McIntosh, A. R.; Weedon, A. C.; Bolton, J. R.; Connolly, J. S.; Hurley, J. K.; Wasielewski, M. R. *J. Am. Chem. Soc.* **1988**, *110*, 1733.
7. Schmidt, J. A.; Liu, J.-Y.; Bolton, J. R.; Archer, M. D.; Gadzekpo, V. P. Y. *J. Chem. Soc., Faraday Trans. 1* **1989**, *85*, 1027.
8. Schmidt, J. A., Ph.D. Thesis, University of Western Ontario, 1986.
9. Ho, T.-F.; Bolton, J. R., unpublished work.
10. Liu, J.-Y.; Bolton, J. R., unpublished work.
11. Roach, K. J.; Bolton, J. R.; Weedon, A. C., unpublished work.
12. Siemiarczuk, A.; McIntosh, A. R.; Ho, T.-F.; Stillman, M. J.; Roach, K. J.; Weedon, A. C.; Bolton, J. R.; Connolly, J. S. *J. Am. Chem. Soc.* **1983**, *105*, 7224.
13. Gouterman, M.; Khalil, G.-E. *J. Mol. Spectrosc.* **1974**, *53*, 88.
14. Marcus, R. A.; Sutin, N. *Biochim. Biophys. Acta* **1985**, *811*, 265.
15. Gunner, M. R.; Robertson, D. E.; Dutton, P. L. *J. Phys. Chem.* **1986**, *90*, 3783.
16. Isied, S. S.; Vassilian, A. *J. Am. Chem. Soc.* **1984**, *106*, 1726 and 1732.
17. Isied, S. S. *Prog. Inorg. Chem.* **1984**, *32*, 443.
18. Isied, S. S. In *Electron Transfer in Biology and the Solid State*; Johnson, M. K.; King, R. B.; Kurtz, D. M., Jr.; Kutal, C.; Norton, M. L.; Scott, R. A., Eds.; Advances in Chemistry 226; American Chemical Society: Washington, DC, 1990; p 91.
19. Schanze, K. S.; Sauer, K. *J. Am. Chem. Soc.* **1988**, *110*, 1180.
20. Cabana, L. A.; Schanze, K. S. In *Electron Transfer in Biology and the Solid State*; Johnson, M. K.; King, R. B.; Kurtz, D. M., Jr.; Kutal, C.; Norton, M. L.; Scott, R. A., Eds.; Advances in Chemistry 226; American Chemical Society: Washington, DC, 1990; p 101.
21. Sakata, Y.; Nakashima, S.; Goto, Y.; Tatemitsu, H.; Misumi, S.; Asahi, T.; Hagihara, M.; Nishikawa, S.; Okada, T.; Mataga, N. *J. Am. Chem. Soc.* **1989**, *111*, 8979.
22. Onuchic, J. N.; Beratan, D. N. *J. Am. Chem. Soc.* **1987**, *109*, 6771.

RECEIVED for review April 27, 1990. ACCEPTED revised manuscript September 11, 1990.

Solvent-Dependent Photophysics of Fixed-Distance Chlorophyll– Porphyrin Molecules

The Possible Role of Low-Lying Charge-Transfer States

Michael R. Wasielewski[1], Douglas G. Johnson[1], Mark P. Niemczyk[1], George L. Gaines III[2], Michael P. O'Neil[1], and Walter A. Svec[1]

[1]Chemistry Division and [2]Biological, Environmental, and Medical Research Division, Argonne National Laboratory, Argonne, IL 60439

The properties of a series of fixed-distance chlorophyll–porphyrin molecules are described. These molecules consist of a methyl pyrochlorophyllide a moiety that is directly bonded at its 2-position to the 5-position of a 2,8,12,18-tetraethyl-3,7,13,17-tetramethyl-15-(p-tolyl)porphyrin. Steric hindrance between adjacent substituents rigidly positions the π systems of both macrocycles perpendicular to each other. The macrocycles were selectively metalated with zinc to give the four possible derivatives, HCHP, ZCHP, HCZP, and ZCZP (where H, Z, C, and P denote free base, Zn derivative, chlorin, and porphyrin, respectively). The lowest excited singlet states of HCHP and ZCHP, which are localized on HC and ZC, respectively, exhibit lifetimes and fluorescence quantum yields that are solvent-polarity-independent and do not differ significantly from those of chlorophyll itself. ZCZP and HCZP, however, display solvent-polarity-dependent photophysics. HCZP forms an ion-pair state following excitation in polar media, although ZCZP does not. Nevertheless, nonradiative decay is substantially enhanced in ZCZP as the solvent polarity increases. These effects are discussed in terms of mixing low-lying charge-transfer states of ZCZP into its locally excited singlet state. Enhanced nonradiative decay of excited heterodimers within bacterial reaction centers has recently been observed.

0065–2393/91/0228–0133$06.00/0

In photosynthetic reaction centers the electron donors and acceptors are positioned at precise distances and orientations relative to one another to promote efficient photoinduced charge separation and to impede charge recombination (1–3). However, the nature of the medium that engulfs each donor–acceptor pair is thought to have a large influence on the observed rates of electron transfer. For example, in the bacterial photosynthetic reaction center a bacteriochlorophyll (BChl) molecule lies between the dimeric bacteriochlorophyll donor (BChl$_2$) and the bacteriopheophytin (BPh) acceptor. The π systems of these chromophores lie at large angles relative to one another (about 70°) in an approximate edge-to-edge configuration. It is thought that superexchange, which mixes low-lying, ionic, virtual states involving the intermediate BChl with the locally excited state of the donor, may lead to a greatly increased rate of charge separation (4–7).

Recently, both photochemical hole-burning experiments (8–11) and Stark effect spectroscopy (12, 13) on reaction centers have been reported. The results of these experiments suggest that excitation of BChl$_2$ produces an excited state with significant charge-transfer (CT) character. A recent report on reaction centers from genetically altered *Rhodobacter capsulatus*, in which the BChl$_2$ donor is transformed into a BChl–BPh heterodimer, shows that excitation of this species results in a 50% quantum yield of nonradiative decay to ground state with a 30-ps time constant (14). The majority of femtosecond spectroscopic experiments of native reaction centers show no evidence for intradimer charge-separated intermediates (15–17). However, recent femtosecond transient absorption measurements suggest that a distinct BChl$^-$ intermediate may be involved in the primary charge separation (18).

However, the impact of low-lying virtual states possessing CT character on the photophysics of porphyrin and chlorophyll donor–acceptor molecules remains unclear. We recently demonstrated that a chlorophyll dimer can undergo solvent-induced symmetry breaking within the excited singlet manifold (19). The result of this symmetry breaking is enhanced nonradiative decay of the excited state without the appearance of distinct charge-separated intermediates as monitored by transient absorption spectroscopy. Strictly speaking, CT-state formation within a symmetric dimer is symmetry forbidden. However, by providing different environments for each of the two chromophores, solvent molecules can break the symmetry and thereby allow the formation of CT states. A well-characterized example of a molecule that undergoes solvent-induced symmetry breaking is the symmetric bichromophore 9,9'-bianthryl (20, 21). Because solvation is adequate to destroy the symmetry and allow the formation of CT states in model systems, it is reasonable that the environment of the reaction-center protein surrounding the dimeric donor is sufficiently asymmetric to promote mixing of CT states with the locally excited singlet state of the dimer (1). Mataga and co-workers have recently discussed several interesting features regarding the formation of CT states in symmetric bichromophores (22, 23).

To better understand the influence of low-lying CT states on the observable photophysics of chlorophyll and porphyrin donor–acceptor molecules, we prepared a series of chlorophyll–porphyrin molecules in which the two macrocycles maintain a fixed distance and orientation relative to one another. Molecules HCHP, ZCHP, HCZP, and ZCZP (where H, Z, C, and P denote free base, Zn derivative, chlorin, and porphyrin, respectively) possess a chlorophyll molecule rigidly bound to a porphyrin such that the π systems of both macrocycles are constrained by steric interactions to be perpendicular to each other. This geometry was chosen to mimic approximately the geometric relationship between the $BChl_2$ donor and the intermediary BChl within photosynthetic reaction centers.

Experimental Details

Solvents for all spectroscopic experiments were dried and stored over 3-Å molecular sieves. HPLC-grade toluene was distilled from $LiAlH_4$ (LAH). Butyronitrile was refluxed over $KMnO_4$ and Na_2CO_3, then twice distilled; we retained the middle portion each time. 2-Methyltetrahydrofuran (MTHF) was freshly distilled from LAH before each experiment.

UV–visible absorption spectra were taken on a Shimadzu UV-160 spectrometer. Fluorescence spectra were obtained by using a Perkin–Elmer MPF-2A fluorometer interfaced to a PDP 11/34 computer. All samples for fluorescence were purified by preparative thin-layer chromatography (TLC) on Merck silica gel plates. Samples for fluorescence measurements were 10^{-7} M in 1-cm cuvettes. The emission was measured 90° to the excitation beam. Fluorescence quantum yields were determined by integrating the digitized emission spectra from 600 to 800 nm and referencing the integral to that for chlorophyll a in diethyl ether (24).

Redox potentials for HCHP, HCZP, ZCHP, and ZCZP were determined in butyronitrile containing 0.1 M tetra-n-butylammonium perchlorate by using a Pt disc electrode at 21 °C. These potentials were measured relative to a saturated calomel electrode using ac voltammetry (25). Both one-electron oxidations and reductions of these molecules exhibited good reversibility.

The picosecond transient absorption and emission measurements were obtained by using apparatus described previously (26). The fluorescence lifetimes and error limits given in Table I are the average of at least three measurements. Lifetimes >1 ns were measured with a Hamamatsu R1294U microchannel plate photomultiplier/TEK 7912 digitizer combination. The total system possessed a time response of 0.8 ns. Lifetimes <1 ns were determined with a Hamamatsu C979 streak camera, with repetitive averaging to obtain good data on weakly luminescent samples. The system response time was 10 ps. The lifetimes were obtained from the data and instrument response by using a Levenberg–Marquardt iterative reconvolution technique.

Results

The synthesis of HCHP, HCZP, ZCHP, and ZCZP (*see* structure) will be reported in a later publication. The methyl groups at the 3 and 7 positions of the porphyrin buttress the chlorophyll and fix the plane of the chlorophyll

Table I. Fluorescence Decay Times

Solvent	ZC	HC	ZCZP	HCZP
Dioxane	7.68 ± 0.04	11.5 ± 0.5	4.66 ± 0.03	8.2 ± 0.3
Toluene	3.14 ± 0.01	10.5 ± 0.5	3.43 ± 0.01	8.5 ± 0.3
Ether	5.97 ± 0.01	10.3 ± 0.5	3.69 ± 0.01	11.9 ± 0.09
EtOAc	5.94 ± 0.01	10.2 ± 0.5	2.61 ± 0.01	1.49 ± 0.01
MTHF	5.90 ± 0.01	10.8 ± 0.5	3.22 ± 0.01	2.06 ± 0.01
CH_2Cl_2	3.33 ± 0.01	9.0 ± 0.4	0.87 ± 0.01	0.41 ± 0.01
Pyridine	6.18 ± 0.02	10.7 ± 0.5	0.29 ± 0.01	0.016 ± 0.0006
Butyronitrile	6.00 ± 0.01	9.4 ± 0.4	0.119 ± 0.001	0.0070 ± 0.0006
DMF	6.63 ± 0.01	7.9 ± 0.3	0.052 ± 0.001	0.0049 ± 0.0008

NOTE: Fluorescence was excited at 610 nm and detected between 650 and 700 nm. No wavelength dependences of the lifetimes were observed. All decays are single exponential and recorded in nanoseconds.

macrocycle perpendicular to that of the porphyrin (27). Rotation about the single bond joining the porphyrin to the chlorin is completely restricted.

The ground-state optical absorption spectra of ZCZP and HCZP are shown in Figure 1. It is immediately apparent that the spectra of both molecules are a superposition of the respective spectra of the porphyrin and chlorin moieties, with each macrocycle possessing its own distinct Soret band and Q bands. Because the energy levels of the chlorin and porphyrin macrocycles are nonresonant, exciton splittings of the optical absorption bands are not observed. The Q_y band of the chlorin rings in HCZP and

HCHP: $M_1 = M_2 = 2H$ ZCHP: $M_1 = Zn, M_2 = 2H$

ZCZP: $M_1 = M_2 = Zn$ HCZP: $M_1 = 2H, M_2 = Zn$

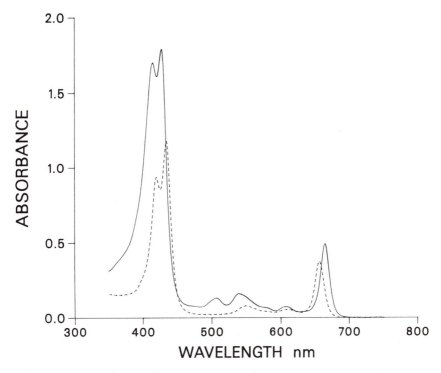

Figure 1. Ground-state absorption spectra of HCZP (—, 10^{-5} M) and ZCZP (---, 6×10^{-6} M) in butyronitrile.

ZCZP occur at 669 and 654 nm, respectively, and are only 1–2 nm blue-shifted from the respective bands in free-base and zinc methyl pyropheophorbide *a*. Similarly, the (0,0) fluorescence emission bands of HCZP and ZCZP that occur at 674 and 660 nm, respectively, are only slightly blue-shifted from the corresponding bands in free-base and zinc methyl pyropheophorbide *a*. Thus, the π systems of the two macrocycles are only weakly coupled. The prominence of the Q bands of the chlorin makes it possible to selectively excite the chlorin in our experiments. At 610 nm, the wavelength of our laser excitation experiments, the ratio of chlorin–porphyrin absorbance is >25 for ZCZP and about 8 for HCZP. In addition, we can use the distinct spectroscopic signature of the Soret bands to detect the involvement of chlorin or porphyrin states in the overall excited-state description of these molecules.

Figure 2 shows the energy levels of the locally excited states of the chromophores within HCHP, ZCHP, HCZP, and ZCZP and several hypothetical CT states. The energies of the excited states are determined from the position of the (0,0) band of the fluorescence emission. The hypothetical CT state energies are estimated from a simple sum of the one-electron

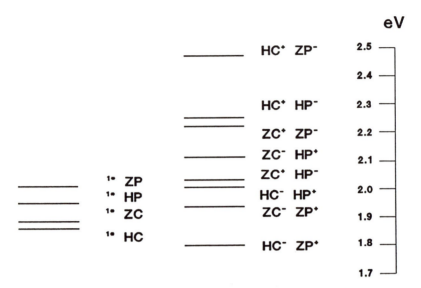

Figure 2. Energy-level diagrams for low-lying locally excited singlet states and charge-transfer states of HCHP, HCZP, ZCHP, and ZCZP in butyronitrile.

oxidation and reduction potentials of the chromophores measured within each molecule in butyronitrile. These CT-state energies are reasonable estimates only in highly polar media, where the ions are strongly solvated and the coulombic interaction between the ions is small. Two important features of the energetics of these molecules can be noted from these data. First, most of the predicted CT states are low-lying (i.e., energetically close to S_1). Second, only HC^-ZP^+ is below the locally excited singlet states of the chromophores. The next lowest-lying state belongs to ZC^-ZP^+. Thus, one might expect HCZP and ZCZP to show the greatest perturbation of their locally excited singlet states by mixing with the CT states.

Figure 3 gives the fluorescence quantum yields of HCHP, ZCHP, HCZP, and ZCZP, along with the zinc and free-base methyl pyropheophorbide *a* reference compounds. We found that energy transfer from the porphyrin to the chlorin in each of these molecules proceeds in times <1 ps; therefore, the locally excited singlet state resides on the chlorin. The only compounds in the series that show solvent-polarity-dependent fluorescence quantum yields are HCZP and ZCZP. The emission from both molecules is strongly quenched in polar media. The static dielectric constant required to diminish the emission yield of ZCZP is somewhat higher than that of HCZP. Table I presents fluorescence lifetime data for ZCZP and HCZP in the same solvents used to obtain the quantum yield data. Figure 3 and Table I show that the reduction in emission quantum yield closely parallels the observed decrease in fluorescence lifetime as the solvent polarity increases. This finding suggests that increasing the solvent polarity results in a new nonradiative chan-

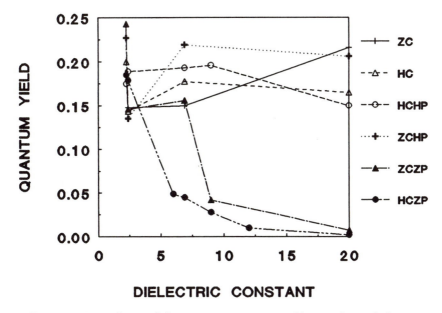

Figure 3. Dependence of fluorescence quantum yields on solvent dielectric constant for the indicated compounds.

nel that competes with radiative decay. The nature of this channel can be examined by looking for transient spectral changes that arise following excitation of HCZP and ZCZP.

Figures 4A and 4B show the transient absorption difference spectrum of ZCZP in toluene (molar absorption ϵ = 2.4) and in butyronitrile (ϵ = 20) obtained 20 ps after a 1-ps laser flash at 610 nm. In both solvents the Soret band due to the chlorin chromophore at 430 nm bleaches within the 1-ps time of the laser flash. There is no evidence for bleaching of the porphyrin Soret band at 414 nm. Similarly, in both solvents (Figure 4B) the Q_y band of the chlorin bleaches immediately upon excitation. There is no evidence of significant absorption in the near-infrared region of the spectrum, where the radical ion states of both porphyrins and chlorins have absorbances (28–30).

Figure 5 shows the recovery of the Q_y band bleach as a function of time. The data show that the excited state recovers much faster in polar media than in nonpolar media, 119 ps vs. 3.4 ns, respectively. The excited-state recovery times obtained via transient absorption agree well with those obtained from fluorescence emission decays, as shown in Table I. Throughout the recovery time of the absorption changes, the spectral features illustrated in Figures 4A and 4B only diminish. No new features or changes in band shape or structure are observed.

Figures 6A and 6B show the corresponding transient absorption changes

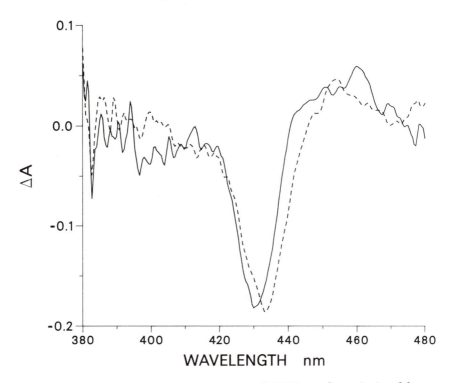

Figure 4A. Transient absorption spectrum of ZCZP in toluene (—) and bu-tyronitrile (---) at 20 ps following a 1-ps laser flash at 610 nm.

for HCZP in both toluene and butyronitrile. In toluene the transient absorption spectra shown in Figures 6A and 6B appear with the laser flash and decay with an 8.5-ns time constant. The spectral features show only involvement of the chlorin ring in the excited state and no evidence of ion-pair intermediates in the near-infrared region of the spectrum. On the other hand, when HCZP is excited in butyronitrile, the Soret bands from both the porphyrin and the chlorin are bleached. The chlorin bleach at 435 nm appears within the 1-ps laser flash and is followed within 5 ps by the appearance of the porphyrin bleach at 415 nm. The transient absorption changes in the Q_y-band region of the spectrum also differ substantially from those observed in toluene. The spectrum of HCZP in toluene is similar to that of ZCZP in Figure 4B and gives no indication of significant ion-pair character. In butyronitrile the broad absorption feature between 625 and 700 nm is characteristic of the porphyrin cation radical (30), and the absorption feature at near 800 nm is characteristic of the chlorophyll anion radical (29). Thus, a real ion-pair intermediate is formed following excitation of HCZP in a polar solvent.

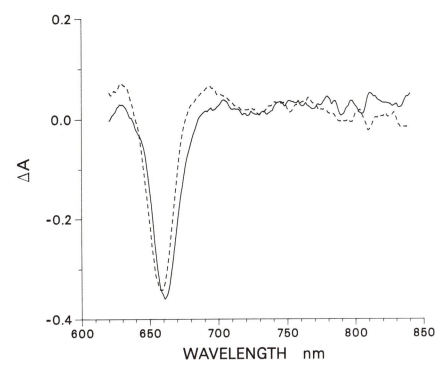

Figure 4B. Transient absorption spectrum of ZCZP in toluene (——) and bu-tyronitrile (---) at 20 ps following a 1-ps laser flash at 610 nm.

Further evidence for the formation of the ion-pair state of HCZP in polar media is obtained by comparing the kinetics of the excited singlet state decay with those for the recovery of the ion pair. The disappearance of the excited state can be obtained from the decay of stimulated emission of the HC chromophore in HCZP (*31*). This result is shown in Figure 7. The weak stimulated emission recovers with a 5($\pm$1)-ps time constant and is within experimental error of the fluorescence lifetime of HCZP under these con-ditions (Table I). This contrasts with the 43-ps decay time for the HCZP ion-pair features (Figure 7).

Discussion

The photophysical data suggest that rapid nonradiative decay pathways for the lowest excited singlet states of HCZP and ZCZP are dominant in polar solvents and relatively unimportant in nonpolar media. In polar solvents, electron transfer from ZP to ^{1*}HC occurs within HCZP. Placing either HCZP or ZCZP in nonpolar media substantially increases the energies of the CT

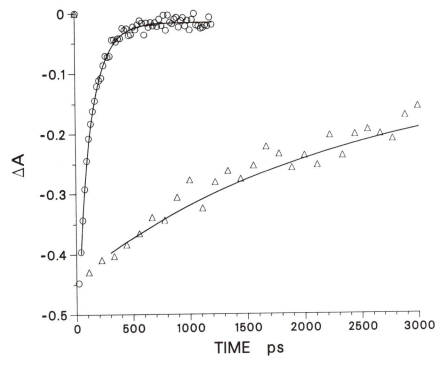

Figure 5. Transient absorption recovery at 660 nm for ZCZP in toluene (△)
and butyronitrile (○) following a 1-ps, 610-nm laser flash.

states. An estimate of this increase can be made by using dielectric continuum theory to obtain the solvation energy of the ions as a function of solvent polarity (32). For two ions with 5-Å radii separated by 10 Å in toluene, the CT-state energies should lie about 0.6 eV higher than those depicted in Figure 1. Thus, in toluene one expects the energies of HC^-ZP^+ and HC^+ZP^- to be at least 0.5 eV above that of $^{1*}HCZP$, so that ion-pair formation is precluded. For ZCZP in both toluene and butyronitrile, the energies of ZC^-ZP^+ and ZC^+ZP^- should be above that of $^{1*}ZCZP$. In toluene the energies of the CT states should be sufficiently above that of the locally excited singlet state to preclude significant interaction between these states. As the solvent polarity increases, mixing of CT character into the locally excited state of ZCZP increases. This change may enhance nonradiative transitions between the mixed excited state and the ground state.

The formation of intramolecular charge-transfer states in directly linked bichromophores is a well-known phenomenon (20, 21). In a directly linked bichromophore, CT-state formation is favored when the two chromophores are unable to adopt a conformation in which their π systems are face-to-face, but are able to twist into a conformation in which the two π systems

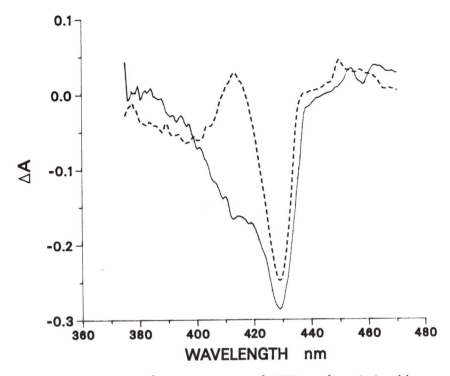

*Figure 6A. Transient absorption spectrum of HCZP in toluene (---) and bu-
tyronitrile (—) at 20 ps following a 1-ps laser flash at 610 nm.*

are orthogonal. This criterion is opposite to that for excimer formation, where
the greatest stabilization is generally obtained when the two chromophores
are able to maximize their π overlap. The term "twisted intramolecular
charge transfer" (TICT) has been used to describe this type of CT state. The
criterion for TICT-state formation within a directly linked bichromophoric
molecule has been described as the rule of minimum overlap (33). The
structures of HCHP, ZCHP, HCZP, and ZCZP are all ideal for TICT-state
formation. It is sterically impossible for the π systems of the chlorophyll and
the porphyrin in these molecules to achieve large overlap. The methyl groups
at the 3 and 7 positions on the porphyrin hold the chlorin ring perpendicular
to the porphyrin.

In a CT or TICT state, the asymmetric charge distribution in the molecule results in a dipole moment that interacts strongly with solvent dipoles.
As solvent polarity increases, the CT or TICT state is stabilized. Because
the λ_{max} of emission from ZCZP is essentially independent of solvent polarity,
the fluorescence of ZCZP at 660 nm is not due to an emissive CT or TICT
state. Yet, as the solvent polarity increases, the quantum yield of emission
from the 660-nm band decreases greatly. Despite intensive efforts, we have

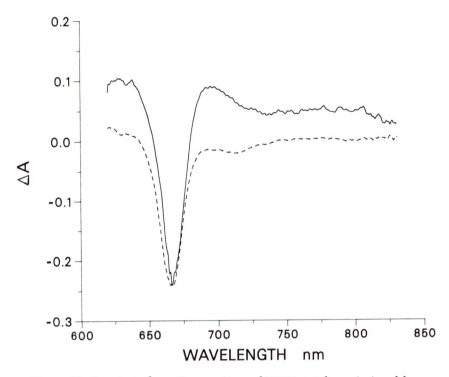

*Figure 6B. Transient absorption spectrum of HCZP in toluene (---) and bu-
tyronitrile (—) at 20 ps following a 1-ps laser flash at 610 nm.*

never observed emission from charge-separated states involving chlorophyll
or porphyrin containing donor–acceptor molecules that are known to form
radical ion pairs in high quantum yield (*34*). Therefore, we hypothesize that
the TICT state of ZCZP decays entirely nonradiatively. However, this hy-
pothesis requires that the energy of the TICT state be below that of the
lowest excited singlet state, S_1. An estimate of the energy of a hypothetical
TICT state in ZCZP can be obtained from our electrochemical data and is
shown in Figure 2. The sum of the one-electron redox potentials of ZCZP
in polar media suggests that the TICT-state energy is probably no lower
than about 1.95 eV. This amount is about 0.15 eV above the observed lowest
excited singlet state energy of ZCZP. Dissolving ZCZP in low-polarity media
such as toluene serves only to increase the energy of the TICT state, thereby
increasing the energy gap between this state and S_1. Placement of the TICT
state of ZCZP above S_1 is consistent with the transient absorption and emission data obtained for ZCZP. The transient spectra of ZCZP shown in Figures
4A and 4B show no evidence of ion pairs as the solvent polarity increases.
 However, nonradiative transitions between S_1 and the ground state, S_0,

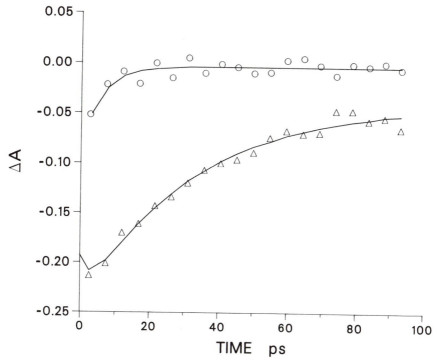

Figure 7. Stimulated emission decay at 685 nm (O) and transient absorption decay at 670 nm (Δ) for HCZP in butyronitrile following a 1-ps, 610-nm laser flash.

in ZCZP may be enhanced by mixing of a low-lying CT (or TICT) state at 1.95 eV into S_1 at 1.87 eV. A small amount of mixing could result in significant enhancement of the nonradiative decay rate in ZCZP without the appearance of large changes in the excited-state absorption spectrum. Nonradiative decay from $S_1 \rightarrow S_0$ in chlorophylls is relatively inefficient. Internal conversion accounts for no more than 15% of excited-state decay. Ab initio self-consistent field–molecular orbital–configuration interaction (SCF–MO–CI) calculations of both S_0 and S_1 for ethyl chlorophyllide a show that the π-electron populations at each atom in the π system do not differ significantly between S_0 and S_1 (35). The similar electron distributions for S_0 and S_1 suggest that the potential surfaces of these states form a nested pair (Figure 8). The absence of a crossing point between the surfaces of S_0 and S_1 results in very small Franck–Condon factors with consequent low rates of internal conversion.

On the other hand, addition or removal of an electron from either the chlorophyll or metalloporphyrin ring in ZCZP results in significant changes

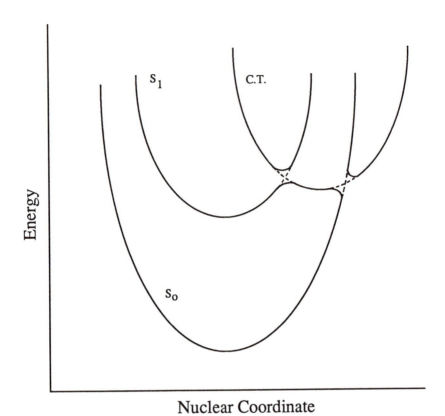

Figure 8. Potential energy surfaces for ZCZP.

in the overall π-electron distribution within these macrocycles (36, 37). Moreover, these changes in π-electron distribution also differ from the π-electron distribution in both S_0 and S_1. In particular, for the CT state ZC^-ZP^+, the π-electron distribution within the reduced chlorophyll ZC^- is very asymmetric, with most of the unpaired electron density concentrated along the ring A–C axis. Rings B and D have relatively little density. Within the oxidized porphyrin ZP^+, the π-electron density is more or less evenly distributed around the periphery of the macrocycle. The changes in electron-density distribution in ZC^-ZP^+ relative to both ZCZP and $^{1*}ZCZP$ should result in a corresponding displacement of the nuclei in the CT state relative to those in the excited and ground states (Figure 8).

If the CT state of ZCZP is low-lying in polar solvents, as is suggested by our electrochemical data, the potential-energy surface for the CT state may intersect the surfaces of both S_1 and S_0. The rapid $(2 \times 10^{11} \text{ s}^{-1})$ electron transfer observed in HCZP in butyronitrile suggests that the electronic coupling between two macrocycles linked in this fashion is on the order of 25–

50 cm^{-1}. Thus, the surface crossings may be avoided; the result would be a relatively smooth pathway for nonradiative deactivation of S_1 to S_0.

Conclusion

The chlorophyll–porphyrin molecules described in this chapter are sensitive probes of their solvation environment. Our data show that two large π macrocycles, which are strongly coupled electronically, may exhibit enhanced rates of nonradiative decay as a function of increasing solvent polarity. Enhanced nonradiative decay need not require the formation of real, spectroscopically observable ion-pair states. It is sufficient to mix CT character into the excited-state description of the macrocycles. This type of behavior can be used to model recent observations of significant nonradiative decay yields in bacterial reaction-center proteins for which the primary $BChl_2$ donor has been genetically altered to yield a BChl–BPh heterodimer (14).

Acknowledgment

The authors acknowledge the support of the Division of Chemical Sciences, Office of Basic Energy Sciences, U.S. Department of Energy, under contract W-31-109-Eng-38.

References

1. Deisenhofer, J.; Epp, O.; Miki, K.; Huber, R.; Michel, H. *J. Mol. Biol.* **1984**, *180*, 385.
2. Chang, C.-H.; Tiede, D.; Tang, J.; Smith, U.; Norris, J.; Schiffer, M. *FEBS Lett.* **1986**, *205*, 82.
3. Allen, J. P.; Feher, G.; Yeates, T.-O.; Komiya, H.; Rees, D. C. *Proc. Natl. Acad. Sci. U.S.A.* **1987**, *84*, 6162.
4. Marcus, R. A. *Chem. Phys. Lett.* **1988**, *133*, 471.
5. Won, Y.; Friesner, R. A. *Biochim. Biophys. Acta* **1988**, *935*, 9.
6. Bixon, M.; Jortner, J.; Plato, M.; Michel-Beyerle, M. E. In *The Bacterial Reaction Center, Structure and Dynamics*; Breton, J.; Vermeglio, A., Eds.; Plenum: New York, 1988; pp 399–419.
7. McConnell, H. M. *J. Chem. Phys.* **1961**, *35*, 508.
8. Boxer, S. G.; Mittendorf, T. R.; Lockhart, D. J. *Chem. Phys. Lett.* **1985**, *123*, 476.
9. Meech, S. R.; Hoff, A. J.; Wiersma, D. A. *Chem. Phys. Lett.* **1985**, *121*, 287.
10. Hayes, J. M.; Small G. J. *J. Phys. Chem.* **1986**, *90*, 4928.
11. Gillie, J. K.; Fearey, B. L.; Hayes, J. M.; Small, G. J.; Golbeck, J. H. *Chem. Phys. Lett.* **1987**, *134*, 316.
12. Lockhart, D. J.; Boxer, S. G. *Biochemistry* **1987**, *26*, 664.
13. Lockhart, D. J.; Boxer, S. G. *Chem. Phys. Lett.* **1988**, *144*, 243.
14. Kirmaier, C.; Holten, D.; Bylina, E. J.; Youvan, D. C. *Proc. Natl. Acad. Sci. U.S.A.* **1988**, *85*, 7562.

15. Wasielewski, M. R.; Tiede, D. *FEBS Lett.* **1986**, *204*, 368.
16. Martin, J.-L.; Breton, J.; Hoff, A.; Migus, A.; Antonetti, A. *Proc. Natl. Acad. Sci. U.S.A.* **1986**, *83*, 957.
17. Breton, J.; Martin, J.-L.; Migus, A.; Antonetti, A.; Orszag, A. *Proc. Natl. Acad. Sci. U.S.A.* **1986**, *83*, 5121.
18. Holzapfel, W.; Finkele, U.; Kaiser, W.; Oesterhelt, D.; Scheer, H.; Stilz, H. U.; Zinth, W. *Chem. Phys. Lett.* **1989**, *160*, 1.
19. Johnson, D. G.; Svec, W. A.; Wasielewski, M. R. *Isr. J. Chem.* **1988**, *28*, 193.
20. Kajimoto, K.; Yamasaki, K.; Arita, K.; Hara, K. *Chem. Phys. Lett.* **1986**, *125*, 184.
21. Nakashima, N.; Murakawa, M.; Mataga, N. *Bull. Chem. Soc. Jpn.* **1976**, *49*, 854.
22. Mataga, N.; Yao, H.; Okada, T.; Rettig, W. *J. Phys. Chem.* **1989**, *93*, 3383.
23. Yao, H.; Okada, T.; Mataga, N. *J. Phys. Chem.* **1989**, *93*, 7388.
24. Latimer, P.; Bannister, T. T.; Rabinowitch, E. *Nature (London)* **1956**, *124*, 585.
25. Wasielewski, M. R.; Smith, R. L.; Kostka, A. G. *J. Am. Chem. Soc.* **1980**, *102*, 6923.
26. Wasielewski, M. R.; Fenton, J. M.; Govindjee *Photosynth. Res.* **1987**, *12*, 181.
27. Wasielewski, M. R.; Niemczyk, M. P.; Svec, W. A.; Pewitt, E. B. *J. Am. Chem. Soc.* **1985**, *107*, 5583.
28. Borg, D. C.; Fajer, J.; Felton, R. H.; Dolphin, D. *Proc. Natl. Acad. Sci. U.S.A.* **1970**, *67*, 813.
29. Fujita, I.; Davis, M. S.; Fajer, J. *J. Am. Chem. Soc.* **1978**, *100*, 6280.
30. Fajer, J.; Borg, D. C.; Forman, A.; Dolphin, D.; Felton, R. H. *J. Am. Chem. Soc.* **1970**, *92*, 3451.
31. Rodriguez, J.; Kirmaier, C.; Holten, D. *J. Am. Chem. Soc.* **1989**, *111*, 6500.
32. Weller, A. *Z. Phys. Chem. Neue Folge* **1982**, *133*, 93.
33. Rettig, W. *Angew. Chem. Int. Ed. Engl.* **1986**, *25*, 971.
34. Wasielewski, M. R.; Johnson, D. G.; Svec, W. A. In *Supramolecular Photochemistry*; Balzani, V., Ed.; Reidel: Dordrecht; p 255.
35. Petke, J. D.; Maggiora, G. M.; Shipman, L. L.; Christoffersen, R. E. *Photochem. Photobiol.* **1979**, *30*, 203.
36. Spangler, D.; Maggiora, G. M.; Shipman, L. L.; Christoffersen, R. E. *J. Mol. Spectrosc.* **1978**, *71*, 64.
37. Petke, J. D.; Maggiora, G. M.; Shipman, L. L.; Christoffersen, R. E. *Photochem. Photobiol.* **1981**, *33*, 663.

RECEIVED for review April 27, 1990. ACCEPTED revised manuscript August 9, 1990.

Manipulation of Electron-Transfer Reaction Rates with Applied Electric Fields

Application to Long-Distance Charge Recombination in Photosynthetic Reaction Centers

Stefan Franzen and Steven G. Boxer

Department of Chemistry, Stanford University, Stanford, CA 94305

The rate of electron-transfer reactions can be manipulated by electric fields. Although there may be several mechanisms by which the electric field influences the rate, the dominant mechanism is a change of free energy for the reaction by changing the energy of dipolar intermediates or products. This change in free energy maps in some nonlinear fashion onto a change in rate. The field may also affect the electronic coupling. The case of long-distance (25 Å) charge recombination in photosynthetic reaction centers was investigated in detail at 80 K for nonoriented samples. The experimental change in the rate as a function of field is inverted to give the relationship between the rate and free energy in the vicinity of the zero-field free energy. The dependence is found to be considerably more shallow than suggested by simplified treatments of the Franck–Condon factor. This discrepancy indicates that coupling to both high- and low-frequency modes is important for this reaction.

THE RATES OF ELECTRON-TRANSFER REACTIONS can be changed by application of external electric fields. During the past few years we have investigated the effects of applied electric fields on the spectral properties (Stark

0065–2393/91/0228–0149$06.00/0

effect spectroscopy) (1, 2) and reaction dynamics (3–5) in photosynthetic reaction centers (RC). In particular, we have investigated the long-distance charge-recombination reaction between a hole on the primary electron donor (P$^+$) and an electron on the secondary quinone electron acceptor (Q$_A^-$) in isotropic samples at low temperature. The purpose of this investigation is to demonstrate that effects can be measured and quantitated, to develop methodology for analyzing the data, and to provide information on this particular electron-transfer reaction. The latter is important because two prior studies of field effects in Langmuir–Blodgett films (6, 7) and in lipid bilayers (8, 9) at room temperature reached very different results.

The relationship among the reactive groups in the RC as determined by X-ray crystallography (10) is shown in Scheme I. Upon excitation, ^{1}P transfers an electron to H within a few picoseconds (11); the electron moves on from H$^-$ to Q$_A$ within a few hundred picoseconds. Of the two C$_2$ symmetry-related halves, L and M, of the RC, only the L side is functional. The intermediate P$^+$Q$_A^-$ is long-lived; it decays on the order of 100 ms at room temperature (about 30 ms at low temperature) back to the neutral initial condition. Because of this long lifetime, it is possible to measure the decay kinetics with excellent signal-to-noise ratio. The distance between P and Q$_A$ is approximately 25 Å (10).

It is natural to consider electric-field effects because the RC spans a membrane that can have a substantial transmembrane potential. For an idealized bilayer 50 Å thick, a transmembrane potential of 500 meV (somewhat larger than is likely to be physiologically relevant) corresponds to an applied electric field of 10^6 V/cm (10 mV/Å). Because the RC is oriented with respect to this applied potential, the energies of dipolar states such as P$^+$H$_L^-$ and P$^+$Q$_A^-$ will be changed: the interaction energy $\Delta U(F) = -\mu \cdot F$, where μ is the dipole of the state and F is the applied electric field. For example, the P$^+$Q$_A^-$ dipole is roughly aligned with the local C$_2$ axis (Scheme I), which is parallel to the membrane normal, and $|\mu(P^+Q_A^-)| \sim 130$ D on the basis of crystal structure (10). Thus, the energy of this dipole will be increased by nearly 300 meV for a field of 10^6 V/cm, a substantial fraction of the free energy ($\Delta G^0 = 520$ meV) (12, 13) for the P$^+$Q$_A^- \rightarrow$ PQ$_A$ recombination reaction (Scheme II).

To obtain an effect of the applied field on the rate of electron transfer, the field must affect some factor that determines the rate. In the simplest picture, the rate is factored into the product of an electronic factor (V^2) and a Franck–Condon (FC) factor, and the field can change the rate by changing either or both factors. As discussed in the previous paragraph, the field will always change the energy of dipolar states. Consequently, a change in the driving force for the reaction is expected (Figure 1A). ΔG^0 shifts to higher or lower values, depending on the orientation of the dipole. For an oriented sample, one predicts that the rate will shift as the field is applied, with the magnitude and direction of the shift depending on the shape of the rate vs. the ΔG^0 curve in the vicinity of the zero-field ΔG^0, the magnitude and

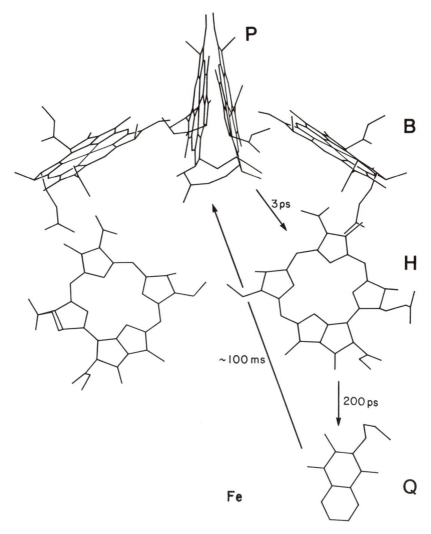

Scheme I. Arrangement of the reactive components taken from the X-ray crystal structure coordinates of the Rps. viridis *reaction center* (10).

direction of the dipole relative to the field, and the field strength. This situation is ideal from an experimental perspective.

If the sample is not oriented, but instead all orientations of dipoles relative to the field are present (an isotropic sample), then the driving force spreads out around the zero-field value: the extrema of the change correspond to those orientational subpopulations whose dipoles are parallel or antiparallel to the field, whereas other subpopulations experience a smaller change. The consequences for the rate are illustrated in Figure 1B, corresponding to the three regions of the rate vs. the ΔG^0 dependence highlighted

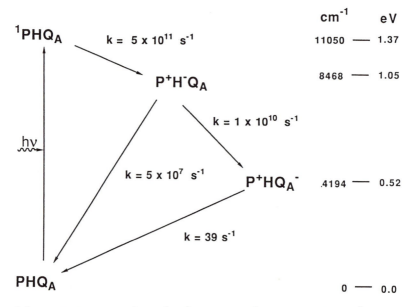

Scheme II. *Reaction scheme for the primary charge separation and recombination steps of bacterial photosynthesis, and approximate rate constants at 80 K. The energies relative to the ground state are given on the right in reciprocal centimeters and electron volts.*

in Figure 1A (*14, 15*). Consider the case in which, at zero applied field, the electron-transfer reaction is well-described by a single exponential. Upon application of the field, the value of ΔG^0 spreads around the zero-field value to produce a spread of rates (i.e., the kinetics becomes highly nonexponential).

This unsavory situation is illustrated in Figure 1B for the three regions in plots of the expected normalized difference in absorbance between field-on and field-off as a function of time (referred to as difference decay for brevity). The total differences are expected to be small, but the shape of the difference decay curve depends sensitively on the shape of the rate vs. the ΔG^0 curve in the vicinity of the zero-field ΔG^0. Furthermore, as the applied field is increased, a wider and wider range of the rate vs. the ΔG^0 curve will be sampled. In the following discussion we will demonstrate that it is possible to reverse the process illustrated in Figure 1A → 1B. By measuring the difference decay as a function of applied field, it is possible to obtain the dependence of the electron-transfer rate on ΔG^0 for the system under investigation. Although we focus our initial study on the $P^+Q_A^-$ recombination reaction in RCs, the method should be entirely general.

In this qualitative discussion we have only considered effects on the FC factors as these are easily illustrated. Nonetheless, the applied field may also

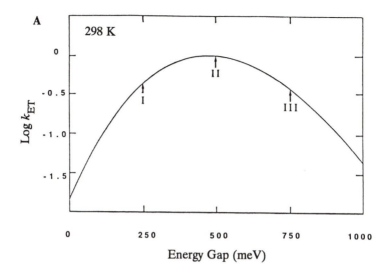

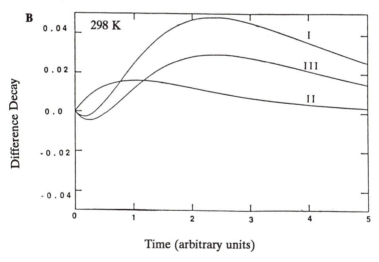

Figure 1. (A) Logarithm of the electron-transfer rate as a function of the free energy change for a typical case. Regions are labeled I for the normal region, II for the optimally exothermic region, and III for the inverted region. (B) Calculated effects of an electric field on the electron-transfer rate for an isotropic sample shown as difference decays (field-on minus field-off). The effect is shown for the three regions illustrated in panel A. For the purpose of calculation, a 50-D dipole and 10^6-V/cm applied field were assumed (possible effects of the field on the electronic coupling matrix element were ignored).

affect the electronic coupling. One possibility is that the field is so large that it distorts the wave functions on the donor and acceptor, and thereby changes their overlap. This direct effect of the field is probably not significant, considering what is known about the polarizability of these molecules and ions; however, definitive experiments need to be developed to prove this. A more subtle issue arises if the electronic coupling involves superexchange coupling via states of species between the donor and acceptor (other chromophores, amino acid residues, etc.). In this case the field might exert a substantial effect on the matrix element under certain conditions. We will argue that superexchange involving P^+H^- may be important in mediating the $P^+Q_A^-$ recombination reaction, but that the field effect is likely not to be very large.

Experimental Details

Reaction centers from *Rhodobacter sphaeroides* R-26 were obtained by standard methods (16) and contain a single ubiquinone. Samples were prepared by spin-coating solutions containing RCs [10 mM Tris, pH 8.0, 0.025% lauryldimethylamine oxide detergent (LDAO) in 18% (w/v) poly(vinyl alcohol) (PVA, avg. MW = 125,000, Aldrich)] onto glass slides coated with indium–tin oxide (ITO) (>85% transparency in the 500–1000-nm range, conductance $>1.25 \times 10^{-3}$ S, 0.17 μm thick). Samples were spun on a photoresist spinner. Different samples had thicknesses ranging from $3.0(\pm0.1)$ to $10.0(\pm0.3)$ μm as measured with a Sloan Dektak IIa thickness-measuring system (precision ±0.1 μm) and had optical densities ranging from 0.02 to 0.1 at 870 nm. The second electrode was prepared by evaporating 0.3 μm of Al onto the polymer film.

    The samples were placed in a closed-cycle helium refrigerator in a copper sample holder designed to allow thermal contact with the cold finger through the aluminum electrode. Data were collected for 3–5-min periods while the compressor was turned off to reduce the noise. The temperature rise during these time periods was less than 2.0 K, as determined by a gold–constantan thermocouple near the sample. The sample was excited by using the frequency-doubled output of a Nd:YAG laser (pulse width ~10 ns); this resulted in a bleach of the 870-nm band that was less than 70% of saturation. The probe beam was broad-band light from a tungsten halogen lamp filtered to produce a beam centered at $880(\pm20)$ nm. The probe beam was reflected off the aluminum electrode at 16.4° incidence, and the transient signal was measured with a silicon photodiode. The high voltage was gated on immediately following the excitation flash with a rise time of <50 μs. Typically about 200 transients were averaged to achieve the necessary signal-to-noise ratio in excess of 1000:1.

Results and Methods of Analysis

The effects of increasing an applied electric field on the $P^+Q_A^-$ decay kinetics at 80 K are shown in Figure 2 as difference decays for field-on minus field-off, which are normalized to the zero-field initial amplitude of the bleach. The decay is slower at all times with the field on than with the field off for all field strengths. The difference reaches a maximum at slightly greater than one $1/e$ time of a single exponential fit to the data. At an applied field of 1.2×10^6 V/cm, differences as large as 7% of the initial amplitude are

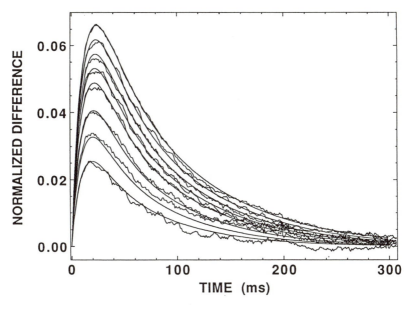

Figure 2. Difference decay curves (field-on minus field-off) for $P^+Q_A^-$ charge recombination at 80 K. Data are shown for external fields 7.61, 8.57, 9.05, 9.52, 10.00, 10.47, 10.96, and 11.43 × 10^5 V/cm, where increasing fields lead to larger differences relative to zero field. The lines through the data were generated from the parameters obtained from the best fit to all the data by using a cumulant expansion in powers of the field. (Reproduced from ref. 24. Copyright 1990 American Chemical Society.)

observed. Even at a qualitative level, these data suggest that this reaction is approximately in the region labeled II in Figure 1. Markedly different results are observed at higher temperatures due to charge recombination via an activated pathway; this will be discussed in detail elsewhere (*17*).

   We shall extract the form of the k_{ET} vs. F_{ext} curve from the raw data (difference decays) by assuming that this curve is well represented by a cumulant expansion and adjusting the cumulants for a best fit. This method is completely general and can be used for any electron-transfer system. To interpret this particular system, we make the following assumptions:

   1. The field affects only the energy of the $P^+Q_A^-$ state dipole; that is, possible effects of the field on the ground state, the reorganization energy, and the zero-order electronic wave functions are ignored. (We briefly address the electric-field dependence of the electronic coupling matrix element due to coupled excited states to second order (superexchange) in the Discussion section.)

2. The interaction energy between the electric field and the $P^+Q_A^-$ dipole moment represents a contribution to the free energy of the charge-separated state $P^+Q_A^-$.

3. There are no other pathways in the decay of $P^+Q_A^-$ to the ground state at 80 K up to the highest fields; that is, we do not consider activated back reactions through other states.

4. Electron transfer is the rate-limiting step in the recombination of $P^+Q_A^-$ at this temperature.

The distribution of electron-transfer rates induced by the field is treated by using a cumulant expansion of each rate in powers of the energy of interaction (ΔU) between the $P^+Q_A^-$ dipole (μ) and the electric field (F_{int}), where $\Delta U = -\mu F_{int}\cos\theta$, and θ is the angle between μ and the internal field F_{int}. The modified rate constant at a given orientation and field is given by:

$$k_{ET}(F_{int},\theta) = \exp\left(\sum_{n=0}^{\infty} P_n \Delta U^n\right) \qquad (1)$$

where the zero-field rate constant is given by $k_{ET}(F_{int} = 0,\theta) = \exp(P_0)$. Equation 1 is simply the representation of the logarithm of the rate constant by a general polynomial. The cumulants P_n can be related to the moments of an expansion of the rate constant on powers of the free energy, and the shape of the curve they describe can be compared with the shape predicted by theory. This approach allows various theories and also levels of approximation within one theory to be tested.

For the case in which the $P^+Q_A^-$ decay-rate constant is well-described by a single exponential at zero field, application of the field shifts the value of the decay-rate constant to a new value depending on the angle θ. The experimentally observed difference decay curves can be fit to the orientation averaged difference decay function $\Delta A(t)$

$$\Delta A(t) = A_0 \left\{ \int_0^1 \exp\left[-\exp\left\{\sum_{n=0}^{\infty} P_n(-\mu F_{int}\cos\theta)^n\right\} t\right] \times \right.$$

$$\left. d(\cos\theta) - \exp\left[-\exp(P_0)t\right]\right\} \qquad (2)$$

by using the Marquardt algorithm for nonlinear least-squares fitting. The parameter P_0 is obtained from the zero-field fit of the data to a single exponential. The parameters P_n resulting from the fit describe the shape of the distribution of rates as a function of energy for the process. The parameters P_n are obtained by fitting a set of difference decay curves obtained at many electric-field values simultaneously and requiring that the parameters be globally valid for all of the fields. As long as the number of field values is large enough, the problem is overdetermined. The determination of the rate vs. interaction energy curve determined in this way is an experimental quantity and is free from theoretical modeling. Starting with the data, which are described by a function of time, $\Delta A(t)$, one obtains the distribution function $k_{ET}(F_{int}, \theta)$, which satisfies eq 2.

Although the $P^+Q_A^-$ recombination kinetics fit well to a single exponential in solution at room temperature, it has been observed in careful measurements that the decay is not single-exponential at low temperature (*18*). Likewise, the fit of the PVA film samples to a single exponential is not good, though the best fit to a single exponential gives a rate constant of 39.0 s^{-1}, which is the value often quoted in the literature at 80 K (*19*). The data can be much better fit by using two exponentials. Thus, following the work of many other investigators (*20–23*), we have analyzed the $P^+Q_A^-$ recombination kinetics as a double exponential and treat the two decays as corresponding to two populations of RCs: population 1 with a faster recombination rate, $k_1 \simeq 75 \text{ s}^{-1}$ and population 2 with a slower rate, $k_2 \simeq 21 \text{ s}^{-1}$. It is possible that differences in the degree of protonation are the origin of these populations (*22*).

By using the result that the zero-field kinetics are biexponential, the method outlined can be extended to allow two k_{ET} vs. applied-field curves to be calculated. The parameters for the faster process (population 1) are denoted P_n and those for the slower process (population 2) are denoted Q_n. The amplitude of population 1 is A_0 and that of population 2 is B_0. Assuming isotropic excitation, the equation used for the biexponential fit is an extension of eq 2.

$$\Delta A(t) = A_0 \left\{ \int_0^1 \exp\left[-\exp\left[\sum_{n=0}^{\infty} P_n(-\mu F_{int} \cos \theta)^n \right] t \right] \times \right.$$

$$\left. \mathrm{d} (\cos \theta) - \exp\left[-\exp (P_0)t \right] \right\} +$$

$$B_0 \left\{ \int_0^1 \exp\left[-\exp\left[\sum_{n=0}^{\infty} Q_n(-\mu F_{int} \cos \theta)^n \right] t \right] \times \right.$$

$$\left. \mathrm{d} (\cos \theta) - \exp\left[-\exp (Q_0)t \right] \right\} \quad (3)$$

Other issues such as the local field correction, complications due to the absorption Stark effect and angle dependence, and details of the error analysis are discussed elsewhere (*24*).

By using eq 3 with the first four cumulants included, the difference decays shown in Figure 2 were fit to give the experimental rate vs. interaction energy curves shown in Figure 3. These curves can be used in turn to regenerate the difference decay curves that are shown in Figure 2 along with the original data. The error curves are shown in Figure 3 as dashed lines above and below the experimental k_{ET} vs. ΔG^0 curve. The errors are smaller near the zero-field rate than at the extremes covered by the field because the effect near zero field is sampled with each application of the field; in contrast, the effect at the extremes is only sampled with the highest applied fields.

Discussion

Several features of the curves in Figure 3 are interesting. The k_{ET} vs. ΔG^0 curves are relatively flat and markedly asymmetrical; they are much flatter for larger values of $-\Delta G^0$ than for smaller values. This result suggests that many modes, including both low- and high-frequency modes, are coupled to the electron-transfer process. The maximum rate is not the zero-field rate. Thus, even though the reaction is essentially activationless, the reorganization energy is not equal to the zero-field free energy, though it is not far removed. There is no evidence for structure in the curves. In contrast to data obtained with model compounds at discrete free-energy points (25), the experimental curves in Figure 3 are continuous because a continuous distribution of ΔG^0 is sampled between $\pm\mu(P^+Q_A^-)F$ (about ±300 meV). Thus, if structure were present it should be seen in this data; none is observed. Finally, the shape of the curves for the two decay processes is quite different, reflecting some difference in the modes that are coupled to the electron-transfer process.

With these experimental curves in hand, it is useful to go further and attempt to fit these curves to a variety of models. We have implemented a nonlinear least-squares fitting program for this purpose, the details of which are described elsewhere (24). The key results are shown in Figure 4, which gives best fits to the models described in the caption. The simplest semiclassical Marcus theory gives a poor fit to the data. By comparison, a model involving linear coupling to two modes compares well with the data for the faster decay process and at least passably with the data for the slower decay process, using the parameter values given in Table I. The values of the mode frequencies coupled to the process are average values, and the coupling constants represent the total coupling of all of the modes with a given average frequency. The product of the electron–phonon coupling S and the frequency ω gives the contribution of a mode to the reorganization energy ($\lambda = \Sigma S_i h\omega_i$, where h is Planck's constant). The clear implication is that coupling to both low- and high-frequency modes is important, and that the total reorganization energy is roughly evenly divided between these modes.

The long-distance charge-recombination reaction between $P^+Q_A^-$ and

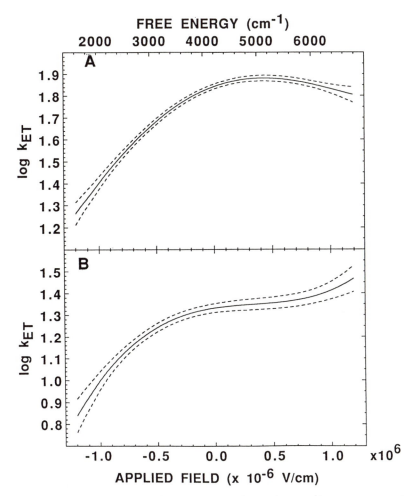

Figure 3. (A) Experimental log k_{ET} vs. ΔG^0 curve for population 1 obtained from the best fit to the electromodulated kinetic data shown in Figure 2. (B) Experimental log k_{ET} vs. ΔG^0 curve for population 2 obtained from the best fit to the electromodulated kinetic data shown in Figure 2. The dotted curves above and below the best fit for each population represent the errors. The relative amounts of populations 1 and 2 are $0.623(\pm 0.064)$ for population 2 and $0.377(\pm 0.064)$ for population 1, as determined from the amplitudes of the fit to two exponentials at zero applied field. The abscissa is the absolute value of free energy $(-\Delta G^0_{et})$. (Reproduced from ref. 24. Copyright 1990 American Chemical Society.)

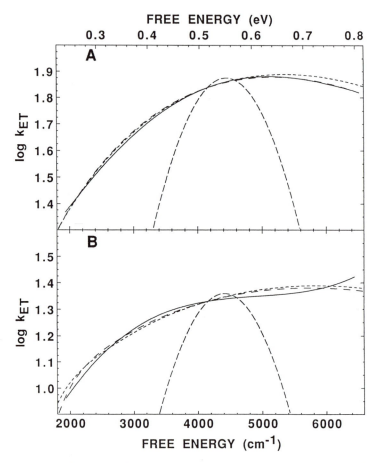

Figure 4. (A) *Fits using theories of electron transfer to the experimental log* k$_{ET}$ *vs.* ΔG^0 *curve for population 1 (—).* (B) *Fits to the experimental log* k$_{ET}$ *vs.* ΔG^0 *curve for population 2 (—). The theoretical curves in both panels are indicated by dielectric continuum theory (– – –), linear coupling model with the saddle-point approximation with two modes (————) and quadratic coupling (- - -). The model used to fit the data is the Franck–Condon factor calculated as a function of the free energy by using the saddle-point approximation. A linear coupling model includes position shifts of the nuclei in a normal mode that is coupled to the reaction. Quadratic coupling includes the frequency shift in a normal mode coupled to the reaction. The details of the saddle-point approximation are discussed in ref. 26. The abscissa is the absolute value of free energy (–ΔG^0). (Reproduced from ref. 24. Copyright 1990 American Chemical Society.)*

Table I. Parameter Values from a Fit to the Linear Coupling Model

Parameter	Population 1 ($k_1 = 75.6\ s^{-1}$)	Population 2 ($k_2 = 21.5\ s^{-1}$)
S_1	2.76 ± 0.10	1.88 ± 0.05
ω_1	$1516.0 \pm 50.0\ cm^{-1}$	$2434.0 \pm 90.0\ cm^{-1}$
S_2	39.4 ± 0.4	13.1 ± 0.10
ω_2	$50.0 \pm 3.1\ cm^{-1}$	$199.0 \pm 6.0\ cm^{-1}$
Reduced χ_2	0.58	0.50

NOTE: Parameters are from a nonlinear least-squares fit of the experimental k_{ET} vs. ΔG^0 curve (Figure 3) to a linear coupling model involving two modes. The saddle-point approximation was used to calculate the k_{ET} vs. ΔG^0 curve that is the fitting function. The electron-phonon coupling constant S_i represents the total coupling for all modes with average frequency ω_i. This fit does not show that only two modes are coupled to the reaction. The standard deviations of the parameters were calculated from the covariance matrix of the fit. All fits assume $\Delta G^0 = 4194\ cm^{-1}$. The fits are illustrated in Figure 4.

the ground state PQ_A may be mediated by higher-lying states such as the intermediate state P^+H^- (24). The electronic factor is calculated by using the ratio $(2V_{12}V_{23}/\Delta E_{21})^2$, where V_{12} is the electronic coupling between the state ^1PH and P^+H^-, V_{23} is the coupling between P^+H^- and $P^+Q_A^-$, and ΔE_{21} is the energy difference between the transition state and the mediating state (P^+H^- in this case). The magnitude of the electronic coupling estimated from the electron-transfer rates is consistent with such a mechanism. Virtual coupling by dipolar states (superexchange) can lead to a new field dependence due to the field-induced change in the energy denominator of the electronic factor (ΔE_{21}). However, the inclusion of even two modes into the description leads to the possibility that the field dependence due to the individual modes may cancel because their contributions to ΔE_{21} cancel. For reasonable values of the parameters obtained by using a two-mode linear coupling model for the Franck–Condon factors (e.g., Figure 4 and Table I) and the known energy of the P^+H^- mediating state, the field dependence of the electronic factor is calculated to be quite small. Thus, the relative contribution of low- and high-frequency modes is not likely to be affected by the presence of superexchange coupling in the electronic factor.

In summary, electric-field modulation of electron-transfer reaction rates has been shown to be a useful approach to providing further information on the factors that determine the rate. Although the application discussed in this chapter has focused on a particular biological example, it is expected that a similar approach can be used for a wide range of reactions. Investigation of model systems is currently in progress.

References

1. Lockhart, D. J.; Boxer, S. G. *Biochemistry* **1987**, *26*, 644–668, 2958.
2. Lockhart, D. J.; Boxer, S. G. *Proc. Natl. Acad. Sci. U.S.A.* **1988**, *85*, 107–111.

3. Lockhart, D. J.; Boxer, S. G. *Chem. Phys. Lett.* **1988**, *144*, 243–250.
4. Lockhart, D. J.; Goldstein, R. F.; Boxer, S. G. *J. Chem. Phys.* **1988**, *89*, 1408–1415.
5. Lockhart, D. J.; Kirmaier, C.; Holten, D.; Boxer, S. G. *J. Phys. Chem.* **1990**, *94*, 6987–6995.
6. Popovic, Z. D.; Kovacs, G. J.; Vincett, P. S.; Dutton, P. L. *Chem. Phys. Lett.* **1985**, *116*, 405–410.
7. Popovic, Z. D.; Kovacs, G. J.; Vincett, P. S.; Alegria, G.; Dutton, P. L. *Chem. Phys.* **1986**, *110*, 227–237.
8. Gopher, A.; Schonfeld, M.; Okamura, M. Y.; Feher, G. *Biophys. J.* **1985**, *48*, 311–320.
9. Feher, G.; Arno, T. R.; Okamura, M. Y. In *The Photosynthetic Bacterial Reaction Center: Structure and Dynamics*; Breton, J.; Vermeglio, A., Eds.; Plenum: New York and London, 1988; pp 271–287.
10. Deisenhofer, J.; Epp, O.; Miki, K.; Huber, R.; Michel, H. *J. Mol. Biol.* **1984**, *180*, 385–398.
11. Kirmaier, C.; Holten, D. *Photosynth. Res.* **1987**, *13*, 225–260.
12. Arata, H.; Parson, W. W. *Biochim. Biophys. Acta* **1981**, *636*, 70–81.
13. Moser C. C.; Alegria, G.; Gunner, M. R.; Dutton, P. L. *Isr. J. Chem.* **1988**, *28*, 133–139.
14. Boxer, S. G.; Goldstein, R. A.; Franzen, S. In *Photoinduced Electron Transfer*; Fox, M. A.; Chanon, M., Eds.; Elsevier Press: New York, 1988; Vol. B, pp 163–215.
15. Boxer, S. G.; Lockhart, D. J.; Franzen, S. In *Photochemical Energy Conversion*; Norris, J. R., Jr.; Meisel, D., Eds.; Elsevier Press: New York, 1989; pp 196–210.
16. Feher, G.; Okamura, M. Y. In *The Photosynthetic Bacteria*; Clayton, R. K.; Sistrom, W. R., Eds.; Plenum: New York, 1978; pp 349–386.
17. Franzen, S.; Boxer, S. G., unpublished results.
18. Kleinfeld, D.; Okamura, M. Y.; Feher, G. *Biochemistry* **1984**, *23*, 5780–5786.
    Gunner, M. R.; Robertson, D. E.; Dutton, P. L. *J. Phys. Chem.* **1986**, *90*, 3783–3795.
20. Parot, P.; Thiery, J.; Vermeglio, A. *Biochim. Biophys. Acta* **1987**, *893*, 534–543.
21. Clayton, R. K.; Yau, H. F. *Biophys. J.* **1972**, *12*, 867–881.
22. Sebban, P.; Wraight, C. A. *Biochim. Biophys. Acta* **1989**, *974*, 54–65.
23. Sebban, P. *Biochim. Biophys. Acta* **1988**, *936*, 124–136.
24. Franzen, S.; Goldstein, R. F.; Boxer, S. G. *J. Phys. Chem.* **1990**, *94*, 5135–5149.
25. Closs, G. L.; Miller, J. R. *Science (Washington, D.C.)* **1988**, *240*, 440–446.
26. Fischer, S. *J. Chem. Phys.* **1970**, *53*, 3195–3207.

Received for review April 27, 1990. Accepted revised manuscript August 29, 1990.

Function of Quinones and Quinonoids in Green-Plant Photosystem I Reaction Center

Masayo Iwaki and Shigeru Itoh

Division of Bioenergetics, National Institute for Basic Biology, 38 Nishigonaka, Myodaijicyo, Okazaki 444, Japan

In photosystem I photosynthetic reaction centers of green plants (spinach), the constituent phylloquinone (2-methyl-3-phytyl-1,4-naphthoquinone, vitamin K_1) that functions as the secondary electron acceptor (A_1) was replaced by various quinones and quinonoid compounds. Most of the quinones and quinonoids tested replaced the function of A_1 and suppressed the charge recombination between the reduced primary electron acceptor chlorophyll a (A_0^-) and the oxidized primary donor ($P700^+$). Redox midpoint potential (E_m) values of the semiquinone$^-$/quinone couple of reconstituted quinones in situ in the photosystem I reaction center were estimated to be about 300 mV more negative than those in dimethylformamide (DMF). This result is different from the situation in the purple bacterial reaction center, in which reconstituted quinones showed E_m values about 400 mV more positive than those in DMF. The variation indicates that there are different quinone environments in these two types of reaction centers.

GREEN PLANTS HAVE TWO TYPES of photosynthetic reaction centers (RC) functioning in series: photosystems I and II. The photosystem (PS) II RC recently isolated (*1*) seems to have a structure essentially similar to that of purple bacterial RC, whose tertiary structure has been determined by X-ray crystallography (*2*, *3*). In both PS II and purple bacterial RCs, four to six chlorophylls, two pheophytins, and two quinones are embedded on two

0065–2393/91/0228–0163$06.00/0

polypeptides of about 30-kDa molecular mass (1–4). Amino acid sequences of the polypeptides of PS II and purple bacterial RCs show only 30% homology to each other, but high homology in essential moieties constituting binding sites for the prosthetic groups (2).

On the other hand, the core part of the PS I RC complex is composed of large polypeptides of about 80-kDa molecular mass, more than 50 chlorophyll a molecules [reaction center chlorophyll (P700), primary electron acceptor chlorophyll (A_0), and mostly antenna chlorophylls; about two phylloquinone (2-methyl-3-phytyl-1,4-naphthoquinone) molecules; and a 4Fe–4S center (F_X, tertiary electron acceptor)] (4–6). A small subunit polypeptide of 9-kDa molecular mass attached to the large-core polypeptides containsj other 4Fe–4S centers F_A and F_B, which accept electrons from F_X (7) (Figure 1).

The amino acid sequences of the PS I polypeptides show almost no homology to those of PS II–purple-bacterial-type RCs (5). This, as well as the difference in prosthetic groups, makes it difficult to consider the evolutionary relationship between these two types of RC complexes. Recent

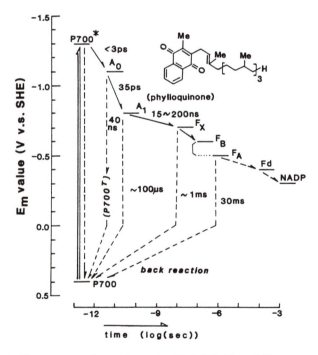

Figure 1. Electron-transfer pathway in PS I RC. Key: P700, ground state; P700*, lowest singlet excited state; P700^T, triplet state of PS I primary electron donor chlorophyll; Fd, ferredoxin. Inset figure shows structure of phylloquinone (A_1) reproduced after refs. 2 and 3. Reaction times are obtained from refs. 4, 6, and 31–35. Continued on next page.

studies, however, indicate that the quinones function as the secondary electron acceptor, both in PS I (*4, 6, 8–20*) [some arguments still remain (*21, 22*)] and PS II–purple-bacterial-type RCs (*23, 24*). Thus, comparison of quinone reactions in these different types of RCs should give new information concerning the electron-transfer mechanism and the evolutionary relationship of photosynthetic RCs.

The energy levels and reaction times of electron-transfer steps in PS I RC and purple bacterial (*Rhodobacter sphaeroides*) RC are compared in Figure 1. Electron transfer in purple bacterial RC is characterized by two ubiquinones (2,3-dimethoxy-5-methyl-6-(prenyl)$_{10}$-1,4-benzoquinone) functioning in series as the secondary (Q_A) and tertiary (Q_B) electron acceptors (*23*). Both of these quinones interact with an Fe atom (*2, 3*). In some purple bacteria such as *Rhodopseudomonas viridis*, Q_A is menaquinone [2-methyl-3-(prenyl)$_n$-1,4-naphthoquinone]. This situation is almost the same in PS II RC, which has plastoquinone [2,3-dimethyl-5-(prenyl)$_9$-1,4-benzoquinone] as Q_A and Q_B (*4, 24*). The large energy gap between the primary electron acceptor pheophytin and Q_A seems to match the reorganization energy of

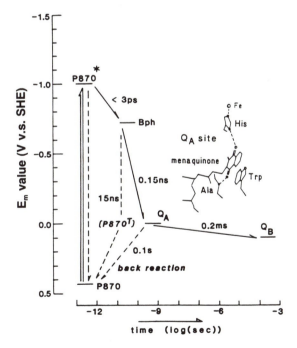

Figure 1. Continued. Electron-transfer pathway in purple bacterial RC. Key: P870, ground state; P870*, lowest singlet excited state; P870^T, triplet state of the primary electron donor bacteriochlorophyll dimer; Bph, bacteriopheophytin. Inset figure shows structure of Q_A site of Rps. viridis RC reproduced after refs. 2 and 3. Reaction times are obtained from refs. 2, 3, 23, and 25.

the reaction (25) and to partially contribute to the activationless behavior of the Q_A reaction in this type of RC. The quinone-extraction–reconstitution studies have been extensively carried out in isolated bacterial RCs to solve the mechanism of efficient electron transfer in RC (25–30). On the other hand, the PS I RC contains phylloquinone as the secondary electron acceptor (A_1), which undergoes a one-electron redox step, does not protonate, and mediates electron transfer between A_0 and F_X (8–20). Reaction of A_1, however, has not been well-characterized yet.

We first reported the extraction and reconstitution method of this phylloquinone in spinach PS I RC (15) and demonstrated the chemical identity of A_1 as phylloquinone (11, 15, 16). This identification was later confirmed in cyanobacterial membranes by another extraction method (9). One molecule of phylloquinone is required to regain the A_1 function (9, 16), and the role of other weakly bound phylloquinone is not known yet. The functional phylloquinone binding site, designated as the Q_ϕ site, also binds herbicides (17, 18) and various artificial quinones (19, 20). The energy diagram of the PS I RC is characterized by a small energy gap for each electron-transfer step, 10-fold faster rate for the quinone reactions, and extremely low redox potentials for components functioning on the reducing side (31–35). The PS I RC thus provides an experimental system to test and refine the electron-transfer mechanism proposed in the purple bacterial RC (25). The reason that PS I RC requires phylloquinone but not plastoquinone and how it distinguishes the former from the latter, which exists in 10 times larger amounts in the same thylakoid membrane (36), are also of interest.

We here summarize and extend the quinone-reconstitution study in PS I RC. Extraction of the original phylloquinone from PS I RC enhances the charge recombination between A_0^- and $P700^+$ [which produces either the triplet $P700^T$ state (9–11) or excited singlet P700*, which produces delayed fluorescence (11)] and results in a decrease in the amount of $P700^+$ detectable in the microsecond-to-millisecond time range. Various quinones or quinonoid compounds, introduced in place of phylloquinone into PS I RC, suppress the charge recombination and mediate the electron transfer between A_0^- and iron–sulfur centers.

Materials and Methods

Lyophilized photosystem I particles, which were obtained by treating spinach chloroplasts with digitonin and contain almost no photosystem II, were twice extracted with a 1:1 mixture of dried and water-saturated diethyl ether, followed by one extraction with dry diethyl ether; this procedure completely removed the two phylloquinone molecules contained in the PS I RC (15). The phylloquinone-depleted particles were also depleted of about 85% of the antenna chlorophyll complement and all carotenoids (15–20, 37, 38). However, P700, A_0, F_X, F_B, and F_A were almost unaffected (15–20). The extracted particles were dispersed in 50 mM glycine OH buffer (pH 10) and then diluted in 50 mM Tris Cl [Tris(hydroxymethyl)aminomethane Cl] buffer (pH 7.5) containing 0.3% (v/v) Triton X-100 (polyethylene glycol *p*-isooctylphenyl ether). After 30 min of incubation, grayish undissolved materials were eliminated by centrifugation. The clear supernatant liquid was diluted about 50 times with 50 mM Tris Cl buffer (pH 7.5), containing 30% (v/v) glycerol to give a final P700 concentration of 0.25 μM, which was used for reconstitution and measurements. All extraction procedure steps were done below 4 °C.

To reconstitute quinones or quinonoids, the suspension of the extracted PS I particles was incubated for a day at 0 °C in the dark with various quinones or quinonoids dissolved in ethyl alcohol, hexyl alcohol, or dimethyl sulfoxide. The quinone-reconstituted RCs were stable when stored below 10 °C. More than 70% of the PS I RCs were reconstituted by this method (*15–20*). Next, 10 mM ascorbate and 0.1 mM dichloroindophenol couple were added to the reaction medium to provide seconds-time-scale reduction of the small amount of P700$^+$ not rapidly reduced by intrinsic components. Quinones and quinonoids used as candidates to reconstitute A_1 were 2,6-diamino-9,10-anthraquinone; fluoren-9-one; anthrone [9(10*H*)-anthracenone] (Aldrich, Milwaukee); 9,10-anthraquinone; menadione (2-methyl-1,4-naphthoquinone) (Katayama, Osaka, Japan); phylloquinone (2-methyl-3-phytyl-1,4-naphthoquinone); menaquinone-4 [2-methyl-3-(prenyl)$_4$-1,4-naphthoquinone] (Sigma, St. Louis); 2,3,5,6-tetramethyl-1,4-benzoquinone; 1,4-naphthoquinone; 1-nitro-9,10-anthraquinone (Tokyokasei, Tokyo, Japan); 1-amino-9,10-anthraquinone; 2,3-dichloro-1,4-naphthoquinone; 2-methyl-9,10-anthraquinone (Wako, Osaka, Japan); 2,3-dimethyl-1,4-naphthoquinone; ubiquinone-10; and plastoquinone-9.

The activity of the reconstituted PS I particles was assayed by measuring the flash-induced absorption change of P700 at 695 nm in a split-beam spectrophotometer (*17*) at 6 °C. The intensity of the actinic flash (532 nm, 10 ns FWHM, 0.7 Hz) from a frequency-doubled Nd-YAG laser (Quanta-Ray, DCR-2-10) was attenuated to excite about a quarter of the RCs to avoid sample damage. Signals were averaged between 32 and 128 scans, as required in each case.

Results

Kinetics of Flash-Induced P700$^+$ in PS I RCs Containing Various Quinones and the Redox Properties of Quinones.

In PS I RCs in which native phylloquinone is extracted with diethyl ether, only a small amount of P700$^+$ was detected in the microsecond-to-millisecond time range after excitation with a laser flash (trace of no additions in Figure 2). This scarcity is due to the rapid return of an electron on A_0^- to P700$^+$ (charge recombination, *see* Figure 1) with a characteristic $t_{1/2}$ of 35 ns (*9–12*). This

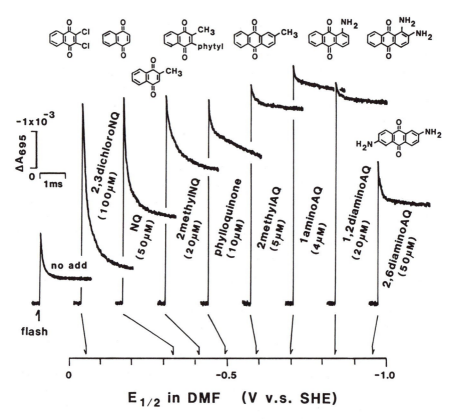

Figure 2. Flash-induced absorption change of P700 in the presence of various naphtho- and anthraquinones in phylloquinone-depleted PS I particles measured at 695 nm. The concentration of added quinone is shown in parentheses. Redox properties of quinones are represented by their $E_{1/2}$ values, referred to the standard hydrogen electrode, measured in DMF (25, 27, 28). NQ and AQ represent 1,4-naphthoquinone and 9,10-anthraquinone, respectively. The figure is redrawn from Figure 1 in ref. 20, with new data added.

reaction produces a small amount of triplet state $P700^T$, which decays with a $t_{1/2}$ of 80 μs [this $t_{1/2}$ is longer than that of a typical few-microsecond decay seen in intact PS I RC, because of the co-extraction of carotenoids (10)] and also is detected at 695 nm (Figure 2). A small amount of $P700^+$ is produced and decays slowly ($t_{1/2}$ = 30–100 ms) through reduction either by an added ascorbate–dichloroindophenol couple or by partially photoreduced iron–sulfur centers.

The flash-induced $P700^+$ extent in the microsecond-to-millisecond range was increased when various naphtho- and anthraquinones, including those known as inhibitors in PS II (39), were substituted for the intrinsic phylloquinone in the PS I RC (Figure 2). This increase indicates that all of these quinones can oxidize A_0^- at a rate rapid enough to compete with the charge recombination with $P700^+$. Benzoquinones were rather poor in this activity except for duroquinone (*19*). The concentration of each quinone shown in Figure 2 was adjusted to provide the maximum level of reconstitution.

   Kinetic patterns of the flash-induced $P700^+$ after the flash, on the other hand, seemed to vary according to the redox potential of the reconstituted quinone given by the polarographically measured potential $E_{1/2}$ value of the semiquinone$^{\cdot -}$/quinone couple in dimethylformamide (DMF). This result suggests that the midpoint potential value of quinone at the Q_ϕ site (E_m in situ with respect to those of A_0 and F_X) is crucial in determining the reaction kinetics of the quinone as recently shown (*19*) and that the E_m in situ is related to $E_{1/2}$ in DMF.

   A group of quinones induced a high extent of flash-oxidized $P700^+$ followed by a slow decay ($t_{1/2}$ = 30–100 ms), as reported for phylloquinone (*16*). The slow decay time indicates that the $P700^+$ is not rereduced by A_0^- or A_1^-, but by $(F_A/F_B)^-$ or by an external reductant (Figure 1) (*16, 19*). Phylloquinone and most anthraquinones (except 1-nitro-9,10-anthraquinone and 2,6-diamino-9,10-anthraquinone) belong to this group. These quinones appear to mediate electron flow from A_0^- to F_X and then to F_A/F_B because the reduction of F_A/F_B can actually be detected at 405 nm, which is an isosbestic wavelength of $P700^+$/P700 (Figure 3) (i.e., they fully reconstituted the function of A_1). The E_m values of these quinones in situ in the Q_ϕ site are, thus, expected to be intermediate between those of A_0 and F_X (*19, 20*) [i.e., between −1000 (*40*) and −720 mV (*41*)].

   With a quinone that has an extremely low potential, 2,6-diamino-9,10-anthraquinone, both the initial extent of flash-induced $P700^+$ and the extent of the slow-decay phase were small even at the saturated concentration (50 μM). This quinone may be incompletely reduced by A_0^-, probably because the E_m of the quinone is comparable to or lower than that of A_0.

   Quinones that have higher potentials [such as 2,3-dichloro-1,4-naphthoquinone, 1,4-naphthoquinone, 2-methyl-1,4-naphthoquinone (menadione), and 1-nitro-9,10-anthraquinone] also promote a high initial extent of the flash-induced $P700^+$, a result indicating that they efficiently accept electrons from A_0^-. However, the decay of $P700^+$ was composed of fast ($t_{1/2}$ = 0.1–1 ms) and slow ($t_{1/2}$ = 30–100 ms) phases. The extent of the latter decreased with the raising of the $E_{1/2}$ value of the tested quinone. This result may be explained if these quinones have E_m values comparable with or more positive than that of F_X so that they cannot fully reduce F_X and therefore, F_A/F_B. In this case, it is expected that the remaining semiquinone$^{\cdot -}$ then will directly reduce $P700^+$ as the fast phase. This mechanism is typified in the quinone with the most positive $E_{1/2}$ value, 2,3-dichloro-1,4-naphthoquinone. It produced a high extent of flash-induced $P700^+$, which decays rap-

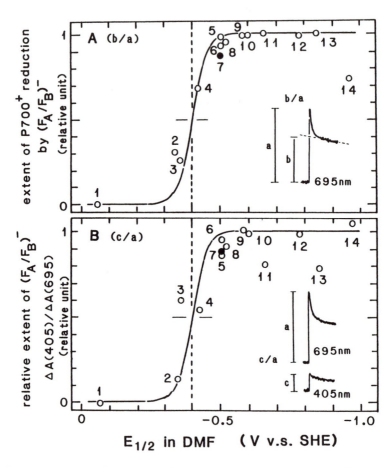

Figure 3. A: Dependence of the relative extent of the quinone-induced increase of the slow-decay phase of P700$^+$ [reduction of P700$^+$ by $(F_A/F_B)^-$] on the $E_{1/2}$ value of added quinone in DMF. B: Dependence of the relative extent of $(F_A/F_B)^-$ measured by the absorption change at 405 nm on the $E_{1/2}$ value of quinone in DMF. Solid lines represent one-electron Nernst's theoretical curve calculated with –0.4-V E_m values. The flash-induced kinetics of P700$^+$ or $(F_A/F_B)^-$ were measured as in Figure 2 at the optimal concentration of quinones and plotted against $E_{1/2}$ value. Quinones used were 1, 2,3-dichloro-1,4-naphthoquinone; 2, 1-nitro-9,10-anthraquinone; 3, 1,4-naphthoquinone; 4, 2-methyl-1,4-naphthoquinone (menadione); 5, 2,3-dimethyl-1,4-naphthoqui-none; 6, menaquinone-4; 7, phylloquinone; 8, 1-chloro-9,10-anthraquinone; 9, 9,10-anthraquinone; 10, 2-methyl-9,10-anthraquinone; 11, 1-amino-9,10-an-thraquinone; 12, 2-amino-9,10-anthraquinone; 13, 1,2-diamino-9,10-anthra-quinone; and 14, 2,6-diamino-9,10-anthraquinone.

idly ($t_{1/2}$ = 100 μs), but almost no increase of the slow decay phase (Figure 2). The semiquinone$^{\cdot-}$/quinone couple of this quinone in the Q_ϕ site is estimated to have an E_m value too positive to reduce F_X and so the semiquinone$^{\cdot-}$ formed will reduce P700$^+$ directly with a fast rate, as observed.

Dependence of P700 kinetics on the redox property of quinones (as shown in Figure 2) was further analyzed with 14 quinones (Figure 3). The initial height of the flash-induced P700$^+$ was almost independent of the redox property of quinones except at extremely negative potentials, as seen in Figure 2. This independence indicates that, for RCs with reconstituted quinones, the efficiency and rate of acceptance of an electron from A_0^- is sufficiently high to compete against the charge-recombination reaction between P700$^+$ and A_0^- for most of quinones tested, except for those with the extremely low redox potentials. On the other hand, decay kinetics of P700$^+$ varied depending on quinone species. Figure 3A indicates the relative extent of the slow-decay phase with respect to the initial extent of P700$^+$ formation. Because the slow phase represents the extent of P700$^+$ that is not rereduced by the flash-reduced quinone, this extent represents the ability of the quinone to reduce F_X.

If the reaction rate between quinone and F_X is assumed to be fast enough to allow equilibrium between them before reaction of F_X with F_A/F_B or P700$^+$, this curve may be used to represent the redox titration of F_X by using quinones of different redox potential. The curve indicates that the quinone with an $E_{1/2}$ value of –0.4 V in DMF gives the half maximal increase of the slow phase. This increase suggests that quinones with this $E_{1/2}$ value show almost the same E_m value as F_X [E_m = –0.72 V (*41*)] in the PS I RC protein. The points with different quinones are fitted with the theoretical one-electron Nernstian curve. This correspondence suggests that E_m values of quinones in situ in PS I RC are shifted about 0.3 V to the negative side from those in DMF. From this figure the E_m in situ value of phylloquinone (closed circles in Figure 3) can be estimated to be 0.1 V more negative than that of F_X (i.e., –0.82 V).

The amount of reduced F_A/F_B was directly measured by monitoring their flash-induced absorption change at an isosbestic wavelength of P700$^+$/P700 (405 nm). Difference spectra of F_A/F_B observed in the PS I RC reconstituted with low-potential quinones (not shown) were similar to that observed in the phylloquinone-reconstituted PS I RC (*16*). The initial amount of flash-induced (F_A/F_B)$^-$ depended on the $E_{1/2}$ value of quinone (Figure 3B). This dependence confirms the indirect estimation from the reaction kinetics of P700 shown in Figure 3A. The plot is also well-fitted with the theoretical curve with a –0.4-V E_m value.

The linear relationship between the E_m value at the Q_ϕ site of quinone and its $E_{1/2}$ in DMF contrasts with the results in the *Rb. sphaeroides* Q_A site at which $E_{1/2}$ in DMF of the reconstituted quinone was a poor indicator of its in situ E_m value (*23, 42, 43*). More work, however, seems to be required before this conclusion can be accepted because the present method of E_m estimation at the Q_ϕ site is not highly accurate for the quinone, for which the E_m significantly differs from that of F_X.

Kinetics of P700$^+$ in PS I RC Containing Quinonoid Compounds. The nonquinone compounds containing one carbonyl group, anthrone, and fluoren-9-one or its derivatives introduced into PS I RC in place of phylloquinone, also function as A_1. They recover the P700$^+$ extent in the microsecond-to-millisecond time range. The kinetics of P700$^+$ varied depending on the $E_{1/2}$ value of the compound used, essentially in a similar way as seen with quinones (not shown). These results indicate that fluorenones or anthrone can also function as the electron acceptor A_1 in PS I RC, as do the quinones, and that they show shifts of redox potential when introduced into the RC protein to an extent similar to that observed with quinones. This will be discussed elsewhere in more detail.

Binding Affinity of Quinones and Quinonoids. The extent of P700$^+$ after the flash excitation depended on the concentration of added quinone. The concentration dependence gives a measure of binding affinity of the quinones to the Q_ϕ site. Dissociation constants (K_d) of quinones were calculated from the dependence of the extent of flash-induced P700$^+$ on the concentration of quinones by fitting with a theoretical curve. Results are summarized in Table I. Increasing the number of aromatic rings of the quinone leads to a lower K_d value and hence to tighter binding (19). The

Table I. Dissociation Constant and Binding Free Energy Between Quinone–Quinonoid and PS I Quinone Binding Q_ϕ Site

Compounds	$log\ [K_d\ in\ M]$	$-\Delta G\ (kJ/mol)$
Quinones[a]		
Duroquinone	-5.2	-28
1,4-Naphthoquinone	-5.7	-31
2-Methyl-1,4-naphthoquinone	-6.5	-35
2,3-Dimethyl-1,4-naphthoquinone	-7.2	-39
Phylloquinone	<-8	<-43
Menaquinone-4	<-8	<-43
9,10-Anthraquinone	-7.7	-41
Quinonoids[b]		
Fluoren-9-one	-6.5	-35
Anthrone	-6.3	-33
Inhibitors[c]		
Atrazine	>-2.0	>-10
4,7-Dimethyl-o-phenanthroline	-4.8	-25
Antimycin A	-5.7	-31
Myxothiazole	-5.2	-28

[a]K_d values were obtained from ref. 19.
[b]K_d values were determined in this work.
[c]K_d values were obtained from ref. 17.

attachment of a hydrocarbon tail seems to result in tighter binding when the K_d values of 1,4-naphthoquinone, 2-methyl-1,4-naphthoquinone (menadione), 2,3-dimethyl-1,4-naphthoquinone, menaquinone, and phylloquinone are compared. The more hydrophobic quinones bind to the reaction center more tightly, as reported in the bacterial Q_A site (*42*). However, plastoquinone-9 and ubiquinone-10, which are known to function in PS II and bacterial RCs (*4*, *23*), respectively, were less effective, although they have long hydrocarbon tails. They gave log K_d values of higher than –4 (M. Iwaki and S. Itoh, unpublished data). These quinones may not properly bind to the Q_ϕ site. Reconstitution of one molecule of phylloquinone per PS I RC fully recovers the extent of P700$^+$ (*16*). This recovery indicates that only one of the two phylloquinone molecules contained in the intact PS I RC functions as A_1. The role of the other phylloquinone molecule with weaker binding affinity (*13*, *14*) is not known yet.

From the difference in K_d values we can estimate the contribution of side groups to the binding energy. The difference of binding energy between 1,4-naphthoquinone and 9,10-anthraquinone suggests that the addition of one aromatic ring increases the binding energy by about 10 kJ/mol. On the other hand, about 4 kJ/mol of stabilization is estimated for the addition of one methyl group to the naphthoquinone ring. These results are qualitatively similar to those shown in the quinone binding to the Q_A site of *Rb. sphaeroides* (*25*, *26*, *42*) and indicate that the increase of hydrophobicity of the quinone molecule increases the binding free energy (*20*).

Deletion of one carbonyl from anthraquinone is expected to increase its hydrophobicity. However, K_d values of the one-carbonyl three-aromatic ring compounds, fluoren-9-one, and anthrone were 1.2–1.4 order higher than that of anthraquinone. Thus, at least one of the carbonyl groups of anthraquinone contributes to the increase of affinity to the Q_ϕ site by 6–8 kJ/mol. This affinity strongly suggests that the carbonyl group is hydrogen bonded to the side chain of RC polypeptide, as shown in purple bacterial Q_A site (*2*, *3*). Lower binding affinity of fluorenones were also reported in *Rb. sphaeroides* RC (*26*).

Compounds known as the inhibitors of quinone reactions in PS II–purple bacterial RC or in cytochrome b–c_1 complex also bind to the Q_ϕ site. Their K_d values were determined by analyzing their competitive inhibition of the quinone (menadione) reconstitution into PS I RC (*17*). 4,7-Dimethyl-*o*-phenanthroline, which strongly binds to the Q_B or Q_A site of purple bacteria or PS II, also binds to PS I RC. However, atrazine, which is known to be a specific inhibitor binding to the Q_B site (*24*), showed only a weak binding (*17*). This variation indicates that the Q_ϕ site has a different structure from that of the Q_A or Q_B site. Antimycin A and myxothiazole, which are known to be inhibitors of the quinone reaction in the cytochrome b–c_1 complex (*46*), showed a strong inhibitory effect and gave a low K_d value. Although the structure of the Q_ϕ site is not yet known, the binding affinities of these compounds in Table I indicate the unique nature of the Q_ϕ site.

Discussion

Electron-transfer reactions in purple bacterial RC containing artificial quinones as Q_A in place of the original ubiquinone have been extensively studied (25–30, 42, 43). Various quinones were shown to be incorporated into the RC, and the relationship between the energy level and the reaction rate has been studied by measurements of reactions of quinones with different E_m values (25, 42, 43). However, a homologous study has never been carried out in green-plant RCs because of the difficulty in extraction and reconstitution of quinones. The study presented here shows that this type of analysis can also be done in green-plant PS I RC.

Redox Properties of Quinones and Their Ability To Function as A_1.   We propose that the thermodynamic relationship between the quinone and its reaction partners depends on the E_m values of the semiquinone$^{\cdot-}$/ quinone couple in situ in the Q_ϕ site with respect to those of A_0 and F_X (Figure 4). The E_m value of A_0 may be similar to that of the chlorophyll a couple, which is about −1.0 V (40). The E_m of F_X is reported to be −0.72 V (41). It is apparent that the E_m values of the quinones in the Q_ϕ site parallel the $E_{1/2}$ values in DMF.

The E_m in situ of phylloquinone (A_1) has never been determined directly because of its extremely negative E_m value and has been estimated to be around −0.9 V (i.e., intermediate between A_0 and F_X) (4, 6). The E_m value can be estimated to be −0.82 V from the results in the present study. This E_m value is about 0.3 V more negative than the $E_{1/2}$ value measured in DMF. Moreover, the E_m value for phylloquinone is 0.7 V lower than that of menaquinone (which differs from phylloquinone only in the structure of the hydrocarbon tail and gives similar $E_{1/2}$ in DMF) in the bacterial Q_A site (23, 43, 45, 46) (Figure 4). Menaquinone functions as A_1 in the PS I Q_ϕ site in almost the same way as phylloquinone does. These observations indicate that the stability of the semiquinone$^{\cdot-}$ is significantly lower at the Q_ϕ site than at the Q_A site.

The estimated E_m value of phylloquinone in situ (A_1) allows us to calculate the energy gaps between A_0 and A_1, and between A_1 and F_X, to be about 0.2 and 0.1 eV, respectively. The A_0–A_1 energy gap is somewhat lower than the corresponding energy gap between bacteriopheophytin and Q_A in *Rb. sphaeroides* RC, which is calculated to be more than 0.5 eV (Figure 1). The reactions between A_0, A_1, and F_X are estimated to proceed even at 4 K (8). Preliminary results in the quinone- or fluoren-9-one-reconstituted system indicate that the reaction between P700$^+$ and A_1^- is temperature-independent (S. Itoh and M. Iwaki, unpublished data). Attempts to obtain the relationship between the energy gap and the reaction rate, as shown in *Rb. sphaeroides* RC (25) or in synthesized organic compounds (47), are now in progress for the reaction of reconstituted quinones in the Q_ϕ site of PS I RC.

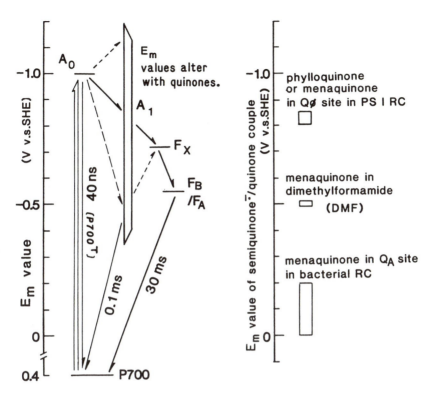

Figure 4. Left: Energy diagram and estimated electron-transfer pathways in PS I RC containing various quinones with different E_m values of semiquinone·⁻/quinone couple in place of intrinsic phylloquinone (see text for details). Right: Comparison of the semiquinone·⁻/quinone E_m values of phylloquinone and menaquinone in different environments.

Quinone Structure and Binding Affinity for the Q_ϕ Site.

Various quinones and quinonoids bind to the PS I Q_ϕ site in place of native phylloquinone and function as A_1. The phytyl tail or naphthoquinone ring of phylloquinone does not seem to be essential in order to function as A_1, although these properties contribute to the tight binding of the phylloquinone to the Q_ϕ site with a significant contribution from hydrophobic interactions, as reported for the bacterial Q_A site (42). Our results suggest that the addition of one methyl group can increase the binding energy by about 4 kJ/mol, and that one aromatic ring can increase the binding energy by about 10 kJ/mol by increasing the hydrophobicity of quinone. This interpretation agrees with the results in *Rb. sphaeroides* RC. On the other hand, deletion of a carbonyl group from anthraquinone is estimated to decrease the binding energy by more than 6–8 kJ/mol. The change of binding free energy seems to be comparable to the energy of hydrogen bonding of water (20 kJ/mol). This comparison indicates that at least one carbonyl group makes a hydrogen bond with a hydrogen atom of an amino acid side chain of the RC polypeptide, as shown in *Rps. viridis* RC (2, 3), although the binding site in PS I is not yet clear.

Our results indicate that the rate of electron transfer at the Q_ϕ site is mainly dependent on the E_m in situ of the semiquinone$^{\cdot-}$/quinone couple, but not significantly dependent on the structure of quinone. This distinction confirms the conclusion of the quinone-reconstitution studies at the Q_A site in the *Rb. sphaeroides* RC (25–30, 42, 43). The quinone reconstitution and replacement studies have given information about the dynamic relationship between the structure and function of the purple bacterial RC (25, 42, 43). This kind of approach can now be adopted for the green-plant PS I RC complex, whose tertiary structure still remains to be characterized. More information is required to establish the precise location of the Q_ϕ site in PS I RC. Studies in photosynthetic RC containing nonquinone compounds seem to open a new field for the study of electron-transfer mechanisms.

List of Symbols

A_0	photosystem I primary electron acceptor chlorophyll a
A_1	photosystem I secondary electron acceptor phylloquinone
DMF	dimethylformamide
$E_{1/2}$	half wave reduction potential, polarographically measured
E_m	midpoint potential
F_X, F_A, and F_B	photosystem I electron acceptor iron–sulfur centers
K_d	dissociation constant
P700	photosystem I primary electron donor chlorophyll a
PS	photosystem
Q_A and Q_B (sites)	primary and secondary electron acceptor quinone (binding sites) in PS II or purple bacterial reaction center
Q_ϕ site	photosystem I phylloquinone binding site
RC	reaction center
$t_{1/2}$	decay half life

Acknowledgments

The authors thank Y. Fujita of NIBB and P. L. Dutton of the University of Pennsylvania for their stimulating discussions; A. Osuka of Kyoto University for a kind gift of 2,3-dimethyl-1,4-naphthoquinone and S. Okayama and Y. Isogai of Kysushu University for their kind gifts of ubiquinone-10 and plastoquinone-9; and J. R. Bolton of the University of Western Ontario for critical reading of the manuscript. The work was supported by Grants-in-Aid for Scientific Research (B) and for Co-operative Research (A) to S. Itoh from the Japanese Ministry of Education, Science, and Culture.

References

1. Nanba, O.; Satoh, K. *Proc. Natl. Acad. Sci. U.S.A.* **1987**, *84*, 109.
2. Deisenhofer, J.; Michel, H. *EMBO J.* **1989**, *8*, 2149.
3. Michel, H.; Epp, O.; Deisenhofer, J. *EMBO J.* **1986**, *5*, 2.
4. Andreasson, L.-E.; Vängard, T. *Annu. Rev. Plant Physiol. Biol.* **1988**, *39*, 379.
5. Kirsch, W.; Seyer, P.; Hermann, R. G. *Curr. Genet.* **1986**, *10*, 843.
6. Golbeck, J. H. *Biochim. Biophys. Acta* **1987**, *895*, 167.
7. Oh-oka, H.; Takahashi, Y.; Wada, K.; Matsubara, H.; Ohyama, K.; Ozeki, H. *FEBS Lett.* **1987**, *218*, 52.
8. Brettel, K. *FEBS Lett.* **1988**, *239*, 93.
9. Biggins, J.; Mathis, P. *Biochemistry* **1988**, *27*, 1494.
10. Ikegami, I.; Sétif, P.; Mathis, P. *Biochim. Biophys. Acta* **1987**, *894*, 414.
11. Itoh, S.; Iwaki, M. *Biochim. Biophys. Acta* **1988**, *934*, 32.
12. Mathis, P.; Ikegami, I.; Sétif, P. *Photosynth. Res.* **1988**, *16*, 203.
13. Takahashi, Y.; Hirota, K.; Katoh, S. *Photosynth. Res.* **1985**, *6*, 183.
14. Schoeder, H.-U.; Lockau, W. *FEBS Lett.* **1986**, *199*, 23.
15. Itoh, S.; Iwaki, M.; Ikegami, I. *Biochim. Biophys. Acta* **1987**, *893*, 508.
16. Itoh, S.; Iwaki, M. *FEBS Lett.* **1989**, *243*, 47.
17. Itoh, S.; Iwaki, M. *FEBS Lett.* **1989**, *250*, 441.
18. Itoh, S.; Iwaki, M. In *Current Research in Photosynthesis*; Baltscheffsky, M., Ed.; Kluwer Academic Publisher: Dordrecht, Netherlands, 1990; Vol. II, pp 651–657.
19. Iwaki, M.; Itoh, S. *FEBS Lett.* **1989**, *256*, 11.
20. Iwaki, M.; Itoh, S. In *Current Research in Photosynthesis*; Baltscheffsky, M., Ed.; Kluwer Academic Publisher: Dordrecht, Netherlands, 1990; Vol. II, pp 647–650.
21. Palace, G. P.; Franke, J. E.; Warden, J. T. *FEBS Lett.* **1987**, *215*, 58.
22. Ziegler, K.; Lockau, W.; Nitschke, W. *FEBS Lett.* **1987**, *217*, 16.
23. Prince, R. C.; Dutton, P. L. In *The Photosynthetic Bacteria*; Clayton, R. K.; Sistrom, W. R., Eds.; Plenum: New York, 1978; pp 439–469.
24. Trebst, A. *Z. Naturforsch.* **1987**, *42C*, 742.
25. Gunner, M. R.; Robertson, E. E.; Dutton, P. L. *J. Phys. Chem.* **1986**, *90*, 3783.
26. Warncke, K.; Dutton, P. L. In *Current Research in Photosynthesis*; Baltscheffsky, M., Ed.; Kluwer Academic Publisher: Dordrecht, Netherlands, 1990; Vol. I, pp 157–160.
27. Gunner, M. R.; Tiede, D. M.; Prince, R. C.; Dutton, P. L. In *Function of Quinone in Energy-Conserving Systems*; Trumpower, B. L., Ed.; Academic Press: New York, 1982; pp 265–269.
28. Dutton, P. L.; Gunner, M. R.; Prince, R. C. In *Trends in Photobiology*; Helene, C.; Chalier, M.; Montenay-Garestier, T.; Laustriat, G., Eds.; Plenum Press: New York, 1982; pp 561–570.
29. Cogdell, R. J.; Brune, C. D.; Clayton, R. K. *FEBS Lett.* **1974**, *45*, 344.
30. Okamura, M. Y.; Isaacson, R. A.; Feher, G. *Proc. Natl. Acad. Sci. U.S.A.* **1975**, *72*, 3491.
31. Owens, T. G.; Webb, S. P.; Mets, L.; Alberte, R. S.; Fleming, G. R. *Proc. Natl. Acad. Sci. U.S.A.* **1987**, *84*, 1532.
32. Shuvalov, V. A.; Nuijs, A. M.; van Gorkom, H. J.; Smit, H. W. J.; Duysens, L. N. M. *Biochim. Biophys. Acta* **1986**, *850*, 319.
33. Wasielewski, M. R.; Fenton, J. M.; Govindjee *Photosynth. Res.* **1987**, *12*, 181.
34. Nuijs, A. M.; Shuvalov, V. A.; van Gorkom, H. J.; Plijter, J. J.; Duysens, L. N. M. *Biochim. Biophys. Acta* **1986**, *850*, 310.
35. Mathis, P.; Sétif, P. *FEBS Lett.* **1988**, *237*, 65.

36. Barr, R.; Crane, F. L. *Methods Enzymol.* **1971**, *23*, 372.
37. Ikegami, I.; Katoh, S. *Biochim. Biophys. Acta* **1975**, *376*, 588.
38. Ikegami, I.; Itoh, S. *Biochim. Biophys. Acta* **1987**, *893*, 517.
39. Oettmeier, W.; Masson, K.; Donner, A. *FEBS Lett.* **1988**, *231*, 259.
40. Fujita, I.; Davis, M. S.; Fajer, J. *J. Am. Chem. Soc.* **1978**, *100*, 6280.
41. Chamorovsky, S. K.; Cammack, R. *Photobiochem. Photobiophys.* **1982**, *4*, 195.
42. Warncke, K.; Gunner, M. R.; Braun, B. S.; Yu, C.-A.; Dutton, P. L. *Prog. Photosynth. Res., Proc. Int. Congr.* **1987**, *1*, 225.
43. Woodbury, N. W.; Parson, W. W.; Gunner, M. R.; Prince, R. C.; Dutton, P. L. *Biochim. Biophys. Acta* **1986**, *851*, 6.
44. Prince, R. C.; Gunner, M. R.; Dutton, P. L. In *Function of Quinones in Energy Conserving Systems*; Trumpower, B. L., Ed.; Academic Press: New York, 1982; pp 29–33.
45. Prince, R. C.; Dutton, P. L.; Bruce, J. M. *FEBS Lett.* **1983**, *160*, 273.
46. Cramer, W. A.; Black, M. T.; Widger, W. R.; Girvin, M. E. In *The Light Reactions*; Barber, J., Ed.; Elsevier: Amsterdam, 1987; pp 447–493.
47. Closs, G. L.; Miller, J. R. *Science (Washington, D.C.)* **1988**, *240*, 440.

RECEIVED for review April 27, 1990. ACCEPTED revised manuscript August 20, 1990.

Effects of Reaction Free Energy in Biological Electron Transfer In Vitro and In Vivo

George McLendon[1], David Hickey[1,2], Albert Berghuis[3], Fred Sherman[2], and Gary Brayer[3]

[1]Department of Chemistry and [2]Department of Biochemistry, University of Rochester, Rochester, NY 14627
[3]Department of Biochemistry, University of British Columbia, Vancouver, British Columbia V6T 1W5, Canada

Although much work detailed in this volume and elsewhere has focused on the determinants of protein electron-transfer rates in vivo, far less is known about the control of electron transfer in vivo. To this end, we here report results concerning the in vitro and in vivo reactivity of cytochrome c and a single-site replacement, N52I. This mutant, despite a significant (50 mV) change in redox potential, surprisingly does not have a deleterious effect on growth in vivo. Thus, a possible mechanism is proposed for regulation of growth of yeast on lactate.

WORK ON BIOLOGICALLY RELEVANT ELECTRON TRANSFER over the last decade has transformed our understanding of the fundamental chemical aspects of this key biochemical reaction. Theoretical advances built on the pioneering work of Marcus and others (1–3) have outlined the effects on rate of donor–acceptor distance, reaction free energy, internal molecular motion and solvent motion, and solvent relaxation dynamics.

Experimental tests of each of these predictions have been made through in vitro experiments (4–8). For example, Gray and co-workers (4, 5) have systematically varied the donor–acceptor distance in proteins by complexing redox-active Ru adducts to specific histidines located on the surface of proteins like cytochrome *c*, myoglobin, and azurin. By looking at several sites,

0065–2393/91/0228–0179$06.00/0

they found exponential dependence of rate (k_{et}) on distance (R), $k_{et} \propto$ $\exp(-0.9\,R)$, in accord with both theory and previous experiments on small-molecule electron transfer.

In complementary work, other groups have examined electron transfer in physiological protein–protein complexes (6–8). By substituting different metal-containing porphyrins into the active sites of proteins like hemoglobin, cytochrome c, and cytochrome c peroxidase, the reaction free energy could be continuously varied. From Marcus theory, the electron-transfer rate constant depends on reaction free energy and reorganization energy, $k_{et} \propto$ $\exp(-\Delta G^{\ddagger}/kT)$

$$\Delta G^{\ddagger} = \frac{(\Delta G^0 - \lambda)^2}{4\lambda}$$

where $\Delta G^{\ddagger}$ is activation free energy, k is the Boltzmann constant, T is absolute temperature, ΔG^0 is reaction free energy, and λ is reorganization energy. An example of this dependence for a biological protein–protein redox couple, the cytochrome c–cytochrome c peroxidase complex, is shown in Figure 1. A fit to the Marcus equation gives an estimate of $\lambda \simeq 1.4$ eV,

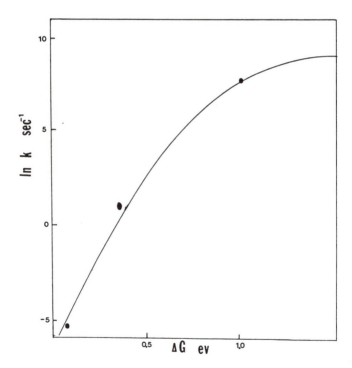

Figure 1. Dependence of the rate constant for oxidation on reaction free energy (ΔG) for the system $[Fe(IV)O]$ cytochrome c peroxidase–(M)cytochrome c [where M is Fe^{II}, Zn^{II}, or $(H^+)_2$]. The solid line is a fit to the Marcus equation, with $\lambda = 1.4$ eV.

which is corroborated by a direct measurement of the activation free energy in this system. Such studies suggested that in protein complexes electron transfer is accompanied by significant reorganization of the protein structure. In Marcus' terms, the reorganization energy, λ, is large; λ ~ 0.8–1.5 V. The limited data for such complexes also support a strong exponential dependence of the electron-transfer rate on the distance between the electron-donor group and the acceptor.

Although such studies have clarified the fundamental parameters that are important in chemically and biochemically relevant electron-transfer reactions, they do not directly test to what extent these physical parameters affect the net metabolic electron-transfer flux (due to electron transport) in vivo. We assume that because these parameters control rates under simple in vitro laboratory conditions, they will exercise similar control in vivo. I call this the first law of biophysics: "If we can observe something, it must be important." Clearly, basic physical and chemical principles operate equally inside and outside living cells. It is less clear how sensitive the net metabolic fluxes (which are the key to cellular growth) are to small changes in the reaction free energy.

We have therefore chosen to examine directly how changes in biological redox potential affect in vivo metabolism. Specifically, the reaction free energy of the key metabolic electron-transfer steps involving cytochrome c have been changed by changing the redox potential of cytochrome c with site-directed mutagenesis.

These inorganic-based redox reactions offer a unique opportunity to explore free-energy effects on metabolism. For metabolic transformations of organic substrates, the reaction free energy is determined by the different bond energies of the reactant(s) and product(s), which in turn are set by the covalent structure. Such transformations cannot be modified without changing the chemical identity of the reactant, a change that would require new metabolic pathways. For redox proteins, however, the redox potentials that set the metabolic free energy are controlled by subtle conformational effects in the surrounding protein, so that the reaction free energy may be altered without requiring the creation of any new metabolic pathways.

One final caveat is necessary. To make meaningful measurements of in vivo activity, the strains compared must be genetically identical in all respects. The exception would be the cytochrome c gene itself, which should be present as a single-copy gene integrated into the correct position on the yeast chromosome. These requirements have not been met in several attempted studies of in vivo function. Unfortunately, it is impossible to obtain any quantitative estimation of in vivo function (the "differential specific activity") in these constructs with multicopy genes, like replicating plasmids.

Factors That Affect the Redox Potential of Cytochrome c

Although cytochrome c serves as a paradigm for many structure–function relationships in proteins (9), the factors controlling the basic redox properties of cytochrome c remain elusive (10, 11). Factors that have been suggested as important include heme "exposure" (i.e., local dielectric environment) (12–14), axial ligation (15), surface-charge distribution (16), and redox state-specific hydrogen-bonding patterns (17, 18) specifically involving the H-bond network between a heme propionate, Asn 57, Arg 38, and an interstitial H_2O molecule. These factors are carefully controlled; in all species, ranging from yeast to plants to humans, the redox potential of cytochrome c is essentially invariant at $E° = 265(±10)$ mV.

With the advent of protein-engineering techniques (12–14, 17–21), it has become possible to investigate the roles played by individual amino acids in controlling the redox potential and redox activity of this protein.

Results and Discussion

Charge Mutants. *Structural Studies.* We first consider the effects of single-charged residues. In early work, Rees (22) suggested that the coulombic potential between a surface-charged residue and the heme could modulate the redox potential and provided some data on chemically modified proteins in support of this suggestion. This analysis was questioned by Moore and co-workers (10, 11), who pointed out several possible pitfalls in the analysis but did not directly address the data of Rees (22). We have isolated and extensively characterized (19–21) single-site replacements in yeast iso-l-cytochrome c for several of the evolutionarily invariant lysines that surround the heme and are thought to be functionally significant. The derivatives investigated here include Lys 32 → Leu, Ile, Arg 18 → Ile, and Lys 77 → Arg, Asp. Each of these replacements support aerobic respiration in vivo and have been extensively characterized in vitro.

The reactivities of these derivatives with various physiological partners of cytochrome c have been reported previously: The Michaelis parameters, K_m (related to binding), and the maximum rate, K_{cat}, vary less than twofold for all these derivatives relative to wild type. Structural characterizations in solution have been carried out by UV–visible spectroscopy, circular dichroism, and NMR spectroscopy. No significant changes in structure can be observed by any of these techniques. Within experimental error, absorption wavelengths and relative extinction coefficients (e.g., Soret, 695) for each of these derivatives are the same as wild type in both the visible spectra and the circular dichroic (CD) spectrum. The positions of the very structure-sensitive hyperfine shifted heme resonances are also unperturbed between wild type and these charge mutants.

Redox Potential. The potentials of these and other mutants have been determined by potentiometric titration with $[Co(II)(terpyridine)_2]Cl_2{}^-$, optically transparent thin-layer electrochemistry, and direct electrochemistry on partially oxidized graphite electrodes. The redox potentials so obtained for single (Lys → neutral) or double (Lys → Asp) charge changes at individual residues are listed in Table I. Only very small changes in potential, which can lead to either a decrease or an increase in $E°$, are observed. These results support the analysis by Moore and co-workers (*10, 11*) of charge effects. Indirectly, the data imply that any charge effects are screened by a high effective dielectric medium between the lysine Nε and heme iron.

Table I. Cyclic Voltammetric Results of the Cyt *c* Mutants at Modified Gold and Edge-Plane Graphite Electrodes

Cyt c Mutant	$i_{p,a}$ $(\mu A)^a$	ΔE_p $(\pm 5\ mV)^b$	$E_{1/2}$ $(\pm 5\ mV)$
Cys 102 Thr	0.22^c	70	+41
(70 μM)	0.53^d	80	+39
Asn 52 Ile–Cys 102 Thr	0.36^c	67	−10
(99 μM)	0.73^d	68	−10
Asn 52 Ala–Cys 102 Thr	0.31^c	70	+18
(80 μM)	0.47^d	152	+12
Lys 72 Asp	0.44^c	66	+31
(143 μM)	$—^{d,e}$	—	—
Lys 27 Gln	0.31^c	74	+33
(110 μM)	$—^{d,e}$	—	—
Arg 13 Ile	0.49^c	68	+62
(181 μM)	$—^{d,e}$	—	—

NOTE: Results for cyt *c* are in 100 mM KCl/10 mM HEPES, pH 7.4; at modified gold, 0.125 cm²; and edge-plane graphite (EPG) electrodes, 0.148 cm².
[a]Anodic peak current measured at 20 mV s⁻¹.
[b]Peak separation measured at 100 mV s⁻¹.
[c]Associated with modified gold electrodes.
[d]Associated with EPG electrodes.
[e]Deterioration of the voltammograms prevented precise measurements.

H-Bonding Mutants: Asn 57 → Ile, Ala.   The second class of mutants examined are involved in a complex hydrogen-bond network that includes the heme propionate and an intrinsic water molecule (Figure 2). Several residues participate in this network, including Arg 38, Asn 52, Tyr 67, and Thr 78. Recently, Cutler et al. (*17*) reported that mutation of Arg 38 can result in significant shifts in $E°$ (up to 50 mV) in the direction predicted by an electrostatic model developed by Cutler et al. (*17*). Subsequently, Luntz et al. (*23*) demonstrated that the replacement of Tyr 67 by Phe leads to a 40-mV decrease in $E°$. We have found that replacements at position 52 lead to similar changes.

Structural Studies.   Solution structural characterization of the Asn 52 → (Ile, Ala) mutants shows that there are no significant changes (<2%) in the visible or CD spectra between the wild-type protein and the mutants (*19–21*). This finding is strongly underscored by the recent solution of the structure of the Asn 57 → Ile mutant using X-ray diffraction techniques,

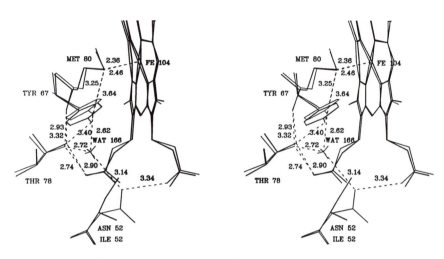

Figure 2. Hydrogen-bond network in left, native (N 52) and right, the low-potential mutant (I 52).

which show that there are no significant changes in the main-chain positions between the mutant and wild-type proteins (24). However, one important change that has occurred is the displacement of an internal H_2O by the substituted Ile side chain. Water displacement has also led to significant alterations of the hydrogen-bonding pattern in this region (Figure 2). Clearly the heme (electronic) environment is altered by the Asn 57 → Ile, Ala replacements, as indicated by these structural analyses and significant shifts in the positions of the heme hyperfine shifted resonances, which are exceedingly sensitive to small perturbations in the electron-density distribution of the heme.

Redox Potentials. The Asn → Ile replacement leads to a significant 50-mV shift in $E°$, from 270 to 220($\pm$5) mV. A similar shift was reported by Luntz et al. (23) for the replacement of Tyr 67 → Phe. They used this shift to predict that loss of any of the critical H-bonding triad (Tyr 67, Asn 52, or Thr 78) could lead to loss of the internal water molecule, with consequent effects on protein stability (which increases when the internal H_2O is replaced) and redox potential (which decreases). The results from the Ile 52 mutant, in which the internal water is known to be absent, support this prediction. However, key structural studies by Hickey et al. (24) of the Tyr 67 → Phe mutant do not show water displacement; in fact, an extra water is added to the cavity to replace the lost Tyr OH. A similar shift is found for the Ala 57 mutant, for which $E° = 230(\pm10)$ mV. Thus, regardless of the steric bulk of the replacement, a similar redox potential is observed. This similarity suggests that the internal water is indeed a key determinant of cytochrome c redox potentials.

In Vivo Effects of Redox Potential on Growth. With the combined structural and redox data in hand, the Ile 52 mutants were chosen as the best candidates for testing the effect of redox potential on redox metabolic flux (and, thereby, cell viability and growth). Given the evolutionary invariance of redox potential over all eukaryotes, it is logical to anticipate that a 40-mV decrease in cytochrome *c* potential might adversely affect electron-transport rate, although the magnitude of such an effect is difficult to anticipate. For any effect to be observed, metabolic conditions are required for which cytochrome *c* is involved in one or more steps that are rate-determining in overall metabolism. Such conditions are unusual. However, detailed experiments on the genetic regulation of cytochrome *c* have shown that when yeast is grown on a synthetic medium that contains lactate as the sole nonfermentable carbon source, then net metabolism and growth directly depend on cytochrome *c* (*25*). By controlling the gene sequence upstream of cytochrome *c*, it is possible to regulate the transcription of the cytochrome *c* gene and thereby to regulate the amount of cytochrome *c* produced in vivo. The amount of cytochrome *c* produced can be quantitatively assayed in intact yeast cells by microspectrophotometric measurements of the 500-nm absorbance due to reduced cytochrome *c*. Experiments have shown that when the total cellular level of cytochrome *c* is modified by transcriptional control, the growth on lactate is coregulated in a monotonic fashion. Figure 3 shows such results. These results suggest that, under specific metabolic conditions, cytochrome *c* is involved in a metabolic step that is partially rate-determining for overall growth. With these key controls in hand, it remained only to compare the growth on lactate of essentially identical yeast strains, which differ only in the sequence of cytochrome *c*:wild type (N 52) and mutant (I 52).

The initial results are quite surprising (Figure 4). The evolutionary invariance of the cytochrome *c* redox potential has been used to argue that the value of 270 mV critically regulates cellular electron transport. The 50-mV shift of the Ile 52 mutant should lead to a significant change in the effective population of (oxidized/reduced) cytochrome *c* and thus might diminish the growth rate. This change is not observed. The strain that contains the I 52 mutant does not grow poorly; it actually appears to grow at a rate that equals or slightly exceeds that of the wild-type strain!

Apparently, the standard redox potentials determined in vitro by conventional electrochemical methods do not control the rates of electron flow in vivo under metabolic conditions in which cytochrome *c* is involved in a rate-determining metabolic step. Furthermore, extensive experiments by Vanderkooi (*26*), Magner (*27*), Chance and Williams (*28*), and others strongly indicate that this surprising result is not due to a simple modulation of the cytochrome redox potential under cellular conditions. Indeed, in a classic paper, Chance and Williams (*28*) showed that the in vitro potential of cytochrome *c* closely agrees with the potential measured for cytochrome *c* in

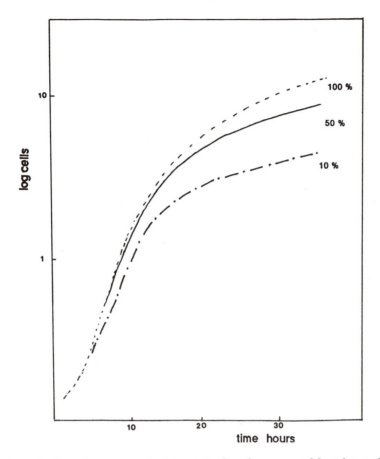

Figure 3. *Growth curves on lactate minimal media measured by Klett techniques for* Saccharomyces *yeast containing different intracellular cytochrome c concentrations (controlled genetically). The monotonic dependence of growth on cytochrome c suggests "autogenous regulation". In chemical terms, cytochrome c is involved in one of the rate-determining steps in overall metabolism.*

intact mitochondria. We are also rather certain that this result is not unique to the Ile 52 variant, on the basis of limited qualitative observations in our laboratory and others for other cytochrome *c* mutants of altered redox potentials (e.g., Ala 52, Phe 67, and Ala 38).

Discussion

Is there any simple chemical explanation for the counterintuitive result that changing the evolutionarily invariant redox potential of cytochrome *c* by a single-site mutation does not diminish its in vivo activity? We suggest two possible explanations that depend on simple chemical kinetics arguments. The first (and the most obvious) is that, under these specific and unusual

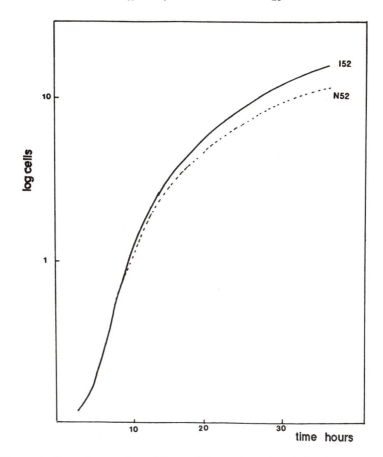

Figure 4. Growth rates for wild type (E° = 270 mV) *and the Ile 52 mutant* (E° = 220 mV). *The mutant reproducibly grows better than the wild type under these conditions.*

metabolic conditions, the effective redox potential of the electron-transport system is controlled "upstream" of cytochrome c, so that, for example, cytochrome c_1 is >99% reduced under steady-state metabolic conditions. If so, the redox potential of cytochrome c will be less important in determining the net flux of the electrons. The flux would still be sensitive to the total amount of cytochrome c, as observed. A second, more specific explanation is possible. For yeast to grow on lactate, lactate must be oxidized to pyruvate by using the yeast lactate dehydrogenase (cytochrome b_2). This enzyme uses cytochrome c as a specific "cofactor" for oxidation:

$$2Fe^{3+} + \text{cyt } c \text{ cyt } b_{2(red)} \rightarrow \text{cyt } b_{2(ox)} + 2Fe^{II} \text{ cyt } c$$

$$\text{cyt } b_{2(ox)} + \text{lactate} \rightarrow \text{cyt } b_{2(red)} + \text{pyruvate}$$

where red is reduced and ox is oxidized. In earlier work (29), we studied

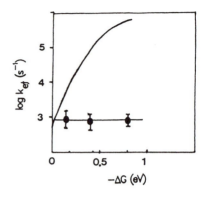

Figure 5. Dependence on ΔG of the rate constant for the reaction of (M)cytochrome c [M is Fe^{III}, Zn^{II}, or (H$^+$)$_2$] with cytochrome b$_2$ (lactate dehydrogenase). The top line shows the dependence predicted by Marcus theory for λ = 1 eV (as in Figure 1). The essential free energy independence of rate suggests conformational gating. The rate-determining step is not electron transfer, but a prior conformational change.

the dependence of the electron-transfer rate on redox potential within the cytochrome c–cytochrome b$_2$ complex (Figure 5). On the basis of Marcus theory and results for the analogous cytochrome c–cytochrome b$_5$ complex, we expected to see a strong dependence of rate on ΔG^0. Instead, we found that the electron-transfer rate was independent of reaction free energy. The simplest explanation for this initially surprising result is that the electron-transfer step, per se, is not the rate-determining step of the reaction. Instead, a slow conformational change in the protein complex precedes the fast electron-transfer step. Such conformationally controlled reaction rates, dubbed "conformational gating" by Hoffman and Ratner (30), have been extensively discussed theoretically. Consider the possibility that the cytochrome c–cytochrome b$_2$ reaction is partially rate-determining for growth in lactate. Then the growth rate can depend only on the total cytochrome c concentration. Equally clearly, on the basis of in vitro studies, a change in the redox potential of cytochrome c will not diminish the rate of this key step because the rate is controlled not by an electron-transfer barrier associated with E° but by a conformational barrier.

At present these suggestions must be considered rather speculative. However, they do suggest further experiments. For example, either of these explanations predict that other metabolic pathways (e.g., growth on glucose) should result in different growth-limiting steps, corresponding to different metabolic pathways with correspondingly different steady-state redox levels. These common metabolic states presumably favor the (higher) cytochrome c potential found in the wild-type protein. This prediction is being tested.

Acknowledgments

We gratefully acknowledge continuing interactions with A. G. Mauk, M. Smith, and E. Margoliash. This work was supported by National Science Foundation grants to George McLendon and Fred Sherman, the National Institutes of Health (GM 33881 GM), and a grant to Gary Brayer from the Medical Research Council of Canada.

References

1. Marcus, R.; Sutin, N. *Biochim. Biophys. Acta* **1985**, *811*, 265.
2. *Tunneling in Biological Systems;* Chance, B., Ed.; Plenum Press: New York, 1980.
3. Hopfield, J. *Proc. Natl. Acad. Sci. U.S.A.* **1974**, *71*, 3640.
4. Gray, H.; Malmstrom, B. *Biochemistry* **1989**, *28*, 7499.
5. Lieber, C.; Karas, J.; Mayo, S.; Albin, M.; Gray, H. *Proceedings, Welch Conference on Chemistry;* Welch Foundation: Houston, TX, 1989; pp 889–893.
6. Wang, N.; Mauk, A.; Pielak, G.; Johnson, J.; Smith, M.; Hoffman, B. *Science (Washington, D.C.)* **1988**, *240*, 311.
7. Peterson, S., Kennedy, S. E.; McGourty, J. M.; Kalwert, J.; Hoffman, B. *J. Am. Chem. Soc.* **1986**, *108*, 1739.
8. McLendon, G. *Acc. Chem. Res.* **1988**, *21*, 160.
9. Moore, G.; Williams, R. J. P. *Coord. Chem. Rev.* **1976**, *18*, 125.
10. Williams, G.; Moore, G.; Williams, R. J. P. *Comments Inorg. Chem.* **1985**, *7*, 55.
11. Moore, G.; Pettigrew, G.; Rogers, N. *Proc. Natl. Acad. Sci. U.S.A.* **1986**, *83*, 4998.
12. Kassner, R. *Proc. Natl. Acad. Sci. U.S.A.* **1972**, *69*, 2263.
13. Louie, G. V.; Pielak, G. J.; Smith, M.; Brayer, G. D. *Biochemistry* **1988**, *27*, 7870.
14. Louie, G. V.; Brayer, G. D. *J. Mol. Biol.* **1989**, *210*, 313.
15. Harbury, H.; Cronen, J.; Fanger, M.; Hettinger, T.; Murphy, A.; Meyer, Y.; Vinograd, S. *Proc. Natl. Acad. Sci. U.S.A.* **1965**, *54*, 1658.
16. Churg, A. K.; Warshel, A. *Biochemistry* **1986**, *25*, 1675.
17. Cutler, R. L.; Davies, A. M.; Creighton, S.; Warshel, A.; Moore, G.; Smith, M.; Mauk, A. G. *Biochemistry* **1989**, *28*, 3188.
18. Bushnell, G. W.; Louie, G. V.; Brayer, G. D. *J. Mol. Biol.* **1990**, *214*, 585.
19. Das, G.; McLendon, G.; McLendon, D.; Sherman, F. *Proc. Natl. Acad. Sci. U.S.A.* **1989**, *86*, 496.
20. Hazzard, J.; McLendon, G.; Tollin, G.; Sherman, F. *Biochemistry* **1988**, *27*, 4445.
21. Principio, L.; Conklin, K.; Short, J.; McLendon, G.; Das, G.; Hickey, D.; Sherman, F. *J. Biol. Chem.* **1988**, *263*, 18290.
22. Rees, D. C. *Proc. Natl. Acad. Sci. U.S.A.* **1985**, *82*, 3082.
23. Luntz, T.; Schejter, A.; Garber, A.; Margoliash, E. *Proc. Natl. Acad. Sci. U.S.A.* **1989**, *86*, 3524.
24. Hickey, D.; Berghuis, A.; Brayer, G.; McLendon, G.; Sherman, F. *J. Biol. Chem.*, accepted.
25. Hampsey, D.; Das, G.; Sherman, F. *J. Biol. Chem.* **1986**, *261*, 3259.
26. Vanderkooi, J.; Ercalinska, M. *Arch. Biochem. Biophys.* **1973**, *162*, 385.
27. Magner, E. Ph.D. Thesis, University of Rochester, 1988.
28. Chance, B.; Williams, R. *Nature (London)* **1962**, 395.
29. Pardue, K.; McLendon, G.; Bak, C. *J. Am. Chem. Soc.* **1987**, *109*, 7540.
30. Hoffman, B.; Ratner, M. *J. Am. Chem. Soc.* **1987**, *109*, 6237.

RECEIVED for review April 27, 1990. ACCEPTED revised manuscript February 4, 1991.

Long-Range Electron Transfer in Heme Proteins

Porphyrin–Ruthenium Electronic Couplings in Three Ru(His)Cytochromes c

Michael J. Therien[1], Bruce E. Bowler[1], Mary A. Selman[1], Harry B. Gray[1], I-Jy Chang[2], and Jay R. Winkler[2]

[1]Arthur Amos Noyes Laboratory of Chemical Physics, California Institute of Technology, Pasadena, CA 91125
[2]Chemistry Department, Brookhaven National Laboratory, Upton, NY 11973

The kinetics of long-range electron transfer (ET) have been measured in Ru(NH₃)₄L(His 39) derivatives (L is NH₃, pyridine, or isonicotinamide) of Zn-substituted Candida krusei cytochrome c *and Ru(NH₃)₄L(His 62) derivatives (L is NH₃ or pyridine) of Zn-substituted* Saccharomyces cerevisiae cytochrome c. *The rates of both excited-state electron transfer and thermal recombination are approximately 3 times greater in Ru(His 39)cytochrome c (Zn) than the rates of the corresponding reactions in Ru(His 33)cytochrome c (Zn), but analogous ET reactions in Ru(His 62)cytochrome c (Zn) are roughly 2 orders of magnitude slower than in the His 33-modified protein. Analysis of driving-force dependences establishes that the large variations in the ET rates are due to differences in donor–acceptor electronic couplings. Examination of potential ET pathways indicates that hydrogen bonds could be responsible for the enhanced electronic couplings in the Ru(His 39) and Ru(His 33) proteins.*

ELECTRON TRANSFER (ET) CAN TAKE PLACE at appreciable rates over long distances (>10 Å) in organic and inorganic molecules (*1–6*) and in proteins (*6–15*). In nonprotein systems, the evidence suggests that ET rates depend

0065–2393/91/0228–0191$06.00/0

upon the number of covalent bonds separating the donor and the acceptor, rather than upon their direct separation distance (1, 2). There are many potential ET pathways in proteins (16–19); the through-peptide routes generally involve so many bonds that they cannot possibly account for the observed rates (20, 21). Pathways that include ionic contacts (e.g., hydrogen bonds) or small through-space jumps often can be found, and it has been postulated that such shortcuts greatly enhance the donor–acceptor electronic coupling (16, 22). Our work on ruthenium-modified cytochromes c is focused in part on understanding how variations in electron-tunneling pathways affect long-range donor–acceptor electronic couplings (17, 21, 23–25). Here we examine three modified proteins: horse heart Ru(His 33)cytochrome c (Zn) (21), *Candida krusei* Ru(His 39)cytochrome c (Zn) (23), and *Saccharomyces cerevisiae* Ru(His 62)cytochrome c (Zn) (24).

Ruthenium [Ru(NH$_3$)$_4$L(His 33; His 39) with L as NH$_3$, pyridine, or isonicotinamide, and Ru(NH$_3$)$_4$L(His 62) with L as NH$_3$ or pyridine] derivatives of Zn-substituted cyt c [Ru(His X)Zn cyt c] were prepared by standard procedures (9, 23–26). Intramolecular ET can be initiated in these protein derivatives by photoexcitation of the Zn–porphyrin (ZnP) to its strongly reducing triplet excited state (21). In addition to its intrinsic radiative and nonradiative decay pathways, this triplet can decay by ET to a histidine-bound Ru(III)–ammine complex (ET*). The metastable product of the ET* reaction, Ru(II)–ZnP$^+$, relaxes via a thermal ET process (ET^b) to reform the ground-state complex, Ru(III)–ZnP.

Kinetics of Electron-Transfer Reactions

The kinetics of the ET reactions of the Ru(His 33)Zn cyt c (21), Ru(His 39)Zn cyt c (23), and Ru(His 62)Zn cyt c (24) derivatives have been measured by laser flash photolysis (9). ET rates and activation parameters are set out in Table I. Although the ET rates in Ru(His 39)Zn cyt c and Ru(His 33)Zn cyt c are not very different, it is striking that ET in Ru(His 62)Zn cyt c is roughly 2 orders of magnitude slower than in the His 33-modified protein.

Semiclassical theories describe ET rates as the product of three factors: nuclear frequency, electronic coupling, and nuclear reorientation. A nonadiabatic expression (eq 1) is appropriate for long-range ET in derivatized proteins (27).

$$k_{ET} = \frac{2\pi(H_{rp})^2}{\hbar(4\pi\lambda kT)^{1/2}} \exp\left[\frac{-(\Delta G^0 + \lambda)^2}{4\lambda kT}\right] \tag{1}$$

The parameter H_{rp} in eq 1 is the electronic coupling matrix element for the ET reaction, $\hbar$ is Planck's constant, ΔG^0 is the reaction free energy, k is the Boltzmann constant, T is absolute temperature, and λ is the nuclear reor-

Table I. Rate Constants and Activation Parameters

Reaction	$-\Delta G^0$ (eV)	k_{ET} (s^{-1})	$\Delta H^{\ddagger}$ (kcal mol^{-1})	$\Delta S^{\ddagger}$ (eu)
		Ru(His 33)Zn cyt c[a]		
Rua$_4$(isn)(His)$^{2+}$ → ZnP$^+$	0.66	2.0 × 10^5	<0.5	−35
ZnP* → Rua$_5$(His)$^{3+}$	0.70	7.7 × 10^5	1.7	−27
Rua$_4$(py)(His)$^{2+}$ → ZnP$^+$	0.74	3.5 × 10^5	<0.5	−34
ZnP* → Rua$_4$(py)(His)$^{3+}$	0.97	3.3 × 10^6	2.2	−22
Rua$_5$(His)$^{2+}$ → ZnP$^+$	1.01	1.6 × 10^6	—	—
ZnP* → Rua$_4$(isn)(His)$^{3+}$	1.05	2.9 × 10^6	<0.5	−30
		Ru(His 39)Zn cyt c[b]		
Rua$_4$(isn)(His)$^{2+}$ → ZnP$^+$	0.66	6.5 × 10^5	−1.7	−39
ZnP* → Rua$_5$(His)$^{3+}$	0.70	1.5 × 10^6	1.3	−27
Rua$_4$(py)(His)$^{2+}$ → ZnP$^+$	0.74	1.5 × 10^6	−1.8	−37
ZnP* → Rua$_4$(py)(His)$^{3+}$	0.97	8.9 × 10^6	0.2	−27
Rua$_5$(His)$^{2+}$ → ZnP$^+$	1.01	5.7 × 10^6	−0.2	−29
ZnP* → Rua$_4$(isn)(His)$^{3+}$	1.05	1.0 × 10^7	0.2	−27
		Ru(His 62)Zn cyt c[c]		
ZnP* → Rua$_5$(His)$^{3+}$	0.70	6.5 × 10^3	1.4	−37
Rua$_4$(py)(His)$^{2+}$ → ZnP$^+$	0.74	8.1 × 10^3	—	—
ZnP* → Rua$_4$(py)(His)$^{3+}$	0.97	3.6 × 10^4	—	—
Rua$_5$(His)$^{2+}$ → ZnP$^+$	1.01	2.0 × 10^4	0.7	−37

NOTE: a is NH$_3$, py is pyridine, isn is isonicotinamide, $\Delta H^{\ddagger}$ is the activation enthalpy, and $\Delta S^{\ddagger}$ is the activation entropy.
[a]Data are from ref. 21.
[b]Data are from ref. 23.
[c]Data are from ref. 24.

ganization energy, which comprises inner-sphere (λ_{in}) and outer-sphere (λ_{out}) contributions. Plots of the Ru(His 33)Zn cyt c, Ru(His 39)Zn cyt c, and Ru(His 62)Zn cyt c data, along with the corresponding fits to eq 1, are shown in Figure 1. Although the reorganization energy is nearly the same for the ET reactions in the three proteins (~1.2 eV), the H_{rp} value for Ru(His 39)Zn cyt c (0.21 cm^{-1}) (23) is almost twice as large as that for Ru(His 33)Zn cyt c (0.12 cm^{-1}) (21) and over 20 times larger than the H_{rp} for Ru(His 62)Zn cyt c (0.01 cm^{-1}) (24).

Both the electronic coupling matrix element and the outer-sphere component of the nuclear reorganization energy are thought to vary with donor–acceptor separation and orientation (27, 28). Studies of Os– and Ru–ammines bridged by polyproline spacers show that the distance dependence of λ can even be greater than that of H_{rp} (4). Dielectric continuum models of solvent reorganization predict that λ_{out} will increase with donor–acceptor separation (r_{DA}). Models that describe charge transfer within low-dielectric spheres or ellipsoids embedded in dielectric continua exhibit

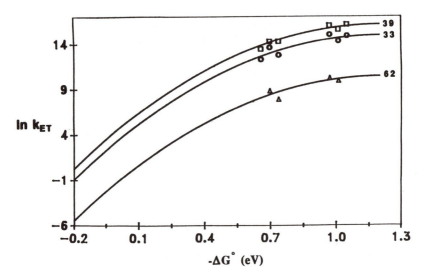

Figure 1. Plots of ln k_{ET} vs. $-\Delta G^0$ for the Ru(His X)cytochrome c ET reactions (Table I). Boxes (His 39), circles (His 33), and triangles (His 62) represent the experimental data (ET* and ET[b]). Solid lines are the best fits to eq 1: λ = 1.21, H_{rp} = 0.21 (His 39); λ = 1.20, H_{rp} = 0.12 (His 33); λ = 1.20 eV, H_{rp} = 0.01 cm⁻¹ (His 62).

a dependence upon r_{DA} as well as upon the positions of the redox sites inside the sphere or ellipsoid (29). Modeling the Ru–Zn–cyt c systems as single spheres suggests, however, that variations in λ_{out} for the Ru(His 33)Zn cyt c, Ru(His 39)Zn cyt c, and Ru(His 62)Zn cyt c ET reactions will not be significant (0.57, 0.60, and 0.63 eV, respectively).

In the calculations of λ_{out}, the cyt c molecule was represented as a 34-Å sphere enclosing 90–95% of the nonhydrogen atoms in the protein. The Ru–ammine group was taken as a 6-Å sphere. These two interpenetrating spheres were enclosed by a third sphere of radius 17.1 Å for Ru(His 33)Zn cyt c, 18.0 Å for Ru(His 39)Zn cyt c, and 17.9 Å for Ru(His 62)Zn cyt c. The Zn and Ru redox centers were placed 5.8 and 14.1 Å from the center of the sphere, respectively, and separated from one another by 17.6 Å in Ru(His 33)Zn cyt c. The corresponding distances were 6.2, 15.0, and 19.9 Å for the Ru(His 39)Zn cyt c model, and 7.0, 14.9, and 21.8 Å for the Ru(His 62)Zn cyt c model. The dielectric constant of the sphere was taken as 1.8; the solvent was assigned a static dielectric constant of 78.54 and an optical dielectric constant of 1.78.

Models of Electron Transfer

In an extension of these calculations, we examined the maximum variation in λ_{out} predicted by the single-sphere continuum model (34-Å-radius sphere for cyt c, with its metal center 5.8 Å from the origin). The Ru–ammine complex was taken as a 6-Å sphere centered on the Ru atom that was assumed to be fixed 16 Å from the center of the cyt c sphere. The small sphere can occupy any position on the large sphere, with values of r_{DA} varying from 10.2 to 21.8 Å. A third sphere then encloses the two other spheres, and λ_{out} for electron transfer between the two metals was calculated by treating the solvent as a dielectric continuum. The value of λ_{out} varies from 0.38 to 0.63 eV almost linearly as r_{DA} increases from 10.2 to 21.8 Å. The total variation of 0.25 eV is only slightly larger than the uncertainty range in our estimates of λ (± 0.1 eV). The invariance of λ found among the different ET reactions is, therefore, consistent with theoretical considerations.

The differences in ET rates among the Ru(His 33), Ru(His 39), and Ru(His 62) derivatives arise from variations in donor–acceptor electronic coupling. The shortest direct distances between the porphyrin and imidazole carbon atoms of His 33 (13.2 Å), His 39 (13.0 Å), and His 62 (15.5 Å) are much too long for any direct donor–acceptor interaction (*16, 30*). Because virtually the same donor and acceptor electronic states are found in the three proteins, the differences in H_{rp} must arise from the manner in which the intervening atoms couple the two states. If a homogeneous medium of constant tunneling-barrier height separated the donor and the acceptor in the three systems, then H_{rp} would depend primarily on r_{DA}. It would be nearly the same for Ru(His 33)Zn cyt c and Ru(His 39)Zn cyt c and decrease only slightly for Ru(His 62)Zn cyt c relative to Ru(His 39)Zn cyt c. This prediction clearly is not in accord with experiment, so it is logical to conclude that the inhomogeneous nature of the polypeptide medium separating the Ru–ammine and metalloporphyrin sites is responsible for the differential electronic coupling in these ruthenium-modified Zn cyt c derivatives.

Bridge-Mediated Electron Transfer

Bridge-mediated ET involves a superexchange mechanism in which electronic states of the intervening medium mix with localized donor states to produce a nonzero H_{rp} (*31, 32*). Beratan and co-workers (*16, 33*) developed a simple model to describe the contribution of the polypeptide bridge to the electronic coupling in long-range ET in protein systems. The essence of the model is that H_{rp} decreases from its maximal value (at van der Waals contact of donor and acceptor) by a constant factor for each covalent bond in the ET pathway. Ionic contacts (H bonds and salt bridges) and through-space jumps decrease H_{rp} by somewhat larger factors. For Ru(His X)cyt c, every potential tunneling pathway is taken to originate at a carbon atom of the relevant histidine and defined to terminate at the first point of contact with the porphyrin. The optimum pathway is that with the combination of through-bond, ionic, and through-space contacts that yields the smallest diminution of H_{rp}.

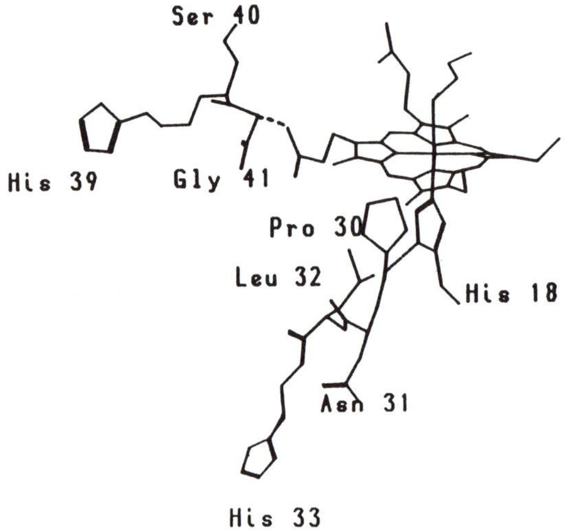

Figure 2. Possible ET pathways from His 33 and His 39 to the heme in cyto-chrome c (33). Edge–edge distances are as follows: His 39 to the heme, 13.0 Å; His 33 to His 18, 11.7 Å; His 33 to the heme, 13.2 Å. Calculations of distances were made by using Biograf/III version 1.34 (Biodesign, Inc.). The structures of Candida krusei and horse heart cytochromes were generated from the structure of the tuna protein by standard methods (23). In both Candida krusei and horse heart proteins, an imidazole carbon on His 33 is 11.7 Å from an imidazole carbon of His 18, an axial ligand of the metallo-porphyrin. This value has been used as the edge-to-edge distance in previous studies (9, 21). His 18 is not likely to be as strongly coupled to the porphyrin-localized donor and acceptor states as are carbon atoms of the porphyrin ring. Hence, in comparing donor–acceptor coupling in Ru(His 33)Zn cyt c and Ru(His 39)Zn cyt c, distances to porphyrin carbon atoms have been used.

The ET pathways in Ru(His 33)Zn cyt c and Ru(His 39)Zn cyt c generated (33) by applying the Beratan–Onuchic criteria (16) are shown in Figure 2. The best pathway from His 33 to the metalloporphyrin is a 15-bond route to the Zn atom through His 18 that includes a 1.85-Å hydrogen bond between the Pro 30 carboxyl oxygen and the proton on the His 18 nitrogen. The shortest pathway from His 39 is a 12-bond route that includes a 2.4-Å H bond between the α-amino hydrogen atom of Gly 41 and the carboxyl oxygen of a propionate side chain on the porphyrin. The key difference between these two pathways is that the His 39 pathway is built from 11 covalent bonds and 1 H bond; the His 33 pathway includes 14 covalent bonds and 1 H bond (30). Hence, the experimental observation that the electronic coupling is stronger in the His 39 derivative than in the His 33-modified protein (even though the edge–edge distances in the two modified proteins are roughly the same) is consistent with the Beratan–Onuchic pathway anal-ysis.

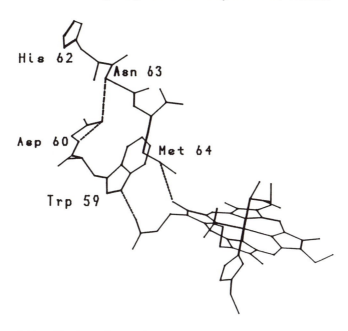

Figure 3. Possible ET pathways from His 62 to the heme in cytochrome c (30). The His 62–heme edge–edge distance is 15.5 Å (see Figure 2 legend).

Tunneling Pathways

The analysis suggests that there are two comparable tunneling pathways for Ru(His 62)Zn cyt *c* (Figure 3). One is a 17-bond route with 14 covalent bonds and 3 H bonds (the third of which connects the Trp 59 nitrogen atom to the carbonyl oxygen of the heme propionate side chain); the other is a 13-bond route with 12 covalent bonds and a through-space interaction between the sulfur atom of Met 64 and the heme edge (30). The sharply lower electronic coupling in the His 62 protein relative to both the His 33 and His 39 systems indicates that neither the 17-bond nor the 13-bond pathway is very good. The 13-bond pathway is the most direct route, a finding that suggests that the Met 64–heme through-space interaction is a poor shortcut. An amino acid with an aromatic group or a sulfur atom in the pathway does not necessarily enhance the donor–acceptor electronic coupling. Although the decay of a tunneling electron wave function might be slow across an aromatic group, mixing of the wave function onto and off of this group may in fact be quite unfavorable (30).

Experiments with systematic variations in tunneling pathways and tunneling medium energetics should further clarify the possible roles that through-space interactions and hydrogen bonds play in biological ET reactions.

Acknowledgments

We thank David Beratan for helpful discussions. Both M. J. Therien (National Institutes of Health) and B. E. Bowler (Medical Research Council of Canada) acknowledge postdoctoral fellowships. Research at the California Institute of Technology was supported by the National Science Foundation (CHE88–22988) and the National Institutes of Health (DK19038). Research performed at Brookhaven National Laboratory was carried out under Contract DE–AC02–CH00016 with the U.S. Department of Energy and supported by its Division of Chemical Sciences, Office of Basic Energy Sciences.

References

1. Closs, G. L.; Miller, J. R. *Science (Washington, D.C.)* **1988**, *240*, 440–447.
2. Oevering, H.; Paddon-Row, M. N.; Heppener, M.; Oliver, A. M.; Cotsaris, E.; Verhoeven, J. W.; Hush, N. S. *J. Am. Chem. Soc.* **1987**, *109*, 3258–3289.
3. Schanze, K. S.; Sauer, K. *J. Am. Chem. Soc.* **1988**, *110*, 1180–1188.
4. Isied, S. S.; Vassilian, A.; Wishart, J. W.; Creutz, C.; Schwarz, H.; Sutin, N. *J. Am. Chem. Soc.* **1988**, *110*, 835–837.
5. Faraggi, M.; DeFelippis, M. R.; Klapper, M. H. *J. Am. Chem. Soc.* **1989**, *111*, 5141–5145.
6. McLendon, G. *Acc. Chem. Res.* **1988**, *21*, 180–187.
7. Gray, H. B.; Malmström, B. G. *Biochemistry* **1989**, *28*, 7499–7505.
8. Liang, N.; Mauk, A. G.; Pielak, G. J.; Johnson, J. A.; Smith, M.; Hoffman, B. M. *Science (Washington, D.C.)* **1988**, *240*, 311–313.
9. Elias, H.; Chou, M. H.; Winkler, J. R. *J. Am. Chem. Soc.* **1988**, *110*, 429–434.
10. Conrad, D. W.; Scott, R. A. *J. Am. Chem. Soc.* **1989**, *111*, 3481–3463.
11. Pan, L. P.; Durham, B.; Wolinska, J.; Millett, F. *Biochemistry* **1988**, *27*, 7180–7184.
12. Farver, O.; Pecht, I. *FEBS Lett.* **1989**, *244*, 379–382.
13. Jackman, M. P.; McGinnis, J.; Powls, R.; Salmon, G. A.; Sykes, A. G. *J. Am. Chem. Soc.* **1988**, *110*, 5880–5887.
14. Faraggi, M.; Klapper, M. H. *J. Am. Chem. Soc.* **1988**, *110*, 5753–5758.
15. Cusanovich, M. A.; Meyer, T. E.; Tollin, G. *Adv. Inorg. Biochem.* **1987**, *7*, 37–91.
16. Beratan, D. N.; Onuchic, J. N. *Photosynth. Res.* **1989**, *22*, 173–188.
17. Bowler, B. E.; Meade, T. J.; Mayo, S. L.; Richards, J. H.; Gray, H. B. *J. Am. Chem. Soc.* **1989**, *111*, 8757–8759.
18. Kuki, A.; Wolynes, P. G. *Science (Washington, D.C.)* **1987**, *236*, 1847–1652.
19. Larsson, S. *J. Chem. Soc., Faraday Trans. 2* **1983**, *79*, 1375–1388.
20. Cowan, J. A.; Upmacis, R. K.; Beratan, D. N.; Onuchic, J. N.; Gray, H. B. *Ann. N.Y. Acad. Sci.* **1989**, *550*, 88–84.
21. Meade, T. J.; Gray, H. B.; Winkler, J. R. *J. Am. Chem. Soc.* **1989**, *111*, 4353–4358.
22. Beratan, D. N.; Onuchic, J. N.; Hopfield, J. J. *J. Chem. Phys.* **1987**, *88*, 4488–4498.
23. Therien, M. J.; Selman, M. A.; Gray, H. B.; Chang, I.-J.; Winkler, J. R. *J. Am. Chem. Soc.* **1990**, *112*, 2420–2422.
24. Therien, M. J.; Bowler, B. E.; Gray, H. B.; Chang, I.-J.; Winkler, J. R., unpublished results.
25. Selman, M. A. Ph.D. Thesis, California Institute of Technology, 1989.

26. Yocom, K. M.; Shelton, J. B.; Shelton, J. R.; Schroeder, W. A.; Worosila, G.; Isied, S. S.; Bordignon, E.; Gray, H. B. *Proc. Natl. Acad. Sci. U.S.A.* **1982**, *79*, 7052–7055.
27. Marcus, R. A.; Sutin, N. *Biochim. Biophys. Acta* **1985**, *811*, 285–322.
28. Brunschwig, B. S.; Ehrenson, S.; Sutin, N. *J. Am. Chem. Soc.* **1984**, *106*, 8858–8859.
29. Brunschwig, B. S.; Ehrenson, S.; Sutin, N. *J. Phys. Chem.* **1988**, *90*, 3857–3888.
30. Therien, M. J.; Bowler, B. E.; Beratan, D. N.; Gray, H. B., unpublished results.
31. McConnell, H. M. *J. Chem. Phys.* **1981**, *36*, 508–515.
32. Marcus, R. A. *Chem. Phys. Lett.* **1987**, *133*, 472–477.
33. Beratan, D. N.; Onuchic, J. N.; Betts, J.; Bowler, B. E.; Gray, H. B. *J. Am. Chem. Soc.* **1990**, *112*, 7915–7921.

RECEIVED for review April 27, 1990. ACCEPTED revised manuscript August 29, 1990.

Long-Range Electron Transfer Within Mixed-Metal Hemoglobin Hybrids

Michael J. Natan, Wade W. Baxter, Debasish Kuila, David J. Gingrich, Gregory S. Martin, and Brian M. Hoffman

Department of Chemistry, Northwestern University, Evanston, IL 60208–3113

Studies of long-range electron transfer (ET) within mixed-metal hemoglobin (Hb) hybrids [MP, $Fe^{3+}(L)P$, where M is Mg or Zn; P is protoporphyrin IX; and L is H_2O, imidazole, N_3^-, F^-, or CN^-] are discussed. Because the structure of Hb is crystallographically known, ET occurs between redox centers held at known distance and orientation. The ET energetics are easily manipulated through variation of M and L. In these systems, cyclic ET is initiated through photoproduction of the strong reductant $^3(MP)$. ET quenching yields the charge-separated intermediate, $[(MP)^+, Fe^{2+}(L)P]$, which returns to the ground state by thermal ET. Direct spectroscopic observation of $[(MP)^+, Fe^{2+}(L)P]$ confirms the cyclic ET scheme. Comparison of rate constants for photoinitiated and thermally activated ET within various [MP, $Fe^{3+}(L)P$] hybrids indicates that ET is direct and not "gated" by either protein conformational changes or ligand loss.

ELECTRON-TRANSFER REACTIONS ARE CENTRAL to biology and chemistry (*1, 2*), but only recently have techniques been developed to study long-range interprotein electron transfer (ET) (*3, 4*) without the complication of second-order processes through use of modified proteins that hold an electron donor–acceptor redox pair at fixed distance. In one approach, several groups have developed techniques for studying ET within proteins modified by covalent attachment of redox-active inorganic complexes to surface amino acid residues (*5–11*). For example, $[(L)_5Ru]^{2+}$, when bound to a histidine residue on the outside of proteins such as cytochrome c or myoglobin, can

0065–2393/91/0228–0201$06.00/0

exchange an electron with a metal-containing redox center located on the inside of the protein.

In parallel with McLendon (3), we focused on studies of long-range ET within protein–protein complexes, such as the physiologically important complex of cytochrome c peroxidase with cytochrome c (12–14) or the $[\alpha_1,\beta_2]$ complex of the hemoglobin tetramer (15–17). Our approach involves replacing the heme (FeP) of one protein partner with a closed-shell porphyrin MP (M is Zn or Mg; P is protoporphyrin IX) and studying ET between the MP and FeP groups (12–17). Reversible ET within such metal-substituted ET complexes (Scheme I) is initiated by laser flash photoproduction of the

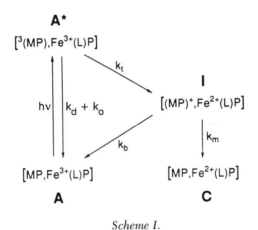

Scheme I.

slowly decaying ^{3}(MP) triplet state (**A***). The ^{3}(MP) is a good reductant and can reduce the ferriheme partner (Fe^{3+}P) by long-range ET with a photoinitiated ET rate constant k_t (eq 1).

$$^3(MP) + Fe^{3+}P \xrightarrow{k_t} (MP)^+ + Fe^{2+}P \qquad (1)$$

The resulting charge-separated intermediate, $[(MP)^+, Fe^{2+}P]$ (**I**), returns to the ground state by thermal ET from Fe^{2+}P to the cation radical $(MP)^+$ (eq 2) with rate constant k_b.

$$(MP)^+ + Fe^{2+}P \xrightarrow{k_b} MP + Fe^{3+}P \qquad (2)$$

In our studies we used transient absorption and emission techniques to monitor A* and I, thereby allowing us to measure both k_t and k_b. The key benefit to studying long-range ET processes within hemoglobin hybrids (Hb) is that, under the conditions of our experiments, the hemoglobin tetramers in solution adopt deoxy-Hb (T-state) geometry with a crystallographically known structure. Thus, electron transfer occurs between redox centers held at known distance and orientation.

Preparation, Structure, and Characterization of Mixed-Metal Hemoglobin Hybrids

Preparation of mixed-metal hemoglobin hybrids is achieved by separation of $[2\alpha,2\beta]$ hemoglobin into its constituent α and β chains, followed by demetalation of one of the chains, metalation with MP, and chain recombination, to yield the tetrameric $[2\alpha(MP), 2\beta(Fe^{3+}P)]$ or $[2\alpha(Fe^{3+}P),2\beta(MP)]$ species (18). Thus, MP $\rightarrow$ FeP ET might in principle occur between α_1–β_1 or α_1–β_2 subunits. However, the distance between α_1 and β_1 hemes is more than 10 Å greater than that between α_1 and β_2. This extra distance is expected and indeed is found to reduce ET rates by several orders of magnitude. Hence, for all practical purposes we may treat the $[2\alpha,2\beta]$ tetramer in terms of two independent $[\alpha_1,\beta_2]$ ET complexes.

If the geometry of mixed-metal hemoglobin hybrids is fixed in the known structure of T-state (deoxy) Hb, metal replacement should not perturb that structure. The first structural issue that must be considered is local: Does substitution of Zn or Mg cause significant perturbations? The structure of MgHb, in which all four prosthetic groups are MgP, has recently been crystallographically determined at 2.2-Å resolution (19). With the atomic model of deoxy Hb as the starting point in a least-squares refinement, only trivially small structural differences were noted. Therefore, replacement of Fe^{2+} with Mg^{2+} in hemoglobin does not significantly alter the structure.

The second structural issue is global: Is the quaternary structure of a mixed-metal hybrid significantly different from that of T-state Hb? Again the answer is no, as indicated by an X-ray structure of the $[\alpha(FeCO),\beta(Mn)]$ hybrid (20). Thus, the distances and geometric relationships of the heme groups involved in ET within the $[\alpha_1,\beta_2]$ ET complex are preserved in the metal-substituted species. With this corroborative structural data, it is possible to discuss structure of the $[\alpha_1,\beta_2]$ complex with high precision.

As an example we considered the [Fe,M] hybrids, where M is Mg or Zn. In the $[\alpha_1,\beta_2]$ ET complex (Figure 1), the $\beta_2(MP)$ and $\alpha_1(FeP)$ are roughly parallel, with distances of 25 Å between metals and about 17 Å edge-to-edge (21). This structurally defined but chemically manipulable system offers many avenues for study. Recently we focused on the effects of changing ET energetics. One means to do this is to vary the closed-shell MP. The $(MgP)^+$–MgP reduction potential is about 100 mV lower than the ZnP^+–ZnP reduction potential (15, 16). Consequently, the free energy change, $-\Delta G$, for photoinitiated $A^* \rightarrow I$ process is ~1.0 eV for $[^3(ZnP), Fe^{3+}P]$, and ~1.1 eV for $[^3(MgP), Fe^{3+}P]$. For the I $\rightarrow$ A ET, $-\Delta G$ is ~0.8 eV for $[(ZnP)^+, Fe^{2+}P]$ and ~0.7 eV for $[(MgP)^+, Fe^{2+}P]$ (15, 16).

Heme ligand variation provides an even more effective means of altering

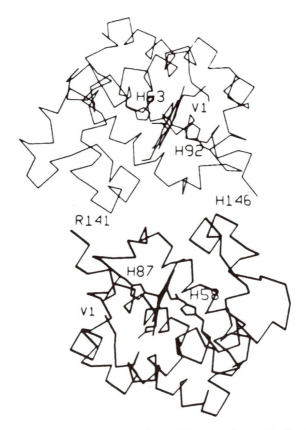

Figure 1. α-Carbon backbone of an [α₁,β₂] ET complex within [2α(Fe(CO)P, 2β(MnP)]. (Reproduced with permission from ref. 20. Copyright 1986 Academic Press.)

the energetics for ET without changing the structure of the ET complex. At neutral pH, the Fe^{3+}P in methemoglobin has H_2O in the distal coordination site, with the remaining five sites taken by the nitrogens of P and of the proximal histidine (22). The coordinated H_2O can be replaced by other ligands L, both neutral (L = L^0 is imidazole) and anionic (L = X^- is CN^-, F^-, or N_3^-). As the Fe^{3+}P–Fe^{2+}P redox potential depends on L, the driving force for ET changes correspondingly. X-ray crystallographic measurements of liganded hemoglobins show negligible changes upon ligand variation, and thus the heme geometry is retained (23).

The combination of metal–ligand variation in these hybrids not only allows alteration of the ET energetics but also provides a means to study mechanistic questions. It could be used to determine whether ET is direct or involves a hopping mechanism and whether ET is "gated" (24–27) by linkage to protein conformational change or heme ligand dissociation.

Kinetics by Triplet Decay

Reversible ET within Mg- and Zn-substituted hemoglobin hybrids is initiated by flash photoproduction of the long-lived triplet state ^{3}(MP). Figure 2 shows the progress curves for triplet decay in [Mg, Fe^{2+}], [Mg, $Fe^{3+}(H_2O)$], [Zn, Fe^{2+}], and [Zn, $Fe^{3+}(H_2O)$], as monitored by ^{3}(MP)–MP absorbance difference spectra. The data are shown normalized to unit triplet population. The triplet decay for both reduced hybrids, [Mg, Fe^{2+}] and [Zn, Fe^{2+}], is exponential over five half-lives. The rate constant for this intrinsic triplet decay, k_d, is 20($\pm$2) s^{-1} at 25 °C when M is Mg and 53($\pm$5) s^{-1} at 25 °C when M is Zn. Identical rate constants are obtained by following triplet-state emission.

Triplet decay in [Mg, $Fe^{3+}(H_2O)$] and [Zn, $Fe^{3+}(H_2O)$] monitored at λ = 415 nm, the $Fe^{3+/2+}$P isosbestic point, or at 475 nm, where contributions from the charge-separated intermediate are minimal, remains exponential. However, the decay rate in the oxidized hybrids, k_p, is increased to 55($\pm$5) s^{-1} when M is Mg and to 138($\pm$7) s^{-1} when M is Zn. Two additional quenching processes can contribute to deactivation of ^{3}(MP) when the iron-containing chain of the hybrid is oxidized to the Fe^{3+}P state: ET quenching as in eq 1 (with rate constant k_t) and Forster energy transfer (with rate constant k_o). ET quenching is not possible in the Fe^{2+}P hybrid, and Forster energy transfer also is unimportant because spectral overlap is minimal (*17*). The

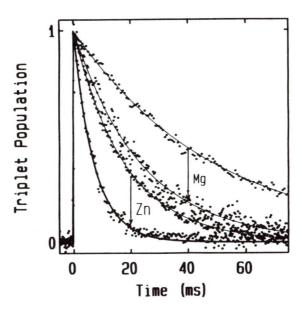

Figure 2. Normalized triplet-decay curves for [M(P), Fe(P)] hybrids. For a given M (M is Zn or Mg), the arrow is directed from the curve for the Fe^{2+}P state toward that for the Fe^{3+}P state.

net rate of triplet disappearance in oxidized hybrids is thus the sum of three terms: $k_p = k_d + k_t + k_o$. The difference in triplet-decay rate constants for the oxidized and reduced hybrids gives the quenching rate constant, $k_q = k_p - k_d = k_t + k_o$, which is thus an upper bound to k_t. Subtraction yields $k_q = 85(\pm 10)$ s^{-1} for [ZnP, $Fe^{3+}(H_2O)P$] and $k_q = 35(\pm 5)$ s^{-1} for [MgP, $Fe^{3+}(H_2O)P$].

By definition, the intrinsic triplet-decay rate constant, k_d, is independent of heme ligation. Therefore, differences in kinetic progress curves for [^{3}(MP), $Fe^{3+}(L)P$] reflect inequivalent values of k_q for the various ligands. The data for M = Zn clearly fall into two classes: Hybrids with ferriheme coordinated to the neutral ligands H_2O and imidazole give $k_q \approx 80$ s^{-1}, but those with bound anionic ligands give dramatically reduced values [$3(\pm 2)$ $s^{-1} < k_q < 12(\pm 3)$ s^{-1}]. The data for [MgP, $Fe^{3+}(L)P$] hybrids show a similar grouping.

In our initial studies, which were solely limited to measurements of triplet quenching, we considered whether energy transfer could be contributing to k_q. Forster energy transfer would be proportional to spectral overlap between the ^{3}(MP) emission spectrum and the $Fe^{3+}(L)P$ absorption spectrum. Thus, a lack of correlation between the ligated heme optical spectra and the observed k_q indicated that for L = L^0 = H_2O (and imidazole), most, if not all, triplet quenching is associated with ET (17). However, with the smaller rate constants for triplet quenching by the anion-ligated hemes, it was by no means clear whether ET was the predominant quenching mechanism. That is, a small value for k_o would have minimal consequence for L = L^0, but could account for much or all of k_q in the case where L = X^-. Thus, direct observation of the charge-separated intermediate I, in addition to yielding k_b, the rate constant for the thermal process, is required to confirm the very existence of long-range ET in cases where k_q is small.

Direct Observation of Charge-Separated ET Intermediates

The time course of the charge-separated intermediate I can be measured in a flash photolysis experiment that monitors the I–A transient absorbance difference at a ground state–triplet state isosbestic point (e.g., $\lambda = 432$ nm when M is Mg and 435 nm when M is Zn). We have observed this intermediate for the [MP, $Fe^{3+}P$] hybrids when M is Mg or Zn; representative kinetic progress curves are shown in Figure 3 (15). In a kinetic scheme that includes eqs 1 and 2 as the only ET processes, when $k_b > k_p$, as is the case here, I appears exponentially with rate constant k_b and disappears completely in an exponential fall with rate constant k_p.

However, the occurrence of a persistent absorbance change (ΔA_∞) for the [M, Fe] hybrids requires an extended kinetic model (Scheme I). In this model, $(MP)^+$ is reduced not only by $Fe^{2+}P$ with rate constant k_b (regenerating the [MP, $Fe^{3+}P$] state), but also by an as-yet-unidentified amino acid

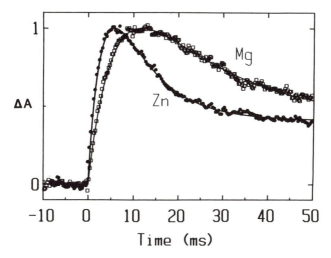

Figure 3. Normalized kinetic progress curves at 5 °C for ET intermediate (I) plus photoproduct (C) (see Scheme I) formed upon flash photolysis of the mixed-metal Hb hybrids: [β(MgP), α(Fe³⁺P)] (λ = 432 nm); [β(ZnP), α(Fe³⁺P)] (λ = 435 nm). Solid lines are nonlinear least-squares fits to the equations in ref. 15. For [Mg, Fe], k_b = 155(±15) s⁻¹, k_p = 47(±5) s⁻¹, and k_m = 20(±5) s⁻¹; for [Zn, Fe], k_b = 350(±35) s⁻¹, k_p = 122(±10) s⁻¹, and k_m = 40(±8) s⁻¹. Buffer: 0.01 M KP_i, pH 7.0.

residue X and/or solution impurities with rate constant k_m, leading to [MP, $Fe^{2+}P$] (eq 3):

$$[(MP^+), Fe^{2+}P] + X \xrightarrow{k_m} [MP, Fe^{2+}P] + X^+ \tag{3}$$

Solution of the kinetic equations implicit in Scheme I indicates that the magnitude of ΔA_∞ is proportional to k_m, and that I appears exponentially with rate constant $k_x = k_b + k_m$. Figure 3 shows that the kinetic progress curves for I for the Zn- and Mg-substituted hybrids are well-described by nonlinear least-squares fits to these kinetic equations (15). The data in Figure 3 show that the time course of the intermediate [(MP)⁺, $Fe^{2+}P$] (I) strongly depends on M. At 5 °C when M is Mg, k_b = 155(±15) s⁻¹, k_p = 47(±5) s⁻¹, and k_m = 20(±5) s⁻¹; when M is Zn, k_b = 350(±35) s⁻¹, k_p = 112(±10) s⁻¹, and k_m = 40(±8) s⁻¹.

Direct observation of I, the charge-separated intermediate, has verified the occurrence of long-range ET within [³(MP), Fe³⁺(L)P] for both M and all L. Figure 4 shows a comparison of the kinetic progress curves obtained for [β(ZnP), α(Fe³⁺(H₂O)P] and [β(ZnP), α(Fe³⁺(CN⁻)P] (16). The long-time exponential fall for the latter, k_p = 65(±8) s⁻¹, is in agreement with that

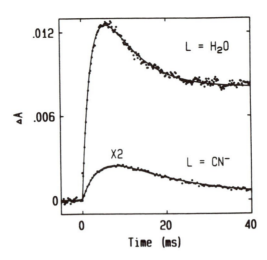

Figure 4. Kinetic progress curves as in Figure 3 at 435 nm for [β(ZnP), α(Fe³⁺(L)P)]. Experimental points and nonlinear least-squares fits for L = H₂O and L = CN⁻ are shown, with absorbance changes normalized to a zero-time triplet concentration (A_0) = 10^{-6} M. For [β(ZnP), α(Fe³⁺(H₂O)P)], k_b = 345(±45) s^{-1} and k_p = 134(±15) s^{-1}; for [β(ZnP), α(Fe³⁺(CN⁻)P)], k_b = 240(±30) s^{-1} and k_p = 65(±8) s^{-1}. Buffer: 0.01 M KP$_i$, pH 7.0.*

observed in triplet-decay data. Absorbance changes resulting from formation of the charge-separated intermediate I are proportional to the rate constant k_t. Thus k_t can be calculated independently of any other contributions to triplet-state quenching if the quantum yield for the formation of I can be determined.

With this procedure, analysis of the relatively large absorbance changes observed with the intermediate in the M = Zn, L = H₂O hybrid gives $k_t^{Zn}(H_2O)$ = 90(±30) s^{-1} and thus confirms the previous assignment k_t = $k_p - k_d$ for L = H₂O. On the other hand, k_q = 14(±4) s^{-1} for [β(ZnP), α(Fe³⁺(CN⁻)P], but analysis of absorbance changes associated with [β(ZnP)⁺, α(Fe²⁺(CN⁻)P] gives an even smaller value, $k_t^{Zn}(CN^-)$ = 6(±3) s^{-1}. Thus, replacement of H₂O by CN⁻ in the heme coordination sphere reduces k_t by over an order of magnitude. Quantitation of I when M is Mg or Zn, and with all the ligands studied so far, gives the $k_t^M(L)$ shown in Figure 5. Replacement of H₂O with another neutral ligand, imidazole, does not significantly alter k_t. Replacement with other anions (N₃⁻, F⁻) affects k_t in the same fashion as CN⁻, a 10-fold reduction in rate.

The effect of anion binding on k_b is not nearly as great. The data in Figure 5 show that, for both metals, a less than 50% reduction in thermally activated ET rate constant is observed between hybrids containing neutral and anionic ligands.

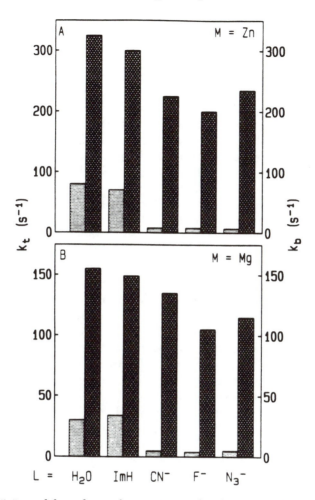

Figure 5. Ligand dependence of rate constants for photoinitiated ET (k_t, light, speckled) and thermal ET (k_b, dark, crosshatched) within the hybrids (A) [ZnP, $Fe^{3+}(L)P$] and (B) [MgP, $Fe^{3+}(L)P$]. When M is Zn, the data refer to 1-methylimidazole rather than imidazole.

Mechanistic Aspects of Electron Transfer within [MP, $Fe^{3+}(L)P$]

The single most important mechanistic question concerning long-range ET in hemoglobin hybrids is whether the ET reactions under consideration are single-step events or multiple-step processes with one or more real intermediate states (such as one with an oxidized or reduced amino acid residue).

A second key issue in mechanistic studies of long-range electron transfer in proteins is the role of gating. If photoinitiated or thermal ET requires a protein conformational change, then k_{obs}, the observed rate constant, may actually be measuring a conformational rate rather than an ET rate (24–27). One aspect of this issue involves the fate of the heme ligand. Reduction of $Fe^{3+}(H_2O)P$ promptly yields the unliganded ferroheme $Fe^{2+}P$, but the fate of anionic ligands upon reduction of $Fe^{3+}(X^-)P$ is not clear. In some cases involving exogenous reductants, it has been shown that ligand dissociation is a prerequisite to reduction (28, 29). Thus, the possibility of "ligand gating" must also be considered. The ability to alter the redox potentials of both MP and Fe(L)P allows us to address these questions directly, and our data show that k_t and k_b represent rate constants for direct, ungated electron transfer between the MP and FeP.

Photoinitiated ET is easily proved to be direct. Indirect hopping of an electron from $^3(MP)$ to $Fe^{3+}(L)P$ would correspond to oxidative triplet quenching by an amino acid, with subsequent reduction of $Fe^{3+}P$ by the amino acid anion. However, any such quenching of $^3(MP)$ would occur equally in the reduced $(Fe^{2+}P)$ hybrids because this process does not involve the FeP. Consequently, ET by this mechanism would not give rise to an increase in triplet decay, and the increased triplet quenching in the $Fe^{3+}P$ hybrids must be associated with a direct ET process.

Our data also indicate that the $I \rightarrow A$ ET process is direct. If it were not, then there must exist an amino acid, y, that mediates electron flow from $Fe^{2+}P$ to $(MP)^+$ via an internal ET. This is equivalent to postulating a thermodynamically accessible discrete intermediate, $[(MP), y^+, (Fe^{2+}P)]$, which would decay by a second ET process back to the ground state. If the $(MP)^+ \rightarrow y$ ET process were rapid and $Fe^{2+}P \rightarrow y^+$ ET were rate limiting, changing M would not affect the observed rate constant and k_b^{Mg} would equal k_b^{Zn}. If a rate-limiting $(MP)^+ \rightarrow y$ ET were succeeded by rapid $Fe^{2+}P \rightarrow y^+$ ET, changing the heme ligand L would not affect the ET rate and $k_b^{Zn}(H_2O)$ would equal $k_b^{Zn}(CN^-)$. However, $k_b^{Mg}(H_2O) \neq k_b^{Zn}(H_2O) \neq k_b^{Zn}(CN^-)$. We conclude that a two-step electron hopping mechanism does not obtain and that k_b describes direct $Fe^{2+}P \rightarrow (MP)^+$ ET.

Comparison of rates for the [M, Fe] hybrids where M is Mg and Zn provides a test of whether ET is gated (i.e., controlled by a slow conformational transformation) to an "ET-active" state in which ET is presumed to be rapid. The rate of such a conformational transformation would not change because of the alteration in driving force caused by the Zn–Mg switch. Because k_t is indeed different when M is Mg and Zn, it cannot represent a rate constant for conformational interconversion.

What is the fate of the heme-ligand, L, during the ET cycle of Scheme I? When L is H_2O, reduction of the aquo-bound heme yields the five-coordinate ferroheme, $Fe^{2+}P$. For CN^-, ligand dissociation from $Fe^{2+}(CN^-)P$ is slow; this fact indicates that the $I \rightarrow A$ process involves reoxidation of the CN^--bound species. The data in Figure 5 show that N_3^- and F^- also remain bound on the ET time scale. If ET-induced ligand loss were rapid compared to the I → A ET process, one would predict that for a given metal, $k_b{}^M(X^-)$ = $k_b{}^M(H_2O)$. This prediction is contrary to observation with both metals. The data also show that for a given anion, $k_b{}^{Mg}(X^-) < k_b{}^{Zn}(X^-)$. This dependence on metal ion indicates that $k_b{}^M(X^-)$ cannot represent a rate-limiting ligand dissociation from the $Fe^{2+}(X^-)P$ partner of I, followed by fast ET (i.e., ligand gating of I → A). Thus the variation in the thermal ET rate, $k_b{}^M(L)$, as a function of ligand and metal indicates that all anionic ligands remain bound during the entire ET cycle of Scheme I. In contrast, the extremely slow reduction of ligated ferrimyoglobin [$Mb^{3+}(L)$] by exogenous $S_2O_4{}^{2-}$ requires anionic dissociation for most ligands, notably fluoride (28, 29). Taken together, these data suggest a difference in the mechanisms for reduction by exogenous dithionite and by an internal $^3(MP)$.

In summary, the simple variation of ligands and metals within [MP,$Fe^{3+}(L)P$] allows a heretofore unparalleled view of the mechanistic aspects of long-range ET between proteins.

Conclusions and Prospectus

Metal-substituted hemoglobin hybrids, [MP, $Fe^{3+}(H_2O)P$] are well-suited to the study of long-range ET within protein complexes. Both photoinitiated and thermally activated ET can be studied by flash excitation of Zn- or Mg-substituted complexes. Direct spectroscopic observation of the charge-separated intermediate, [$(MP)^+$, $Fe^{2+}P$], unambiguously demonstrates photoinitiated ET, and the time course of this ET product indicates the presence of thermal ET. Replacement of the coordinated H_2O by anionic ligands (CN^-, F^-, or $N_3{}^-$) in the ferriheme subunit dramatically lowers the photoinitiated rate constant, k_t, but has a relatively minor effect on the thermal rate, k_b.

Because metal substitution and ligand variation can be effected without structural perturbation of the ET complex, such changes can be used to probe mechanistic aspects of ET. The data show that both photoinitiated and thermal ET are direct processes. Furthermore, ET is not gated either by protein conformational changes or by ligand on–off processes. The stability of hemoglobin tetramers in cryosolvent, coupled with the absence of gating in these systems, has allowed observation of long-range ET at temperatures near 77 K (30), where quantum mechanical tunneling is operative.

The ease with which ET can be studied in mixed-metal hemoglobin hybrids suggests that this system will be of value in addressing several long-standing problems in this field. For example, mixed-metal mutant hemoglobins, in which amino acids between porphyrins have been changed from aliphatic to aromatic and vice versa, are being used to assess the role of ET pathways and of hole superexchange in long-range electron transfer. We are extending our studies of ET within [MP, $Fe^{3+}(L)P$] to liquid helium (4 K) temperatures. Finally, a more complete picture of the role of energetics in long-range ET is being realized through expanded metal substitution and ligand variation. By altering the protein environment, the solvent, the temperature, and the ET sites themselves, we hope to greatly add to the understanding of this important biological process.

Acknowledgment

This research was supported by National Institutes of Health Grants HL 13531 and HL 40453, and by National Science Foundation Grant DMB 8907559 to Brian M. Hoffman, and by National Institutes of Health National Research Service Award postdoctoral fellowship HL07531 to Michael J. Natan.

References

1. Gray, H. B.; Malmström, B. G. *Biochemistry* **1989**, *28*, 7499.
2. Marcus, R. A.; Sutin, N. *Biochim. Biophys. Acta* **1985**, *811*, 265.
3. McLendon, G. *Acc. Chem. Res.* **1988**, *21*, 160.
4. Mayo, S. L.; Ellis, W. R., Jr.; Crutchley, R. J.; Gray, H. B. *Science (Washington, D.C.)* **1986**, *233*, 948.
5. Bowler, B. E.; Meade, T. J.; Mayo, S. L.; Richards, J. H.; Gray, H. B. *J. Am. Chem. Soc.* **1989**, *111*, 8757.
6. Meade, T. J.; Gray, H. B.; Winkler, J. R. *J. Am. Chem. Soc.* **1989**, *111*, 4353.
7. Elias, H.; Chou, M. H.; Winkler, J. R. *J. Am. Chem. Soc.* **1988**, *110*, 429.
8. Cowan, J. A.; Upmacis, R. K.; Beratan, D. N.; Onuchic, J. N.; Gray, H. B. *Ann. N.Y. Acad. Sci.* **1988**, *550*, 68.
9. Axup, A. W.; Albin, M.; Mayo, S. L.; Crutchley, R. J.; Gray, H. B. *J. Am. Chem. Soc.* **1988**, *110*, 435.
10. Jackman, M. P.; McGinnis, J.; Powls, R.; Salmon, G. A.; Sykes, A. G. *J. Am. Chem. Soc.* **1988**, *110*, 5880.
11. Osvath, P.; Salmon, G. A.; Sykes, A. G. *J. Am. Chem. Soc.* **1988**, *110*, 7114.
12. Nocek, J. M.; Liang, N.; Wallin, S. A.; Mauk, A. G.; Hoffman, B. M. *J. Am. Chem. Soc.* **1990**, *112*, 1623.
13. Liang, N.; Mauk, A. G.; Pielak, G. J.; Johnson, J. A.; Smith, M.; Hoffman, B. M. *Science (Washington, D.C.)* **1988**, *240*, 311.
14. Liang, N.; Pielak, G. J.; Mauk, A. G.; Smith, M.; Hoffman, B. M. *Proc. Natl. Acad. Sci. U.S.A.* **1987**, *84*, 1249.
15. Natan, M. J.; Hoffman, B. M. *J. Am. Chem. Soc.* **1989**, *111*, 6468.
16. Natan, M. J.; Kuila, D.; Baxter, W. W.; King, B. C.; Hawkridge, F. M.; Hoffman, B. M. *J. Am. Chem. Soc.* **1990**, *112*, 4081.
17. McGourty, J. M.; Peterson-Kennedy, S. E.; Ruo, W. Y.; Hoffman, B. M. *Biochemistry* **1987**, *26*, 8302.
18. Gingrich, D. J.; Hoffman, B. M., unpublished results.
19. Kuila, D.; Natan, M. J.; Rogers, P.; Gingrich, D. J.; Baxter, W.; Arnone, A.; Hoffman, B. M. *J. Am. Chem. Soc.*, unpublished results.
20. Arnone, A.; Rogers, P.; Blough, N. V.; McGourty, J. L.; Hoffman, B. M. *J. Mol. Biol.* **1986**, *188*, 693.
21. Gingrich, D. J.; Nocek, J. M.; Natan, M. J.; Hoffman, B. M. *J. Am. Chem. Soc.* **1987**, *109*, 7533.
22. Antonini, E.; Brunori, M. *Hemoglobin and Myoglobin in Their Reactions with Ligands*; North Holland Publishing Co.: Amsterdam, Netherlands, 1971.

23. Moffat, J. K.; Deatherage, J. F.; Seybert, D. W. *Science (Washington, D.C.)* **1979**, *206*, 1035.
24. Hoffman, B. M.; Ratner, M. R. *J. Am. Chem. Soc.* **1987**, *109*, 6237.
25. Hoffman, B. M.; Ratner, M. A.; Wallin, S. A. In *Electron Transfer in Biology and the Solid State*; Johnson, M. K.; King, R. B.; Kurtz, D. M., Jr.; Kutal, C.; Norton, M. L.; Scott, R. A., Eds.; Advances in Chemistry Series 226; American Chemical Society: Washington, DC, 1990; pp 125–146.
26. Brunschwig, B. S.; Sutin, N. *J. Am. Chem. Soc.* **1989**, *111*, 7454.
27. McLendon, G.; Pardue, K.; Bak, P. *J. Am. Chem. Soc.* **1987**, *109*, 7540.
28. Cox, R. P.; Holloway, M. R. *Eur. J. Biochem.* **1977**, *74*, 575.
29. Olivas, E.; deWaal, D. J. A.; Wilkins, R. G. *J. Biol. Chem.* **1977**, *252*, 4038.
30. Kuila, D.; Baxter, W. W.; Natan, M. J.; Hoffman, B. M. *J. Phys. Chem.* **1991**, *95*, 1.

RECEIVED for review April 27, 1990. ACCEPTED revised manuscript August 17, 1990.

Electron Transfer, Energy Transfer, and Excited-State Annihilation in Binuclear Compounds of Ruthenium(II)

Takeshi Ohno[1], Koichi Nozaki[1], Noriaki Ikeda[1], and Masa-aki Haga[2]

[1]Department of Chemistry, College of General Education, Osaka University, Toyonaka, Osaka 560, Japan
[2]Department of Chemistry, Faculty of Education, Mie University, Tsu, Mie 514, Japan

Electronic excited states of binuclear compounds of ruthenium(II) bridged by bis-2,2'-(2"-pyridyl)bibenzimidazole (bpbimH$_2$) were studied by means of laser photolysis kinetic spectroscopy. Excitation of RuL$_2$(bpbimH$_2$)RuL$_2^{4+}$ [L is 2,2'-bipyridine (bpy), 4,4'-dimethyl-2,2'-bipyridine (dmbpy), or 1,10-phenanthroline (phen)] into the metal-to-ligand charge-transfer (MLCT) triplet state gives rise to a transient absorption spectrum revealing electron occupation on both L and bpbimH$_2$ in CH$_3$CN. In an asymmetric binuclear compound, excitation-energy transfer takes place from the higher energy site to the other site. Production of the MLCT triplet excited state in a symmetric binuclear compound, Ru(bpy)$_2$(bpbimH$_2$)Ru(bpy)$_2^{4+}$, is compared with that of the corresponding mononuclear compound, Ru(bpy)$_2$(bpbimH$_2$)$^{2+}$. Smaller production of the excited triplet state in the binuclear compound is ascribed to a rapid triplet–triplet annihilation process. A decay rate of the excited Ru(dmbpy)$_2$-(bpbimH$_2$)$^{2+}$ linked to Rh(phen)$_2^{3+}$ in butyronitrile was obtained by extrapolation of rates measured at lower temperatures. Mechanisms of the intramolecular reaction are discussed.

THE RATES OF MANY NONADIABATIC electron-transfer reactions are controlled by both thermally averaged Franck–Condon integrals and electronic

0065–2393/91/0228–0215$06.00/0
© 1991 American Chemical Society

coupling between reactants. The quantum theory of electron transfer predicts a bell-shaped energy-gap dependence on the Franck–Condon integral (1, 2). Because of this expectation, the energy-gap dependencies of electron-transfer rates have been analyzed mostly in terms of the thermally averaged Franck–Condon integral (3–9).

Sometimes electron-transfer rates are either slower than predicted by the Franck–Condon integral or weakly dependent on the energy gap of the process (6, 10–13). In these cases weak electronic coupling between reactants has been regarded as responsible for the slow electron-transfer rates. Electronic coupling has been assumed to be weak for both long-range and spin-inverted electron-transfer processes. However, the intermolecular electronic coupling in transition-state electron transfer has seldom been estimated quantitatively because distance and orientation between transition-state reactants are both unknown (14).

To investigate the electronic-coupling term independently in the electron-transfer rate, intramolecular electron transfers occurring in bichromophoric compounds should be studied. Electronic coupling between the chromophores in these compounds can be estimated by spectroscopic methods (15–20). Accordingly, electronic coupling between Ru(II) and Ru(III) in mixed-valence binuclear complexes has been examined (15–19) to determine a correlation between the electron-transfer rate and electronic coupling.

Metal–metal interaction in the photoexcited states of binuclear Ru(II) compounds has attracted much attention in recent years. If the metal–metal interaction exceeds 10 cm^{-1}, rates of electron transfer, energy transfer, and triplet–triplet (T–T) annihilation can be studied. Electron transfer takes place between an excited-state Ru(II) site and a ground-state metal site if it is energetically feasible (21–23). The rate of metal–metal energy transfer (24–27) depends on the excitation-energy difference between the acceptor and donor. When a laser pulse is strong enough to excite the Ru(II) sites of a binuclear compound, such an excited site will undergo an annihilation process with a neighboring site.

Intramolecular reactions (electron transfer, energy transfer, and T–T annihilation) in binuclear Ru(II)–Ru(II) and Ru(II)–Rh(III) compounds have been examined by means of laser flash kinetic spectroscopy. An intervening tetradentate ligand, bis-2,2'-(2''-pyridyl)bibenzimidazole (bpbimH$_2$) (28), and 2,2'-bibenzimidazole (29, 30) have a strong σ-donor and weak π-acceptor property in comparison with 2,2'-bipyridine (bpy). Recently a binuclear compound, Ru(bpy)$_2$(bpbimH$_2$)Ru(bpy)$_2$$^{4+}$, has been shown to behave as a dibasic acid by using stepwise deprotonation from the imino N–H groups on the bridging bpbimH$_2$. The pK_{a1} and pK_{a2} are 5.61 and 7.12, respectively, in CH$_3$CN buffer (1:1 v/v). The pK_{a1} for the mixed-valence compounds [Ru(bpy)$_2$(bpbimH$_2$)Ru(bpy)$_2$]$^{5+}$ and [Ru(dmbpy)$_2$(bpbimH$_2$)Rh(phen)$_2$]$^{5+}$ (where dmbpy is 4,4'-dimethyl-2,2'-bipyridine and phen is 1,10-phenanthroline) are considerably reduced to 1.2 (28) and 2.89 (31), respectively. Ruthenium–ruthenium interaction in the mixed-valence Ru(II)–Ru(III) compound was estimated on the basis of a weak intervalence transition to be as small as 0.01 eV (28). The structure of bis-2,2'-(2"-pyridyl)bibenzimidazole is

bpbimH$_2$

Experimental Details

Compounds.   Mononuclear ruthenium(II) compounds, $Ru(L)_2(bpbimH_2)^{2+}$ (L is bpy, dmbpy, and phen), and binuclear ruthenium(II–II), $Ru(bpy)_2(bpbimH_2)$-$Ru(bpy)_2^{4+}$, were prepared as described elsewhere (28). An asymmetric binuclear ruthenium compound, $[Ru(dmbpy)_2(bpbimH_2)Ru(phen)_2](ClO_4)_4 \cdot 5H_2O$, was prepared from $Ru(phen)_2Cl_2$ (0.15 g, 0.28 mmol) with $[Ru(dmbpy)_2(bpbimH_2)](ClO_4)_2$ (0.3 g, 0.28 mmol) in ethylene glycol (30 mL). The solid sample obtained was purified by recrystallization from methanol–water (4:1 v/v). Yield, 0.28 g (54%). Anal. Calcd. for $C_{72}H_{56}N_{14}O_{16}Cl_4Ru_2 \cdot 5H_2O$: C, 47.85%; H, 3.58%; N, 10.85%. Found: C, 47.68%; H, 3.50%; N, 10.66%.

A heterobinuclear compound, $[Ru(dmbpy)_2(bpbimH_2)Rh(phen)_2](ClO_4)_5$, was prepared by heating $Rh(phen)_2(bpbimH_2)Cl_3$ (0.47 g, 0.41 mmol) and $Ru(dmbpy)_2Cl_2$ (0.26 g, 0.45 mmol) in ethanol–water. The purification was effected by column chromatography on cross-linked dextran polymer beads (Sephadex LH-20) with methanol as eluent. Yield, 0.52 g (70%). Anal. Calcd. for $C_{72}H_{60}N_{14}O_{22}Cl_5RuRh \cdot 2H_2O$: C, 46.63%; H, 3.26%; N, 10.57%. Found: C, 46.76%; H, 3.59%; N, 10.51%.

Apparatus.   A Hitachi spectrofluorometer (MPF-2A) was used for phosphorescence spectra at 77 K. The Q-switched Nd^{3+}–YAG laser (Quantel YG580) and a transmittance-change acquisition system used have been described elsewhere (32). The laser energies for 532- and 355-nm pulses were less than 80 and 40 mJ, respectively. A xenon arc lamp (150 W) was 30× intensified for 2 ms to improve the signal-to-noise ratio of the transmittance changes of the sample solutions. Time evolution of sample-solution transmittance and phosphorescence were recorded on a transient digital (10 bits) memory (Electronica Co., ELK-5120, 10 MHz) or a storagescope (Iwatu Co., 8123, 200 MHz, 8 bits).

Oxidation potentials of ruthenium(II) compounds were measured by means of differential-pulse voltammetry with a direct-current pulse polarograph (HECS-312B, Huso, Japan). All voltammograms were obtained at a platinum disc electrode (d. 0.5 mm) in CH_3CN containing 0.1 M tetrabutylammonium perchlorate. All potentials are referred to the formal potential of the ferrocenium–ferrocene (Fc^+–Fc) system, which is –0.33 V against a saturated calomel electrode (SCE).

Measurements.   The sample solutions of ruthenium(II) compounds dissolved in acetonitrile, in butyronitrile, and in a mixed solvent of ethanol and methanol (4:1 v/v) were deaerated by bubbling with nitrogen more than 12 min. Either $HClO_4$ or

CF_3COOH (1 mM) was added to suppress deprotonation of the intervening ligand, $bpbimH_2$. Production of the excited states of ruthenium(II) compounds on exposure to the second harmonic pulse (532 nm) of the YAG laser was measured by monitoring the absorbance change of the sample solution. Time profiles of the transmittance in a 580–780-nm region were corrected for the strong phosphorescence of the sample. The temperature of the sample solutions (89–300 K) was controlled by using a cryostat (Oxford DN1704) and a controller (Oxford ITC4). The sample temperature was monitored by putting a thermocouple on a copper sample holder.

Results and Discussion

Intramolecular metal–metal interaction in binuclear Ru(II) compounds depends on an intervening ligand and the electronic states of the metal sites. An intervening ligand, $bpbimH_2$, has a biphenyl structure in which the π electrons are weakly conjugated throughout two moieties of 2-pyridyl-2'-benzimidazole (pbimH). The d_π electrons of one metal site are able to mix with those of the other metal site through the conjugated π and π^* electrons of the ligand. Photoexcitation of a binuclear Ru compound to its metal-to-ligand charge-transfer (MLCT) state is manifested through photophysical and photochemical processes such as energy transfer from one Ru(II) site to another in an asymmetric binuclear compound, electron transfer from a Ru(II) site to a Rh(III) site in a heterobinuclear compound, and T–T annihilation in a symmetrical binuclear Ru(II) compound.

MLCT Excited State and Energy Transfer.   Lowest excited states of many ruthenium(II) polypyridine compounds are described as a phosphorescent state of Ru $\rightarrow$ ligand charge transfer (33, 34). A transferred electron in this localized model occupies an orbital on the most easily reduced ligand. When the redox potentials of the adjacent ligands are close to that of the most easily reduced ligand, electron hopping takes place among the ligands (34–39).

The most easily reduced ligand can be assigned on the basis of redox potentials $[E° (L/L^-)]$ of RuL_3^{2+}, unless the difference in the redox potentials between ligands is subtle (38). The assignment of the electron-occupied orbital (ligand) is feasible by using emission from and absorption of excited states. Emission spectra at 77 K, in this case, are of no use for the assignment of the electron-occupied orbital because the vibronic progressions of phosphorescences are not distinct.

Transient absorption spectra following laser excitation of $Ru(bpy)_2$-$(bpbimH_2)^{2+}$, $Ru(dmbpy)_2(bpbimH_2)^{2+}$, and $Ru(phen)_2(bpbimH_2)^{2+}$ exhibit the π–π^* band of a reduced ligand as a component, and this band is strong evidence for the identification of a reduced ligand. Bands in the ~370–420- and ~500–600-nm regions in Figures 1a and 1b can be assigned to a π–π^* transition of $bpbimH_2^-$ and/or a red-shifted π–π^* of $bpbimH_2$ coordinating to Ru(III). These bands were not evident in the excited-state absorption

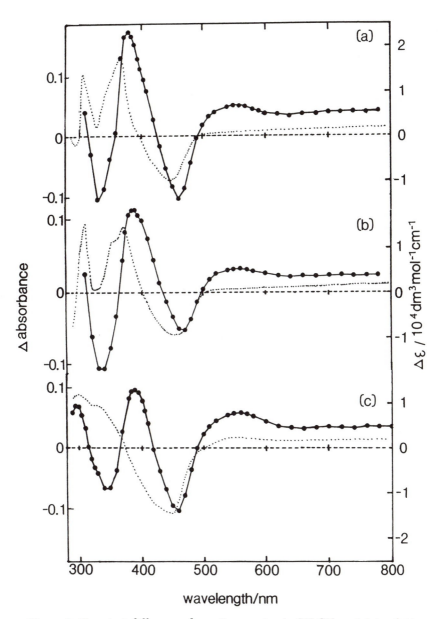

Figure 1. Transient difference absorption spectra in CH_3CN containing 1×10^{-3} M $HClO_4$ at 100 ns after laser excitation. $\Delta\epsilon$ on the right ordinate denotes the difference in molar absorption coefficient between the ground and the MLCT states of $RuL_3{}^{2+}$ (32, 44). a: Solid line, $Ru(bpy)_2(bpimH_2)^{2+}$ (20 μM); dotted line, $Ru(bpy)_3{}^{2+}$. b: Solid line, $Ru(dmbpy)_2(bpbimH_2)^{2+}$ (20 μM); dotted line, $Ru(dmbpy)_3{}^{2+}$. c: Solid line, $Ru(phen)_2(bpbimH_2)^{2+}$ (20 μM); dotted line, $Ru(phen)_3{}^{2+}$.

(ESA) spectra of $Ru(bpy)_3^{2+}$ (32, 40–43) and $Ru(dmbpy)_3^{2+}$ (44). Strong bleaching of π–π^* transition of $bpbimH_2$ at 330 and 350 nm in Figures 1a and 1b also demonstrates the reduction of $bpbimH_2$. The band of bpy⁻ or dmbpy⁻ at 370 nm [molar absorptivity $\epsilon = 20 \times 10^3$ dm³cm⁻¹mol⁻¹ (32, 40–44)], which partly overlapped with the $bpbimH_2^-$ band at 390 nm, was masked by the bleaching of the strong π–π^* band of $bpbimH_2$ ($\epsilon \sim 40 \times 10^3$ dm³cm⁻¹mol⁻¹). A wide band in a region longer than 600 nm (Figure 1a) can be ascribed to π–π^* of bpy⁻; in addition, both ligand-to-metal charge transfer (LMCT) of bpy → Ru(III) and $bpbimH_2^-$ exhibit some intensity in this region.

A sharp band at 310 nm is ascribed to a red-shifted π–π^* of bpy coordinating to Ru(III). This red shift results in the intensity reduction of π–π^* at 290 nm, as has been reported in the case of $Ru(bpy)_3^{2+}$ (40, 42, 43). An enormous bleaching of the π–π^* band (bpy) at 290 nm, which was observed for the dilute (5 μM) solution, may be caused by both the reduction of bpy and the coordination of bpy to Ru(III). The assumption that the extent of bleaching at 330–350 nm is proportional to the formation of $bpbimH_2^-$ implies that more $bpbimH_2^-$ will be formed in $Ru(dmbpy)_2(bpbimH_2)^{2+}$ than in $Ru(bpy)_2(bpbimH_2)^{2+}$.

Excitation of $Ru(phen)_2(bpbimH_2)^{2+}$ produced weak bands at 300, 390, and 580 nm with the bleaching of the MLCT band ($\epsilon = 17,000$ dm³cm⁻¹mol⁻¹ at 460 nm). The bleaching of the 330–350-nm band of $bpbimH_2$ is small compared to the formation of a wide band in a red region. The band at 300 nm is mainly ascribed to phen⁻, because it is evident for the MLCT state of $Ru(phen)_3^{2+}$, as shown in Figure 1c.

When Ru(II) compounds, $[Ru(bpy)_2]_2(bpbimH_2)^{4+}$ and $Ru(phen)_2$-$(bpbimH_2)^{2+}$, were cooled to 83 K in a mixed solvent of ethanol and methanol, the production of the 370-nm band ascribed to bpy⁻ or the 300-nm band ascribed to phen⁻ was more distinct, as Figures 2a and 2c show. On the other hand, the bleaching of the 330–350-nm band at 89 K was reduced. The variation of the transient absorption spectra on cooling demonstrates that the energy level of Ru → $bpbimH_2$ charge transfer (CT) is a little higher than those of Ru → bpy and Ru → phen. The bleaching of the 350-nm band was found to be unchanged for $[Ru(dmbpy)_2]_2(bpbimH_2)^{4+}$ at 83 K. The cooling effect on the transient absorption profiles suggests an order of the ability of the ligand to accept an electron in the MLCT state: bpy > phen > $bpbimH_2$ ~ dmbpy. The decreasing order of ligand effectiveness as an electron acceptor in the MLCT state is consistent with the decreasing order of the redox potential of $Ru(bpy)_3^{2+}$ (−1.34 V vs. SCE), $Ru(phen)_3^{2+}$ (−1.35 V vs. SCE), and $Ru(dmbpy)_3^{2+}$ (−1.45 V vs. SCE) (46).

Transient absorption spectra for the binuclear compounds $[Ru(bpy)_2]_2$-$(bpbimH_2)^{4+}$ and $[Ru(dmbpy)_2]_2(bpbimH_2)^{4+}$ are almost identical to those of the corresponding mononuclear compounds. Bimetalation is accompanied by a small red shift of the band peak from 380–385 to 390 nm; this change

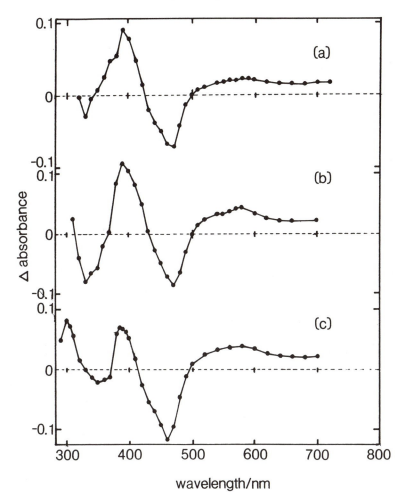

Figure 2. Transient difference absorption spectra in a mixed solvent of ethanol and methanol (4:1 v/v) containing 1 × 10⁻³ M HClO₄ or CF₃COOH at 100 ns after laser excitation at 89 K. a: [Ru(bpy)₂]₂(bpbimH₂)⁴⁺. b: [Ru(dmbpy)₂]₂(bpbimH₂)⁴⁺. c: Ru(phen)₂(bpbimH₂)²⁺.

suggests that less bpy⁻ is formed in the MLCT state. Ru(III) in the singly excited MLCT state of the binuclear compounds probably interacts with unexcited Ru(II) to stabilize the phosphorescent state.

The following list shows the phosphorescence energies (E in reciprocal centimeters) of the Ru compounds.

$Ru(bpy)_2(bpbimH_2)^{2+}$ 16,500

$[Ru(bpy)_2]_2(bpbimH_2)^{4+}$ 16,370

Ru(dmbpy)$_2$(bpbimH$_2$)$^{2+}$ 16,300

[Ru(dmbpy)$_2$]$_2$(bpbimH$_2$)$^{4+}$ 16,070

Ru(phen)$_2$(bpbimH$_2$)$^{2+}$ 16,700

Ru(phen)$_2$(bpbimH$_2$)Ru(dmbpy)$_2$$^{4+}$ 16,100

The energy shift of emission observed for [Ru(bpy)$_2$]$_2$(bpbimH$_2$)$^{4+}$ (130 cm^{-1}) is as small as the Ru(III)–Ru(II) interaction ($\sim$80 cm^{-1}) estimated from the intensity of intervalence transition (28). The result of differential-pulse voltammetry on [Ru(bpy)$_2$]$_2$(bpbimH$_2$)$^{4+}$ confirms that the Ru(III)–Ru(II) interaction is less than 0.040 eV (320 cm^{-1}).

The weak effect of dimetalation in the bpbimH$_2$ compounds is related to weak electronic coupling between the pbimH moieties of bpbimH$_2$, which are not coplanar owing to proton–proton repulsion. In bpbimH$_2$ compounds the dimetalation effects on emission energy are much smaller than those obtained for [Ru(bpy)$_2$]$_2$(bpym)$^{4+}$ (47–49) and [Ru(bpy)$_2$]$_2$(dpp)$^{4+}$ (47–51), where bpym and dpp are 2,2'-bipyrimidine and 2,3-bis(pyridyl)pyrazine, respectively. In the latter cases, the lower energy emission was ascribed to the reduction potentials of bpym or dpp, which are 0.4 V less negative than those of the corresponding mononuclear compounds (50–52).

To see whether the weak metal–metal interaction allows hopping of MLCT between the Ru(II) sites, we examined an asymmetric binuclear compound, Ru(dmbpy)$_2$(bpbimH$_2$)Ru(phen)$_2$$^{4+}$. Excitation-energy transfer from the Ru–phen site to the Ru–dmbpy site is energetically possible; the assigned excitation energies are 16,700 cm^{-1} for Ru(phen)$_2$(bpbimH$_2$)$^{2+}$ and 16,300 cm^{-1} for Ru(dmbpy)$_2$(bpbimH$_2$)$^{2+}$. The phosphorescence of Ru(dmbpy)$_2$(bpbimH$_2$)Ru(phen)$_2$$^{4+}$ observed at 77 K (16,100 cm^{-1}) is emitted from the Ru–dmbpy site. Because the emission lifetime of Ru(phen)$_2$-(bpbimH$_2$)$^{2+}$ is close to 4 μs at 90 K, the energy-transfer rate is estimated as larger than 3 $\times$ 10^7 s^{-1}.

The ESA spectrum of Ru(dmbpy)$_2$(bpbimH$_2$)Ru(phen)$_2$$^{4+}$ immediately after laser excitation at 300 K completely agrees with that of Ru(dmbpy)$_2$-(bpbimH$_2$)$^{2+}$. The characteristic bleaching of the excited Ru–phen in the 420–430-nm region was not observed at all during the laser excitation, a result indicating that the energy transfer at 300 K took place during laser excitation ($k > 10^8$ s^{-1}). The energy transfer from the Ru–phen site to the Ru–dmbpy site occurs via a consecutive process, an electron-energy transfer "cascade" (22, 25, 47). Electron transfer from phen to bpbimH$_2$ generates Ru $\rightarrow$ bpbimH$_2$ CT in one Ru site, and hole transfer from Ru(III) to Ru(II) generates Ru $\rightarrow$ bpbimH$_2$ CT, which is in equilibrium with Ru $\rightarrow$ dmbpy CT.

Electron transfer between adjacent ligands has been interpreted as exciton hopping, with an activation energy equal to the energy difference between the MLCT states (38, 39). A small energy difference between Ru → phen CT and Ru → bpbimH$_2$ CT may not suppress the electron-transfer rate in the Ru(dmbpy)$_2$(bpbimH$_2$)Ru(phen)$_2^{4+}$. Hole transfer, which is exergonic (~0.1 eV), is the rate-determining process. The hole-transfer rate will be discussed in conjunction with a rate of an intramolecular electron transfer occurring in Ru(dmbpy)$_2$(bpbimH$_2$)Rh(phen)$_2^{5+}$.

An alternative mechanism of energy transfer, dipole–dipole interaction mechanism, is improbable because the energy-matching requirement is not fulfilled between the Ru–phen emission and the Ru–dmbpy absorption. If the spin-forbidden MLCT absorption of the acceptor chromophore were in a region of the donor emission, the dipole–dipole interaction mechanism could be competitive with the exchange mechanism (24–27).

Intramolecular Electron Transfer in Ru(dmbpy)$_2$(bpbimH$_2$)-Rh(phen)$_2^{5+}$.

This compound exhibits the sum of the absorption spectra of the component compounds. The phosphorescence and ESA of Ru → dmbpy CT were completely quenched in the 1 mM HClO$_4$ in CH$_3$CN. Both the phosphorescence and the ESA of the Ru–dmbpy site were detected for several hundred nanoseconds in the neutral medium, where deprotonation from an imino N–H group of pbimH moiety coordinating to Rh(III) was observed by means of absorption spectroscopy. The deprotonation shifts the reduction potential of the Rh site negatively from −1.15 to −1.45 V vs. Fc$^+$–Fc, but does not change the oxidation potential of the Ru site (0.68 V vs. Fc$^+$–Fc) in Ru(bpy)$_2$(bpbimH$_2$)$_2$Rh(bpy)$_2^{5+}$. The ergonicity of Ru(II) → Rh(III) electron transfer in the excited state of Ru(dmbpy)$_2$-(bpbimH$_2$)Rh(phen)$_2^{5+}$ can be regarded as −0.16 and 0.14 eV for the acidic form and the basic form, respectively, as it is for Ru(bpy)$_2$(bpbimH$_2$)-Rh(bpy)$_2^{5+}$. Therefore, the phosphorescence quenching in the acidic medium is attributed to Ru(II) → Rh(III) electron transfer.

The rate of Ru(II) → Rh(III) electron transfer was measured on cooling the binuclear compound in butyronitrile (mp 126 K). The decay rate of the excited Ru–dmbpy site monitored at 400 and 560 nm was considerably dependent on temperature. The rate constants of electron transfer (k_{ET}) were obtained from the excited-state lifetime of the binuclear compound by subtracting that of Ru(dmbpy)$_2$(bpbimH$_2$)$^{2+}$. The value of k_{ET} at 300 K, estimated by extrapolation to be 2×10^8 s^{-1} (Figure 3), is not very different from the Ru(II)-to-Ru(II) energy-transfer rate ($>10^8$ s^{-1}) of Ru(dmbpy)$_2$-(bpbimH$_2$)Ru(phen)$_2^{4+}$ in CH$_3$CN. In Ru(II) → Rh(III) electron transfer, the electron residing on one pbimH moiety of bpbimH$_2$ moves to Rh(III) through the other pbimH moiety. In the energy transfer from the Ru–phen site to the Ru–dmbpy site, electron transfer from a phen to bpbimH$_2$ was followed by hole transfer from the Ru(III)–phen to the Ru(II)–dmbpy, which

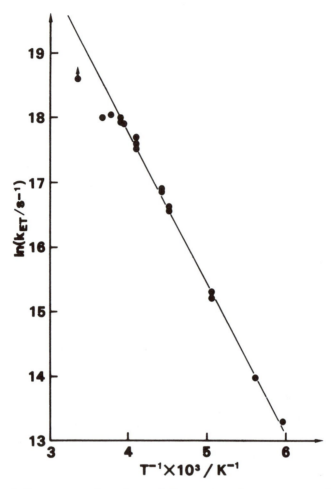

Figure 3. Temperature dependence of electron-transfer rate constant (k_{ET}) of Ru(II) → Rh(III) in the excited MLCT state of Ru(dmbpy)$_2$(bpbimH$_2$)-Rh(phen)$_2$$^{5+}$. [HClO$_4$] is 1 mM in butyronitrile.

takes place via bpbimH$_2$. Therefore, it is reasonable that the rates of the two processes are similar.

A rapid recovery of the MLCT band at 460 nm without delay after the decay of ESA leads to the conclusion that back electron transfer (Rh(II) → Ru(III)) is as rapid as forward electron transfer. Rapid back electron transfer is consistent with the high exergonicity (1.82 eV) involved in the process. The exergonicity of 1.7 eV gives rise to the maximum rate for several electron-transfer reactions within a cage following intermolecular electron-transfer quenching (5, 6, 32). Spin-flip, which is required for the back electron transfer, $^3[^2Ru(III)-^2Rh(II)] \rightarrow {}^1[^1Ru(II)-^1Rh(III)]$, may suppress the rate to some extent. The Ru(II) → Rh(III) electron transfer in Ru(dmbpy)$_2$-(bpbimH$_2$)Rh(phen)$_2{}^{5+}$ is as fast as that in a mixed-valence compound of Ru(II)–Ru(III) bridged by pyrazine (*21*).

Intramolecular Triplet–Triplet Annihilation. Aromatic compounds in the excited state undergo excited-state bimolecular annihilation resulting in the formation of a higher excited state and the ground state (*53*). T–T annihilation of some dye molecules, such as metal–porphyrin (*54*) and lumiflavine (*55*) in the triplet excited state, efficiently occurred to produce a cation radical and an anion radical in polar media.

As for the MLCT triplet excited state of a Ru(II) compound such as Ru(bpy)$_3{}^{2+}$, a channel of T–T annihilation is energetically possible. Electron transfer between excited-state metal sites is strongly exergonic (1.6 eV) because of their excitation energies (2 × 2.1 eV), although disproportionation of two Ru(II) sites in the ground state is endergonic (2.6 eV). Therefore, whether intramolecular T–T annihilation takes place or not is dependent on the density of the excited chromophores in a binuclear compound and Ru(II)–Ru(II) interaction through an intervening ligand.

The intensity of the 532-nm laser was insufficient to convert 40 μmol/dm^3 of Ru(bpy)$_2$(bpbimH$_2$)$^{2+}$ or [Ru(bpy)$_2$]$_2$(bpbimH$_2$)$^{4+}$ to the MLCT excited state. To determine the production of the excited state, energy transfer from the ruthenium(II) compound to anthracene was used. Addition of anthracene (<2 mM) quenched the phosphorescences and ESAs of [Ru(bpy)$_2$(bpbimH$_2$)]$^{2+}$ and [Ru(bpy)$_2$(bpbimH$_2$)Ru(bpy)$_2$]$^{4+}$. An ESA at 421 nm, which is assigned to the excited triplet state of anthracene (^{3}A) (*56*), was produced at the same rate as the decay of phosphorescence of the Ru(II) compound.

^{3}A production was estimated from the absorption change at 421 nm (*56*) by using the molar extinction coefficient [52,000 dm^{3}mol^{-1}cm^{-1} (*57*)]. The ^{3}A production was proportional to the concentration of anthracene added and the quenching efficiency (F_q). The F_q was determined from the decay rates of the excited Ru(II) in the absence and the presence of anthracene.

Although the excitation efficiencies were the same for the mononuclear and the binuclear compounds because of the same absorbances at 532 nm and the same excited-state lifetimes (556 ns), the conversion to the MLCT triplet excited state were 77–83% for the mononuclear compound and 43–45% for the binuclear compound, respectively. The conversion of the binuclear compound, lower than 50%, indicates that some quenching process rapidly occurs between the excited Ru(II) sites. Rapid T–T annihilation between the Ru–bpy sites of the binuclear compound in the excited state is the most probable explanation.

$$Ru^*(bpy)_2(bpbimH_2)Ru^*(bpy)_2{}^{4+} \rightarrow Ru^*(bpy)_2\,(bpbimH_2)Ru(bpy)_2{}^{4+}$$

$$(1)$$

There was no absorption change due to production of Ru(III) [ϵ_{420} = 3300 dm^{3}mol^{-1}cm^{-1} (58)] and bpy$^-$ coordinating to Ru(II) [ϵ_{510} = 12,000 dm^{3}mol^{-1}cm^{-1} (59)] after T–T annihilation, although intramolecular electron transfer is energetically possible.

The following is an alternative explanation for the inefficient excited-state population in the binuclear compound. The excitation of one metal site is delocalized throughout the whole compound so that the MLCT band of the other metal site is also bleached. However, this explanation does not hold true because the molar difference extinction coefficient of the excited binuclear compound at 460 nm obtained from the initial absorption change and the production of ^{3}Ru^{2+} was similar (–9600 dm^{3}mol^{-1}cm^{-1}) to that (–8200 dm^{3}mol^{-1}cm^{-1}) of the excited mononuclear compound. The localization of the excited state in one metal site of the binuclear compound is consistent with weak Ru(II)–Ru(II) interaction, as estimated from the small difference in the emission energy between the mono- and binuclear compounds.

Acknowledgments

M. Haga gratefully acknowledges financial support from the Japanese Ministry of Education for a Grant-in-Aid for Scientific Research, No. 1540508, and the Okasan–Kato Foundation.

References

1. Kestner, R. A.; Jortner, J.; Logan, J. *J. Phys. Chem.* **1974**, *78*, 2148.
2. Ulstrup, J.; Jortner, J. *J. Chem. Phys.* **1976**, *63*, 4358.
3. Miller, J. R.; Beitz, J. V.; Huddleston, R. K. *J. Am. Chem. Soc.* **1984**, *106*, 5057.
4. Closs, G. L.; Calcaterra, L. T.; Greem, M. J.; Penfield, K. W.; Miller, J. R. *J. Phys. Chem.* **1986**, *90*, 3673.
5. Ohno, T.; Yoshimura, A.; Mataga, N. *J. Phys. Chem.* **1986**, *90*, 3295.
6. Ohno, T.; Yoshimura, A.; Shioyama, H.; Mataga, N. *J. Phys. Chem.* **1987**, *91*, 4365.
7. Mataga, N.; Kanda, Y.; Okada, T. *J. Phys. Chem.* **1986**, *90*, 3880.
8. Mataga, N.; Asahi, T.; Kanda, Y.; Okada, T.; Kakitani, T. *J. Chem. Phys.* **1988**, *127*, 249.
9. Levin, P. P.; Pluzhnikov, P. E.; Kuzmin, V. A. *Chem. Phys. Lett.* **1988**, *147*, 283.
10. Chou, M.; Creutz, C.; Sutin, N. *J. Am. Chem. Soc.* **1977**, *99*, 5615.
11. Balzani, V.; Scandola, F.; Orlandi, G.; Sabbatini, N.; Indelli, M. T. *J. Am. Chem. Soc.* **1981**, *103*, 3370.
12. Hoselton, M. A.; Lin, C.-T.; Schwarz, H. A.; Sutin, N. *J. Am. Chem. Soc.* **1978**, *100*, 2383.
13. Mok, C.-Y.; Zanella, A. W.; Creutz, C.; Sutin, N. *Inorg. Chem.* **1984**, *23*, 2381.
14. Ulrich, T.; Steiner, U. E.; Foll, R. E. *J. Phys. Chem.* **1983**, *87*, 1873.
15. Allen, G. C.; Hush, N. S. *Prog. Inorg. Chem.* **1967**, *8*, 357.
16. Hush, N. S. *Prog. Inorg. Chem.* **1967**, *8*, 391.
17. Hush, N. S. *Coord. Chem. Rev.* **1985**, *64*, 135.

18. Creutz, C. *Prog. Inorg. Chem.* **1983**, *30*, 1.
19. Meyer, T. J. *Acc. Chem. Res.* **1978**, *11*, 94.
20. Penfield, K. W.; Miller, J. R.; Paddon-Row, M. N.; Cotsaris, E.; Oliver, A. M.; Hush, N. S. *J. Am. Chem. Soc.* **1987**, *109*, 5061.
21. Creutz, C.; Kroger, P.; Matubara, T.; Netzel, T. L.; Sutin, N. *J. Am. Chem. Soc.* **1979**, *101*, 5442.
22. Curtis, J. C.; Bernstein, J. S.; Meyer, T. J. *Inorg. Chem.* **1985**, *24*, 385.
23. Schanze, K. S.; Neyhart, G. A.; Meyer, T. J. *J. Phys. Chem.* **1986**, *90*, 2182.
24. Furue, M.; Kinosita, S.; Kusida, T. *Chem. Lett.* **1987**, 2355.
25. Schmehl, R. H.; Auerbach, R. A.; Wacholtz, W. F. *J. Phys. Chem.* **1988**, *92*, 6202.
26. Bignozzi, C. A.; Indelli, M. T.; Scandola, F. *J. Am. Chem. Soc.* **1989**, *111*, 5192.
27. Ryo, C. K.; Schmehl, R. H. *J. Phys. Chem.* **1989**, *93*, 7961.
28. Haga, M.-A.; Ano, T.; Kano, K., submitted to *Inorg. Chem.*
29. Haga, M.-A.; Matsumura-Inoue, T.; Yamabe, S. *Inorg. Chem.* **1987**, *26*, 4148.
30. Rillema, D. P.; Sahai, R.; Matthews, P.; Edwards, A. K.; Shaver, R. J. *Inorg. Chem.* **1990**, *29*, 167.
31. Haga, M.-A.; Ishizaki, T.; Nozaki, K.; Ohno, T., unpublished results.
32. Ohno, T.; Yoshimura, A.; Mataga, N. *J. Phys. Chem.* **1990**, *94*, 4871.
33. Dallinger, R. F.; Woodruff, W. *J. Am. Chem. Soc.* **1979**, *101*, 4391.
34. DeArmond, M. K.; Myrick, M. L. *Acc. Chem. Res.* **1989**, *22*, 364.
35. Mabrouk, P. A.; Wrighton, M. S. *Inorg. Chem.* **1986**, *25*, 526.
36. McClanahan, S. F.; Dallinger, R. F.; Holler, F. J.; Kincaid, J. R. *J. Am. Chem. Soc.* **1985**, *107*, 4853.
37. Myrick, M. L.; Blakley, R. L.; DeArmond, M. K.; Arthur, M. L. *J. Am. Chem. Soc.* **1988**, *110*, 1325.
38. Cooley, L. F.; Headford, C. E. L.; Elliott, C. M.; Kelly, D. F. *J. Am. Chem. Soc.* **1988**, *110*, 6673.
39. Cooley, L. F.; Bergquist, P.; Kelley, D. F. *J. Am. Chem. Soc.* **1990**, *112*, 2612.
40. Lachish, U.; Infelta, P. P.; Gratzel, M. *Chem. Phys. Lett.* **1979**, *62*, 317.
41. Creutz, C.; Chou, M.; Netzel, T. L.; Okumura, M.; Sutin, N. *J. Am. Chem. Soc.* **1980**, *102*, 1309.
42. Braterman, P. S.; Harriman, A.; Heath, G. A.; Yellowlees, L. J. *J. Chem. Soc. Dalton Trans.* **1983**, 1801.
43. Hauser, A.; Krausz, E. *Chem. Phys. Lett.* **1987**, *138*, 355.
44. Ohno, T.; Yoshimura, A.; Prasad, D. R.; Hoffman, M. Z. *J. Phys. Chem.* **1991**, in press.
45. Kumar, C. V.; Barton, J. K.; Gould, I. R.; Turro, N. J.; Van Houten, J. *Inorg. Chem.* **1988**, *27*, 648.
46. Kawanishi, Y.; Kitamura, N.; Kim, Y.; Tazuke, S. *Riken Q.* **1984**, *78*, 212.
47. Fuchs, Y.; Lofters, S.; Dieter, T.; Shi, W.; Morgan, R.; Strekas, T. C.; Gafney, H. D.; Baker, A. D. *J. Am. Chem. Soc.* **1987**, *109*, 2691.
48. Sahai, R.; Rillema, D. P.; Shaver, R.; Wallendael, S. V.; Jackman, D. C.; Boldaji, M. *Inorg. Chem.* **1989**, *28*, 1022.
49. Kalyanasundaram, K.; Nazeeruddin, M. K. *Inorg. Chem.* **1990**, *29*, 1888.
50. Braunstein, C. H.; Baker, A. D.; Strekas, T. C.; Gafney, H. D. *Inorg. Chem.* **1984**, *23*, 857.
51. Wallace, A. W.; Murphy, W. R.; Petersen, J. D. *Inorg. Chim. Acta* **1989**, *166*, 47.
52. Tapolsky, G.; Duesing, R.; Meyer, T. J. *J. Phys. Chem.* **1989**, *93*, 3885.
53. Birks, J. B. *Photophysics of Aromatic Molecules*; Wiley: New York, 1970; pp 372–491.
54. Ballard, S. D.; Mauzerall, D. C. *J. Chem. Phys.* **1980**, *72*, 933.

55. Yoshimura, A.; Ohno, T. *Photochem. Photobiol.* **1991,** *53,* 175.
56. Porter, G.; Windsor, M. W. *Discuss. Faraday Soc.* **1954,** *17,* 178.
57. Kikuchi, K.; Kokubun, H.; Koizumi, M. *Bull. Chem. Soc. Jpn.* **1971,** *44,* 1527.
58. Kalyanasundaram, K. *Coord. Chem. Rev.* **1982,** *46,* 159.
59. Mulazzani, Q. G.; Emmi, S.; Fuochi, P. G.; Hoffman, M. Z.; Venturi, M. *J. Am. Chem. Soc.* **1978,** *100,* 981.

RECEIVED for review April 27, 1990. ACCEPTED revised manuscript September 21, 1990.

Electron Transfer Across Model Polypeptide and Protein Bridging Ligands

Distance Dependence, Pathways, and Protein Conformational States

Stephan S. Isied

Department of Chemistry, Rutgers University, New Brunswick, NJ 08904

Studies of intramolecular electron transfer across model peptides and metal-modified proteins are reported in this chapter. Intramolecular electron-transfer studies across polyproline spacers have been compared in the following three different metal donor–acceptor series: $[(NH_3)_5Os\text{-}iso\text{-}(Pro)_n\text{-}Co(NH_3)_5]$, $[(NH_3)_5Os\text{-}iso\text{-}(Pro)_n\text{-}Ru(NH_3)_5]$ (iso is isonicotinyl-), and $[(bpy)_2Ru\text{-}X\text{-}(Pro)_n\text{-}Co(NH_3)_5]$, (bpy is 2,2'-bipyridyl-, X is 4-carboxy-4'-methyl-2,2'-bipyridyl). In these studies the electronic coupling factor β was measured to be ~0.3–0.4 Å^{-1} for the ruthenium–bipyridine series with n = four to six prolines, in contrast to β ~ 0.9 Å^{-1} for the other metal–amine complexes with n = one to four prolines. These experiments suggest the possibility of observing rapid rates of electron transfer across 10 proline residues (~40 Å) in related ruthenium–bipyridine series. In the ruthenium-modified proteins, intramolecular electron-transfer reactions with histidine-33 ruthenium-modified cytochrome c have shown that the rate of intramolecular reduction of the heme of cytochrome c can be changed by more than 5 orders of magnitude for different ruthenium donor complexes. However, oxidation of the heme of cytochrome c by the ruthenium–bipyridine complexes results in a rate that is substantially slower than predicted. Interpretation of these results using a mechanism by which protein conformational change is associated with the rate of electron transfer is proposed.

0065–2393/91/0228–0229$06.00/0

$\mathbf{D}$ONOR–ACCEPTOR MOLECULES separated by synthetic peptides and proteins have contributed significantly to our understanding and analysis of
intramolecular electron-transfer reactions in the past decade. The different
contributions in this volume attest to the variety of elegant experiments that
both demonstrate various aspects of electron-transfer (ET) theory and provide new challenges and questions for the theorist to answer in order to
interpret the new experimental results.

My group's work in this area has centered around metal donor–acceptor
complexes separated by synthetic peptides and electron-transfer proteins
(Structure 1). We have designed simple model systems that emphasize certain properties of the bridging ligands, as well as of the donors and acceptors

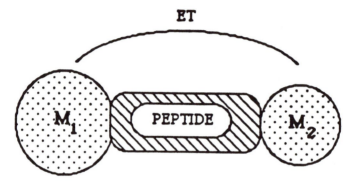

Structure 1. Binuclear metal donor–acceptor complexes.

(1–5). Among the most exciting findings from our current work is the prediction that long-range electron transfer across polypeptides should be observable over 30–40 Å in reasonably short time scales (<1 ms) in related
proline-bridged systems. The role of the peptide in facilitating long-range
ET is also becoming more apparent from our experiments (6). This chapter
reviews the work done in my laboratory (2–11). However, several other
related studies on synthetic peptides and proteins can provide a more comprehensive view. In the field of synthetic peptides, the data of Schanze and
Sauer (12), Sisido et al. (13), and Farraggi et al. (14) are related to the current
results. In the protein area, studies by Gray and co-workers (15–18) on
ruthenium-modified cytochrome c are directly related to the work discussed
in this chapter. Other investigations of protein–protein complexes, including
the results of Hoffman and co-workers (19, 20) and of McLendon et al. (21),
are also relevant.

The first part of this chapter will review results with oligoprolines (2–6)
that suggest that electron transfer can occur across polypeptides at extremely
long distances. The second part reports related experiments with cytochrome
c proteins in which the protein is part of the donor or acceptor bridged

component (*7–11*). The protein work has concentrated on the modification of horse-heart cytochrome *c* and other yeast cytochromes with ruthenium–amine and polypyridine reagents. From this detailed study of ruthenium-modified cytochromes, some aspects of the mechanism of intramolecular electron-transfer reactions and their relationship to protein conformational states in these molecules will be discussed.

Donor–Acceptor Molecules with Peptide Bridges

The versatility of the properties of metal donors and acceptors in controlling the rate of intramolecular electron transfer across bridging ligands has been demonstrated (*1–6*). Table I is a clear illustration of how the rate of intramolecular electron transfer can be changed by over a trillion times through changes in the driving force (ΔG^0) and the reorganization energy of the metal donor and acceptor, with a constant bridging ligand (i.e., the same distance between the metal centers). Similar results have been obtained for longer bridging ligands containing the amino acid proline (*4, 5*).

In addition to driving force and reorganization energy, the distance between the donor and acceptor can also change the rate of the intramolecular electron-transfer reaction by many orders of magnitude. Although the origin of this distance dependence has been assumed for many years to be related mainly to electronic factors, recent work has demonstrated that the outer-sphere reorganization-energy dependence on distance can be sub-

Table I. Effect of Metal Donors and Acceptors on Intramolecular Rates of Electron Transfer

M–M	k (s^{-1})	ΔG^0 (eV)
Os–Ru	$>5 \times 10^9$	-0.25
Os–Co	1.9×10^5	-0.15
Ru–Co	1.2×10^{-2}	$+0.4$

NOTE: In all cases, the M–M distance was 9.0 Å.

stantial and in some instances exceeds the electronic factor (4). Theory predicts that the rate of intramolecular electron transfer (k_{et}) will decrease with distance according to the expression $k_{et} \propto e^{-\beta r}$, where r is the edge-to-edge distance between the donor and acceptor at which the reaction becomes nonadiabatic (22). The electronic coupling factor β is a constant that is characteristic of the electronic interaction between the donor and acceptor across the bridging ligand.

Polyproline Bridging Ligand.   Before introducing our results on the rate of electron transfer, a short introduction to the properties of the polyproline bridging ligand, which has been the cornerstone of my group's studies on the distance dependence of electron transfer across polypeptides, will be

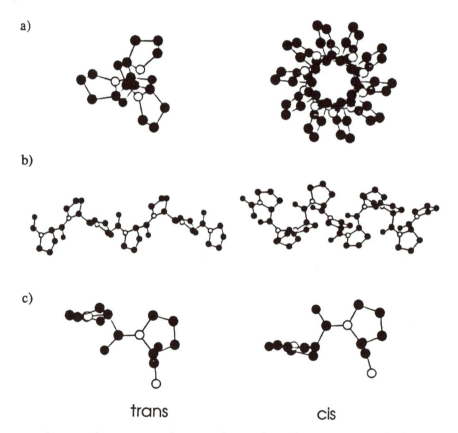

a)

b)

c)

trans   cis

Chart 1. a, b: Structures of trans- *and* cis-*proline polymers, respectively, from fiber X-ray diffraction (29). c: Structures of* trans- *and* cis-*proline monomers. Open circles are nitrogen atoms, closed circles are carbon atoms, and closed circles that are connected by one bond to the main chain are carbonyl oxygen atoms.*

given. The oligoproline peptides (Chart 1, a and b) proved to be reasonably rigid molecules (*23, 24*) for studying long-range intramolecular electron transfer as a function of distance between a donor and acceptor.

The structural rigidity of proline (Pro) oligomers in comparison to other naturally occurring amino acids is due mainly to the cyclic structure of the proline ring. The five-membered ring of the proline side chain restricts rotation around the peptide bond, and this restriction results in *cis–trans* conformational isomerism (*21–32*), as shown in Chart 1, c. Polyproline bridges were used as rigid chemical spacers in early studies of energy transfer between organic donors and acceptors (*23, 24*). In polar solvents the efficiency of energy transfer follows the r^{-6} distance dependence for weak dipolar energy transfer (*33*). The results of these energy-transfer studies show that polyproline can be used as a spectroscopic ruler in the 10–60-Å range (*23, 24*).

Fibers of polyproline crystallized from aqueous solution possess an all-*trans* conformation (>95% *trans*). When the same polyproline is crystallized from solvents of lower polarity, especially aliphatic alcohols, fibers of the *cis* isomer are obtained instead. In Chart 1, a and b show the fiber structures of cis- and *trans*-polyproline (*34, 35*). X-ray diffraction analysis of poly-*l*-proline fibers shows clear structural differences between the *cis* and *trans* forms. As can be seen in Chart 1, b, both proline oligomers are helical in structure, with different unit cell properties. *trans*-Polyproline makes a left-handed helical cycle every three residues, with a 3.12-Å translation per residue along the helical axis. In the *cis* isomer a right-handed helical turn consists of 3 1/3 proline units with a 1.85-Å translation per residue.

One of the interesting features of the fiber structure of these two proline isomers (Chart 1, a) is the fact that the *trans* isomer possesses an extended structure in which polar solvents can hydrate the peptide bonds and stabilize the open structure. In less polar media the *cis* conformation is more compact and is stabilized when the polymer turns the hydrocarbon part of the proline residue to the weakly polar solvents. The conversion between *trans*- and *cis*-polyproline isomers (Chart 1, a) occurs with a half-life of approximately 1–2 min at room temperature (enthalpy change $\Delta H^{\ddagger} \sim 20$ kcal mol^{-1}; entropy change $\Delta S^{\ddagger} \sim 0$) (*36*). (For high-molecular-weight oligomers, several hours are required to complete this isomerization.) The interconversion between the *trans* and *cis* isomers is known to be one of the slowest processes controlling conformational changes in peptides and proteins (*37, 38*). The studies described in this chapter were carried out in aqueous acidic media, under conditions in which the extended *trans* conformation of the proline oligomers is known to predominate (>95%) (*39–43*).

Distance Dependence.   With this introduction to the proline bridging peptide and the sensitivity of inorganic donor–acceptor systems in controlling rates of electron-transfer reactions, the [(NH$_3$)$_5$Os-L-Ru(NH$_3$)$_5$] (Os-

L-Ru) series [where L is iso(Pro)$_n$ and iso is the isonicotinyl group] will be discussed as a function of the number of proline residues separating the donor from the acceptor. Table II shows how the rate of intramolecular electron transfer can be changed by more than 8 orders of magnitude as the distance between the donor and acceptor is increased by the introduction of additional proline residues. In this work the donors and acceptors are kept the same and therefore this substantial change in rate must be attributed to the distance dependence of the rate of intramolecular electron transfer.

For molecules with one, two, or three prolines, the temperature dependence of the rate of intramolecular electron transfer has also been studied, and this information has been used to separate the distance-dependent component from the electronic component of the reorganization energy. This separation was done by using a modified version of the transition-state expression where $\ln k + \Delta H^{\ddagger}/RT$ is plotted vs. distance (Figure 1, Os-L-Ru). From these plots the electronic coupling factor, $\beta \sim 0.6\text{--}0.7 \text{ Å}^{-1}$, can be calculated for the Os-L-Ru series, where L is iso(Pro)$_n$. With data on a similar series of molecules (the Os-L-Co series, where L is iso(Pro)$_n$) that

Table II. Intramolecular Rates of Electron Transfer, Activation Parameters, and Distances for the [(NH$_3$)$_5$Os-L-Ru(NH$_3$)$_5$] Series

n	Os══════════Ru	k (s^{-1})	$\Delta H^{\ddagger}$ (kcal/mol)	$\Delta S^{\ddagger}$ (cal/deg mol)	M–M (Å)
0		$>10^9$	—	—	9.0
1		3.1×10^6	4.2	-15	12.1–12.3
2		3.7×10^4	5.9	-19	14.4–15.1
3		3.2×10^2	7.4	-23	17.8–18.3
4		~ 50	—	—	20.9–21.5

NOTE: L is iso(Pro)$_n$.

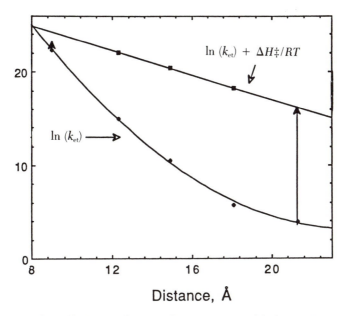

Figure 1. Plot of ln k_{et} vs. distance (lower curve) and ln k_{et} + $\Delta H^{\ddagger}/RT$ vs. distance (upper curve) for the $[(NH_3)_5Os^{III}\text{-}L\text{-}Ru^{II}(NH_3)_5]^{4+}$ series; n = 0–4.

we synthesized and studied earlier, Endicott (*44*) obtained a slightly higher β for the Os-L-Co series of molecules. In the Os-L-Co molecules, the rate of electron transfer occurs at significantly lower time scales because of the larger reorganization energy of the Co(III) metal center. This reorganization energy may be expected to produce more conformational variability of the bridging ligand—especially in the longer proline bridges (*n* ~ 3 or 4). If direct metal–metal overlap or overlap through the solvent in these two systems is negligible, then one would expect similar results to be observed for β in both systems.

A very important assumption involved in this analysis is that the reorganization energy is associated with the activation enthalpy of the reaction with no entropic contributions. This assumption is justified because the donor and acceptor are hydrophilic and similar in nature. The charge on the precursor and successor complex in the electron-transfer reaction is also the same [Os(II)Ru(III) → Os(III)Ru(II)].

For a metal-to-metal distance of 21 Å, a rate of 50 s^{-1} was observed (*5*). The driving force for these reactions in Table II is small; reduction potential $E°$ ~ 250 mV. This observation that rapid rates of electron transfer can be obtained at these long distances (21 Å) even with these small driving forces suggests that rapid electron transfer should be observable at much longer distances with proper control of the driving force and the reorganization energy.

Polyproline Complexes. To achieve these longer-range electron transfers, a new series of polyproline complexes (I) was designed and synthesized (bpy is 2,2'-bipyridine) (6).

$$[(bpy)_2Ru^{II}(bpy)\text{--}\overset{\overset{\displaystyle O}{\parallel}}{C}\text{--}(Pro)_n\text{--}O\text{--}Co^{III}(NH_3)_5]$$

Series I

These complexes have been assembled from the reaction of $[(bpy)_2Ru^{II}(bpy\text{-}COOH)]$ and $[(NH_3)_5Co(Pro)_nOH]$ by using standard peptide synthetic techniques (45). The reason for choosing the ruthenium bipyridine system is the availability of the Ru(I) oxidation state (more accurately described as $Ru^{II}L^{\bullet}$), which is a very strong reductant; $E^{\circ} = -1.3$ V vs. NHE (normal hydrogen electrode). The reduction potential for the $Ru^{I/II}$ couple is expected to be similar to that for $[Ru(bpy)_3]$, $E^{\circ} = -1.2$ V vs. NHE. (*See*, for example, ref. 46.) With such a strong reducing agent, the intramolecular electron-transfer reaction can then take place with much larger driving forces than in the $[(NH_3)_5Os\text{-}iso(Pro)_n\text{-}Ru(NH_3)_5]$ series. Furthermore, the Ru(III) bipyridine can be used as an oxidant when other Ru(II) reductants are used.

The series of complexes (I) were purified and characterized by several physiochemical techniques. The circular dichroic (CD) spectra of series I ($n = 1$–7) are shown in Figure 2. As the number of proline residues increases from 1 to 4, a significant shift to lower energy is observed in the CD spectra. Beyond $n = 4$, no significant changes are observed. This shift is attributed to the formation of the polyproline II left-handed helix (6). Similar results were obtained earlier on related molecules by using ^{13}C and ^1H NMR spectroscopy (5). Other evidence for the secondary structure of the polyproline helix came from early studies by Stryer and Haugland (23) and Gabor (24) on similar polyproline peptides bridged by donor and acceptor molecules of the type, D--$(Pro)_n$--A, where n is 1–12 prolines, A is the dansyl energy acceptor at the amino proline terminal, and D is the naphthyl donor at the carboxyl proline terminal. Their studies showed that for energy transfer across polyprolines ($n = 5$–12), the efficiency of energy transfer decreases with the increasing number of proline residues. This decrease indicates a 50% transfer efficiency at 34.6 Å and shows the r^{-6} dependence predicted by Förster (33) for weak dipole–dipole coupling.

Figure 3 shows how the rate of electron transfer changes with the number of prolines separating the ruthenium and the cobalt. Intramolecular electron transfer does not decrease as fast as expected if the number of proline units between the donor and acceptor adopts the secondary helical structure (Figure 3). Furthermore, the temperature dependence of the reaction (Figure 4) demonstrates that the distance dependence is mainly controlled by electronic effects because the change in temperature does not change the slope of the plot.

An electronic coupling factor, $\beta \sim 0.3$–0.4 Å^{-1}, is calculated for this

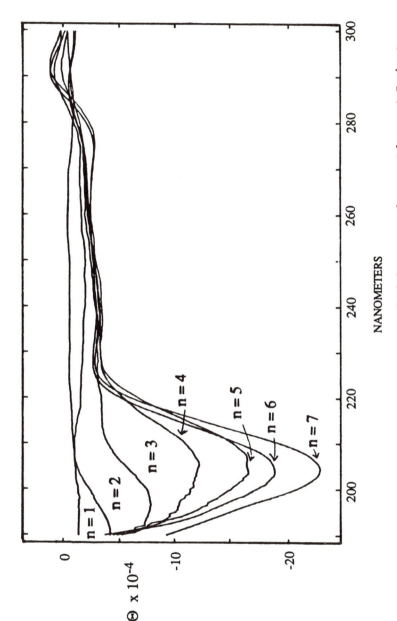

Figure 2. Circular dichroic (CD) spectra of the proline-bridged electron-transfer series I for n = 1–7 taken in aqueous solution. The CD maximum located at ~200 nm (π–π*) is characteristic of the trans-polyproline (II) structure. This feature shifts to longer wavelengths as n is increased from 1 to 4.

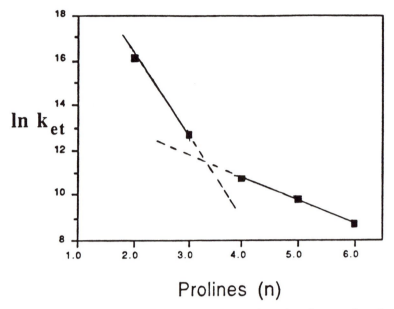

Prolines (n)

Figure 3. A plot of ln k_{et} + $\Delta H^{\ddagger}/RT$ vs. the number of proline residues for n = 2–6 prolines. The smaller slope for n = 4–6 corresponds to the stabilization of the polyproline secondary structure.

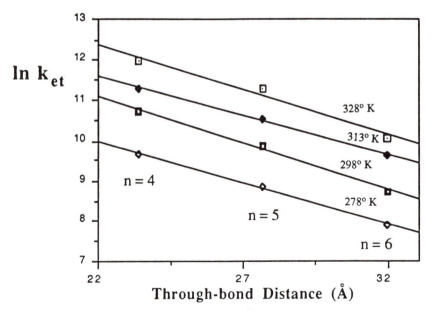

Figure 4. Temperature dependence of the rate of intramolecular electron transfer for the complexes Ru-iso(Pro)$_n$-Co (I), for n = 4–6.

series of molecules. This β can be compared directly with that in the earlier systems. The more facile intramolecular electron-transfer reaction in these longer prolines may be attributed to a better matching of the orbitals of the donor with that of the bridge. This improvement may reflect changes in the electronic structure of the polyproline bridge after it has adopted the helical structure (Structure 2). Regardless of interpretation, the original point made in this chapter that electron transfer can be observed at rapid rates over 40 Å is now further strengthened, because the distance between the Ru and the Co in these series is approximately 30 Å.

Exchanging the Co(III) acceptor with the Ru(III) or other acceptors with much lower inner-sphere reorganization energy is one approach to further extend the distance dependence of intramolecular electron transfer to ~40 Å (i.e., 10 proline residues separating the donor and the acceptor). We are currently carrying out these experiments (6).

Electron transfer across rigid hydrocarbon spacers has been elegantly demonstrated on a variety of systems, starting with the pioneering work of Miller and Closs (47). Different organic hydrocarbon spacers have been investigated by Closs, Dervan, Padden Row, and others; these studies are summarized in ref. 47. It is desirable to compare saturated organic hydrocarbon spacers and peptide spacers. Although the current data is limited, a qualitative comparison indicates that electronic transmission across peptides is more facile than across saturated hydrocarbons (a lower β was observed for the peptide bridging groups).

Protein Donor–Acceptor Complexes

We have shown that it is possible to extend the concept of donor–acceptor complexes to an electron-transfer protein by covalently attaching a well-defined transition metal complex to a specific amino acid side chain in the protein (7–9). Using a variety of ruthenium–amine complexes to modify horse-heart cytochrome *c*, we isolated ruthenium-modified proteins in which the modification by the ruthenium complex occurs at the His-33 position. Variation of the type of Ru(III) complex attached to the cyt *c* can change the protein moiety from an acceptor to a donor.

When the modified protein [cyt *c*-Ru(NH$_3$)$_5$] is prepared in the

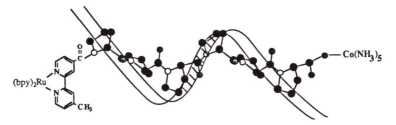

Structure 2. Stabilized secondary structure adopted by type I complexes, Ru-iso(Pro)$_n$-Co, for n = 6. *Similar helices are present for* n = 4 *and* 5.

Ru(III)cyt c(III) state and then reduced with a variety of radicals generated by pulse radiolysis techniques, intramolecular electron transfer from the ruthenium site to the heme site occurs with a rate constant, $k = 53$ s^{-1}, $\Delta E°$ for cyt $c = 0.26$ V, and $\Delta E°$ for $[(NH_3)_5Ru^{II/III}(His)] = 0.10$ V. Similar results have been obtained by flash photolysis techniques (11). The temperature dependence, concentration dependence, and pH dependence of this electron-transfer reaction were investigated. The results of this investigation showed that the rate of electron transfer is independent of concentration and moderately sensitive to temperature ($\Delta H^{\ddagger} \sim 3.5$ kcal M^{-1} and $\Delta S^{\ddagger} \sim -39$ eu). The electron-transfer reaction is independent of pH through pH 5–9; then it increases below pH 5 as the native conformation of the cyt c changes (9).

The observation that intramolecular electron transfer between the ruthenium site and the heme site occurs at distances of 12–15 Å is extremely significant, because it represents the first observation of an intramolecular electron-transfer reaction within an electron-transfer protein. The magnitude of the rate constant, 53 s^{-1}, is similar to other rate constants that are known to occur within the native cyt c molecule ($8, 9$). This finding led us to question whether the unimolecular rate observed is rate-limiting in electron transfer (eq 1) or in a protein-associated conformational change (eqs 2a and 2b).

$$Ru^{II}cyt\ c^{III} \xrightarrow{k_{et}} Ru^{III}cyt\ c^{II} \tag{1}$$

where k_{et} is the rate constant for intramolecular electron transfer

$$Ru^{II}cyt\ c^{III} \xrightarrow{k_{cc}} Ru^{II}*cyt\ c^{III} \tag{2a}$$

$$Ru^{II}*cyt\ c^{III} \xrightarrow{k_{et}} Ru^{III}cyt\ c^{II}\ (fast) \tag{2b}$$

and k_{cc} is the rate constant for a protein conformational change ($8–11$).

To answer this question, a series of related ruthenium molecules that are more oxidizing than cyt c were synthesized and characterized. The redox potentials of the Ru(II)/(III) couple, greater than 0.26 V, allow one to reverse the direction of electron transfer in the modified cyt c such that electron flow in the Ru(III)cyt c(II) is from the heme to the ruthenium. Thus, one can change the heme of cyt c from an electron acceptor to an electron donor. The rationale behind these experiments is rather simple; if the unimolecular rate observed is rate-limiting in electron transfer, then similar variation in rates of electron transfer should be observed for reduction and oxidation (10).

Structure 3 shows the structure of cyt c and the positions of the heme relative to the ruthenium-modified sites. Table III summarizes the rates of intramolecular electron transfer for the reduction and oxidation of cyt c by a number of ruthenium complexes.

The rate of reduction of cyt c can be changed by more than 5 orders of

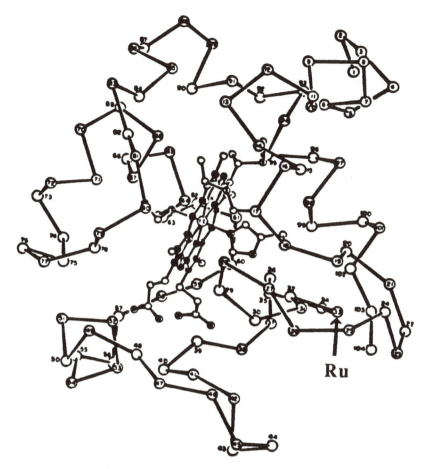

Ru

Structure 3. Ruthenium-modified cytochrome c showing the relative position of the heme and the ruthenium sites.

magnitude, depending on the redox potential and the reorganization energy of the ruthenium-modified species (Table III). Two types of complexes coordinated to His 33 of cytochrome c can be identified. In the first type, electron transfer takes place from (or to) a ruthenium t_{2g} orbital. This condition is satisfied for the oxidation and reduction reactions of cytochrome c by the different ruthenium complexes (I–II and V–VII in Table III). The other type of reactions are those in which the electron is transferred from a radical anion ligand attached to the ruthenium bound at the *cis* or *trans* position to the imidazole moiety of the His 33. This condition is satisfied for the reduction of cyt c with reactions in complexes III–IV (Table III).

The ruthenium-modified proteins constitute an interesting series of modified proteins in which the distance between the ruthenium label and the heme group is relatively well-defined. The variation in the reduction potential of the ruthenium complexes allowed us to study, for the first time,

Table III. Rates of Intramolecular Electron Transfer and Reduction Potential of Ruthenium–Cytochrome c Complexes

Ruthenium-Modified Cyt c	$E^{\circ a}$ (V)	Intramolecular ET, k (s^{-1})	Direction of ET
Native HH cytochrome c	0.26	—	—
(I) c-$[(NH_3)_4Ru(OH)]$-(II/III)	−0.01	5×10^2	Ru → heme
(II) $[(NH_3)_5Ru]$-(II/III)	0.13	55	Ru → heme
(III) $[Ru(bpy)(bpy^-)(py)]$-(II/II•)^b	−1.3	2.8×10^5	Ru → heme
(IV) $[Ru(bpy)(bpy^-)(im)]$-(II/II•)^b	−1.3	2.0×10^5	Ru → heme
(V) $[Ru(bpy)_2(py)]$-(II/III)	0.92	40	Heme → Ru
(VI) $[Ru(bpy)_2(im)]$-(II/III)	0.79	55	Heme → Ru
(VII) $[Ru(bpy)(terpy)]$-(II/III)^b	0.74	40	Heme → Ru
(VIII) $[Ru(bpy)_2OH_2]$-(II/III)	0.65	40	Heme → Ru

$^a E^\circ$ is the reduction potential of the cyt c or substituted cyt c vs. the normal hydrogen electrode (NHE).
^{b}bpy$^-$ is the bipyridine radical anion; bpy is bipyridine; py is pyridine; im is imidazole; and terpy is terpyridine.

the reduction and oxidation of the cyt c from a remote site, His 33, which is approximately 15 Å away from the heme group.

The scheme for studying the intramolecular electron-transfer step by using pulse radiolysis techniques is outlined in Scheme I. In this scheme oxidation of the Ru–cyt c species by $CO_3{}^\bullet$ generates a nonequilibrium distribution between the Ru^{II}cyt c^{III} and Ru^{III}cyt c^{II}. The relaxation to equilibrium distribution is then taken as a measure of the rate of intramolecular electron transfer from the ruthenium site to the heme site or vice versa. Similar reactions can be observed for the reduction of cyt c with $CO_2{}^\bullet$ and

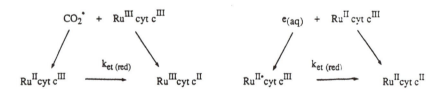

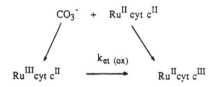

Scheme I. $CO_2{}^\bullet$ and $CO_3{}^-$ are radicals generated from a chemical precursor by using pulse radiolysis techniques; $CO_2{}^\bullet \to k_{et(red)}$; $CO_3{}^- \to k_{et(ox)}$. $CO_2{}^\bullet$ was generated in 0.1 M $NaHCO_2$. $CO_3{}^-$ was generated in 0.1 M $NaHCO_3$. $e_{(aq)}$ was generated in 0.13 M t-BuOH. All experiments were conducted at pH 7–8 in 0.05–0.1 M phosphate buffer.

$e_{(aq)}$ (Scheme I). The first two entries in Table III clearly show that the rate of reduction of cyt c by the ruthenium–ammine complexes decreases with a decrease in driving force. Thus, in going from $[Ru(NH_3)_4OH]$ to $[Ru(NH_3)_5]$, the rate changes by an order of magnitude (*10*). This change is consistent with a simple electron-transfer step that follows Marcus theory. Further increase in driving force can lead to further increase in rate, as is observed for the $[(bpy)_2RuL]$ complexes. (The electron in these complexes is localized on the bpy ligands, and therefore these complexes are more correctly formulated as $Ru^{II}L^{\bullet}$.)

A different type of behavior is observed for the cyt c oxidation with the ruthenium–bipyridine complexes, where the direction of electron transfer is expected to be from the heme to the Ru(III) label. Three related cyt c derivatives in this series have been characterized and studied. The main difference between these ruthenium labels is in the driving force of the reaction (Table III). The rates of oxidation of cyt c in these three complexes are equal within experimental error. The rate constant for this process is $\sim$40–55 s^{-1}. This insensitivity of rate to driving force for these complexes, as well as the magnitude of the observed rate constant, argues against a simple intramolecular electron-transfer step as the rate-limiting step in these reactions. Therefore the oxidation of cyt c by this Ru^{III} label does not seem to be limiting in electron transfer. Earlier we interpreted this phenomenon in terms of a directional electron transfer (*10*). However, these data may be accommodated by changes in the electron-transfer pathway (i.e., different pathways are operational for ruthenium–ammine complexes than for the ruthenium–bipyridine complexes). Further work is required to define the molecular and electronic events that lead to these different rates.

In conclusion, we have shown that in simple donor–acceptor complexes where peptides mediate between the donor and acceptor, rates of electron transfer can vary over many orders of magnitude in a predictable way. In proteins, however, electronic and conformational states may interfere with electron-transfer rates through specific protein dynamical changes that take control of the electron-transfer process. Understanding the elementary steps associated with electron transfer will be one of the future aims that would help in understanding the structure and function of electron-transfer proteins.

Acknowledgments

I thank the graduate students and postdoctoral fellows who participated in this work and whose names are mentioned in the publications, A. Vassilian, M. Ogawa, J. Wishart, R. Bechtold, and B. van Hemelryck. The cytochrome c work was investigated by a number of talented students, including G. Worosila, M. Gardineer, A. Greway, and M. O.K. Cho. Finally I acknowledge the help and continued support of Harold Schwarz of the Brookhaven

National Laboratory, who introduced our group to the pulse radiolysis techniques.

References

1. Isied, S. S. *Prog. Inorg. Chem.* **1984**, *32*, 443–517.
2. Isied, S. S.; Vassilian, A. *J. Am. Chem. Soc.* **1984**, *106*, 1726.
3. Isied, S. S.; Vassilian, A.; Magnuson, R.; Schwarz, H. *J. Am. Chem. Soc.* **1985**, *107*, 7432–7438.
4. Isied, S. S.; Vassilian, A.; Wishart, J.; Creutz, C.; Schwarz, H.; Sutin, N. *J. Am. Chem. Soc.* **1988**, *110*, 635.
5. Vassilian, A.; Wishart, J.; van Hemelryck, B.; Schwarz, H.; Isied, S. S. *J. Am. Chem. Soc.* **1990**, *112*, 7278.
6. Ogawa, M. Y.; Wishart, J. F.; Isied, S. S., unpublished results.
7. Isied, S. S.; Worosila, G.; Atherton, S. J. *J. Am. Chem. Soc.* **1982**, *104*, 7659–7661.
8. Isied, S. S.; Kuehn, C.; Worosila, G. *J. Am. Chem. Soc.* **1984**, *106*, 5145.
9. Bechtold, R.; Gardineer, M. B.; Kazmi, A.; van Hemelryck, B.; Isied, S. S. *J. Phys. Chem.* **1986**, *90*, 3800.
10. Isied, S. S. In *Electron Transfer in Biology and the Solid State: Inorganic Compounds with Unusual Properties*; Johnson, M. K.; King, R. B.; Kurtz, D. M., Jr.; Kutal, C.; Norton, M. L.; Scott, R. A., Eds.; Advances in Chemistry 226; American Chemical Society: Washington, DC, 1990; pp 91–100.
11. Bechtold, R.; Kuehn, C.; Lepre, C.; Isied, S. S. *Nature (London)* **1986**, *322*, 286.
12. Schanze, K.; Sauer, K. *J. Am. Chem. Soc.* **1988**, *110*, 1180.
13. Sisido, M.; Tanaka, R.; Inai, Y.; Imanishi, Y. *J. Am. Chem. Soc.* **1989**, *111*, 6790–6796.
14. Farraggi, M.; DeFelippis, M. R.; Klapper, M. H. *J. Am. Chem. Soc.* **1989**, *111*, 5141.
15. Therien, M. J.; Bowler, B. E.; Selman, M. A.; Gray, H. B.; Chang, I-J.; Winkler, J. R. In *Electron Transfer in Inorganic, Organic, and Biological Systems*; Bolton, J. R.; Mataga, N.; McLendon, G., Eds.; Advances in Chemistry 228; American Chemical Society: Washington, DC, 1991; Chapter 12.
16. Mayo, S. L.; Ellis, W. R., Jr.; Crutchley, R. J.; Gray, H. B. *Science (Washington, D.C.)* **1986**, *233*, 948–952.
17. Yocum, K. M.; Shelton, J. B.; Schroeder, W. A.; Worosila, G.; Isied, S. S.; Bordignon, E.; Gray, H. B. *Proc. Natl. Acad. Sci. U.S.A.* **1982**, *79*, 7052.
18. Nocera, D. G.; Winkler, J. R.; Yocum, K. M.; Bordignon, E.; Gray, H. B. *J. Am. Chem. Soc.* **1984**, *106*, 5145–5150.
19. Natan, M. J.; Baxter, W. W.; Kuila, D.; Gingrich, D. J.; Martin, G. S.; Hoffman, B. M. In *Electron Transfer in Inorganic, Organic, and Biological Systems*; Bolton, J. R.; Mataga, N.; McLendon, G., Eds.; Advances in Chemistry 228; American Chemical Society: Washington, DC, 1991; Chapter 13.
20. Liang, N.; Mauk, A. G.; Pielak, G. J.; Johnson, J. A.; Smith, M.; Hoffman, B. M. *Science (Washington, D.C.)* **1988**, *240*, 311.
21. McLendon, G.; Hickey, D.; Berghuis, A.; Sherman, F.; Brayer, G. In *Electron Transfer in Inorganic, Organic, and Biological Systems*; Bolton, J. R.; Mataga, N.; McLendon, G., Eds.; Advances in Chemistry 228; American Chemical Society: Washington, DC, 1991; Chapter 11.
22. Sutin, N. In *Electron Transfer in Inorganic, Organic, and Biological Systems*;

Bolton, J. R.; Mataga, N.; McLendon, G., Eds.; Advances in Chemistry 228; American Chemical Society: Washington, DC, 1991; Chapter 3.
23. Stryer, L.; Haugland, R. P. *Proc. Natl. Acad. Sci. U.S.A.* **1967**, *58*, 719.
24. Gabor, G. *Biopolymers* **1958**, *6*, 809.
25. Steinberg, I. Z.; Harrington, W. F.; Berger, A.; Sela, M.; Katchalski, E. *J. Am. Chem. Soc.* **1960**, *82*, 5263.
26. Engel, J. *Biopolymers* **1966**, *4*, 945.
27. Schimmel, P. R.; Flory, P. J. *Proc. Natl. Acad. Sci. U.S.A.* **1967**, *58*, 52.
28. Cowan, P. M.; McGavin, S. *Nature (London)* **1955**, *176*, 501.
29. Torchia, P. A.; Bovey, F. A. *Macromolecules* **1971**, *4*, 246.
30. Deber, C. M.; Bovey, F. A.; Carver, J. P.; Blout, E. R. *J. Am. Chem. Soc.* **1970**, *92*, 6191.
31. Chao, Y. H.; Bersohn, R. *Biopolymers* **1978**, *17*, 2761.
32. Chiu, H. C.; Bersohn, R. *Biopolymers* **1977**, *16*, 277.
33. Förster, T. *Ann. Phys. (Leipzig)* **1948**, *2*, 55.
34. Traub, W.; Shmueli, U. *Nature (London)* **1963**, *195*, 1165.
35. Burge, R. E.; Harrison, P. M.; McGavin, S. *Acta Crystallogr.* **1962**, *15*, 914.
36. Cheng, H. N.; Bovey, F. A. *Biopolymers* **1977**, *16*, 1465–1472.
37. Lin, L.; Brandt, J. F. *Biochemistry* **1983**, *22*, 553–559.
38. Brandt, J. F.; Halvorson, H. R.; Brennan, M. *Biochemistry* **1975**, *14*, 4953.
39. Rothe, M.; Theysohn, R.; Steffer, K. D.; Schneider, H.; Amani, M.; Kostrzewa, M. *Angew. Chem. Int. Ed. Engl.* **1969**, *8*, 919.
40. Rothe, M.; Rott, H. *Angew. Chem. Int. Ed. Engl.* **1970**, *15*, 770.
41. Grathwohl, C.; Wuthrich, K. *Biopolymers* **1978**, *17*, 2761.
42. Grathwohl, C.; Wuthrich, K. *Biopolymers* **1976**, *15*, 2025 and 2043.
43. Clark, D. S.; Dechter, J. J.; Mandelkern, L. *Macromolecules* **1979**, *12*, 626.
44. Endicott, J. F. *Acc. Chem. Res.* **1988**, *21*, 59.
45. Stewart, J. M.; Young, J. D. *Solid Phase Peptide Synthesis*, 2nd ed.; Pierce Chemical Co.: Rockford, IL, 1984.
46. Juris, A.; Balzani, V.; Barigelletti, F.; Campagna, S.; Belser, P.; von Zelewsky, A. *Coord. Chem. Rev.* **1988**, *84*, 85.
47. Closs, G. L.; Miller, J. R. *Science (Washington, D.C.)* **1988**, *240*, 440.

RECEIVED for review April 27, 1990. ACCEPTED revised manuscript September 27, 1990.

Solvent Reorganization Energetics and Dynamics in Charge-Transfer Processes of Transition Metal Complexes

Xun Zhang, Mariusz Kozik, Norman Sutin, and Jay R. Winkler[1]

Chemistry Department, Brookhaven National Laboratory, Upton, NY 11973

Time-resolved and steady-state luminescence measurements were used to probe the energetics and dynamics of solvation in two different transition metal complexes. The metal-to-ligand charge-transfer excited state of Ru(bpy)$_2$(CN)$_2$ (bpy is bipyridine) was studied in a series of aliphatic alcohols, and the luminescent excited state of Mo$_2$Cl$_4$[P(CH$_3$)$_3$]$_4$ was studied in aprotic organic solvents. The energetics of excited-state solvation were evaluated from the shapes and positions of the steady-state luminescence spectra recorded at low (~10 K) and room temperatures. The dynamics of excited-state solvation were probed by time-resolved emission spectroscopy.

THE SOLVENT PLAYS A MAJOR ROLE IN GOVERNING the rates of electron-transfer reactions in solution (1, 2). The solvent reorganization associated with electron-transfer reactions in polar solvents is often the major contributor to the total reorganization energy (λ). In recent years, questions have arisen regarding the importance of solvent reorganization dynamics in controlling the rates of fast electron-transfer processes (3–11). Models that treat the solvent as a continuous dielectric medium are often used to describe solvation energetics and dynamics, but the applicability of these models to real chemical systems remains an open question. Electronic absorption and emission spectroscopies are powerful techniques for probing the environments of molecules in solution (12–22). The energies and shapes of absorption

[1]Current address: Beckman Institute, California Institute of Technology, Pasadena, CA 91125

0065–2393/91/0228–0247$06.00/0

and emission profiles correlate with solvent properties and empirical solvent parameters and provide insight into the energetics of solvation. Time-resolved emission spectroscopy (TRES) is an important technique for examining the dynamics of solvation of excited molecules, as well as the solvent dynamics associated with fast electron-transfer processes (23–55).

The equilibrium positions of nuclei in molecules generally shift as a consequence of electronic excitation. Such a shift produces broad spectral profiles and, in some cases, vibrational fine structure. If only solvent nuclei change their equilibrium positions, the energy of the steady-state absorption or emission maximum directly reflects the difference in position along the solvent coordinate of the initial and final states. Similar information is contained in the breadths of the bands. When internal-mode distortions accompany electronic excitation, it is more difficult to extract information about the solvent configuration because the band shape and position also reflect internal-mode rearrangements. To characterize the energetics of solvation of an excited molecule, internal-mode contributions to the absorption or emission profiles must be factored out. Once the solvent contribution to the band shape has been determined, the resulting solvent reorganization energies can be correlated with solvent dielectric properties.

Because electrons move much faster than nuclei, a short laser pulse can be used to prepare a molecule in a nonequilibrium solvation environment. Following excitation, the solvent will rearrange to accommodate the new electron distribution and geometry of the excited molecule. The position and shape of the emission band reflect, in part, the solvation environment; therefore the time evolution of the emission profile can be used to monitor the dynamics of the approach to equilibrium solvation.

Most previous investigations of solvent reorganization dynamics have involved organic probe molecules, especially laser dyes. There are, however, many luminescent metal complexes that can serve as probe molecules in these experiments. One unique feature of metal complexes as compared to organic chromophores is their shape: the organic probes tend to be large flat molecules, but metal complexes can be a variety of shapes (e.g., flat, cylindrical, or spherical). Ruthenium–bipyridine complexes, for example, are roughly spherical and have long-lived (>100 ns) luminescent charge-transfer excited states. The time-resolved emission spectra of one member of this class of molecules, $Ru(bpy)_2(CN)_2$ (bpy is bipyridine), have been examined in alcohols near the glass transition (39). We extended this study to higher temperatures (−20 °C) and faster time scales (>20 ps) in a series of aliphatic alcohols. In addition, we performed a band-shape analysis of the steady-state emission spectra of $Ru(bpy)_2(CN)_2$ in alcohol solvents to characterize the energetics of solvent reorganization about the excited molecule.

$Ru(bpy)_2(CN)_2$ suffers from two shortcomings for solvent dynamics experiments: a low radiative rate constant (the luminescent state is formally a triplet) and low solubility in aprotic solvents. We therefore initiated TRES studies of a second luminescent metal complex, $Mo_2Cl_4[P(CH_3)_3]_4$. This molecule is soluble in most organic solvents (except alcohols), and steady-state spectra suggest that its luminescent state has a respectable dipole moment (vide infra). Furthermore, the fact that the luminescent excited state is a singlet greatly facilitates TRES studies on picosecond time scales. For these reasons we examined the picosecond time-resolved emission spectra of $Mo_2Cl_4[P(CH_3)_3]_4$ in benzonitrile between 20 and −32 °C.

Experimental Details

Materials. All solvents used in this study were HPLC grade. Methanol (MeOH), 1-propanol (PrOH), 1-butanol (BuOH), and absolute ethanol (EtOH) were refluxed over Na, distilled, and stored under Ar over molecular sieves (3 Å for MeOH, 4 Å for higher alcohols). Tetrahydrofuran (THF) was refluxed over Na–benzophenone, distilled, and stored under vacuum over Na–benzophenone. Hexanes (bp 68–69 °C) were refluxed over NaK alloy, distilled, and stored under vacuum over NaK. Methylene chloride (CH_2Cl_2), chloroform ($CHCl_3$), and acetonitrile (CH_3CN) were refluxed over CaH_2, distilled, and stored under vacuum over molecular sieves (4 Å for CH_2Cl_2 and $CHCl_3$, 3 Å for CH_3CN). Ethyl acetate, diethyl ether, butanone, dimethyl sulfoxide (DMSO), and benzonitrile were stored under vacuum over 4-Å molecular sieves.

cis-$Ru(bpy)_2(CN)_2$ was prepared and purified by published procedures (*56, 57*). Purity was determined by TLC on silica, developed with methanol. $Mo_2Cl_4[P(CH_3)_3]_4$ was prepared according to a published procedure (*58*). Sample purity was evaluated by absorption spectroscopy. Samples for steady-state and time-resolved emission spectra were kept under vacuum in sealed fused-silica cuvettes.

Data Collection. *Steady-State Emission Spectra.* Emission spectra were recorded on an instrument constructed at Brookhaven National Laboratory (*59*). Samples for low-temperature spectra were held under vacuum in sealed 4-mm-o.d. fused-silica tubes and mounted on the cold head of a closed-cycle refrigerator. Identical configurations were used for room-temperature and low-temperature spectra.

Time-Resolved Emission Spectra. Picosecond time-resolved emission spectra were recorded following excitation with a vertically polarized 30-ps pulse ($Mo_2Cl_4[P(CH_3)_3]_4$, 532 nm; $Ru(bpy)_2(CN)_2$, 355 nm) from a flashlamp-pumped, actively–passively mode-locked Nd:YAG laser. Emitted light passed through an analyzing polarizer, then was dispersed by a spectrograph and directed to the entrance slit of a streak camera. The instrument response time was 35–40 ps (*59*). Time-resolved emission spectra were recorded with the analyzing polarizer set to the "magic angle" (54.75° from vertical) (*60*). Time-resolved fluorescence depolarization measurements were performed without dispersion of the emission spectrum by using parallel (0°) and perpendicular (90°) orientations of the analyzing polarizer (*61*).

Results

Ru(bpy)$_2$(CN)$_2$. ***Steady-State Emission Spectra.*** The room-temperature and low-temperature emission spectra of Ru(bpy)$_2$(CN)$_2$ in ethanol are shown in Figure 1. At room temperature the emission profile is a broad, asymmetric band. At 12 K a progression in a high-energy vibrational mode can be resolved. The low-temperature spectra were fit to the model described by eqs A3–A6 (*see* Appendix), in which a single quantum mode was included, and the remainder of the internal-mode broadening was treated semiclassically (dashed lines, Figure 1). The frequency of the quantum mode for all of the alcohols, about 1300 cm^{-1}, contributed ~0.16 eV to the inner-shell reorganization parameter (λ_{in}). The resonance Raman spectrum of solid Ru(bpy)$_2$(CN)$_2$ exhibits, in addition to a 1317-cm^{-1} peak, intense features at 366, 663, 1024, 1172, 1485, 1557, and 1601 cm^{-1} indicative of distortions along these vibrational coordinates in the MLCT (metal-to-ligand charge-transfer) excited state (62). The >1400-cm^{-1} vibrations do not clearly contribute to the low-temperature emission spectrum, and their contributions to λ_{in} have been neglected.

Assuming that the contribution to the bandwidth from solvent relaxation during the lifetime of the excited state can be neglected at low temperatures, and that the remaining inner-shell distortions can be represented by an

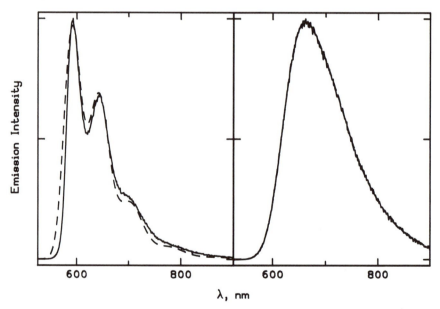

Figure 1. Emission spectra of Ru(bpy)$_2$(CN)$_2$ in EtOH. Left: 12 K. Right: room temperature. Dashed lines are Franck–Condon fits to the spectra using one quantum mode.

average vibrational frequency of 600 cm^{-1}, the semiclassical inner-shell distortion can be calculated ($\lambda_{in,j} \sim 0.05$ eV, eq A5). This calculation may be combined with the quantum-mode reorganization to provide an estimate of $\lambda_{in} \sim 0.21$ eV for the Ru(bpy)$_2$(CN)$_2$ MLCT excited state. The solvent reorganization parameter (λ_{out}) can be determined from the breadth of the room-temperature emission profiles by using the inner-shell distortion parameters obtained from the low-temperature spectra (eq A5). The resulting λ_{out} values are 0.09 eV for MeOH, 0.11 eV for EtOH, 0.09 eV for PrOH, and 0.07 eV for BuOH.

Time-Resolved Spectra.   The luminescence lifetime of Ru(bpy)$_2$(CN)$_2$ is ~400 ns in alcohols at –20 °C. In the four alcohols studied, solvent relaxation at this temperature was complete before there had been any significant excited-state depopulation. The wavelength dependence of the Ru(bpy)$_2$(CN)$_2$ luminescence decays in BuOH (–20 °C) are shown in Figure 2. The most striking aspect of these data is that the form of the emission decay function is wavelength-dependent. At shorter wavelengths (<640 nm), a large initial emission intensity rapidly decays to a smaller, constant value (on the 2-ns time scale). At longer wavelengths (>680 nm), a rapid increase

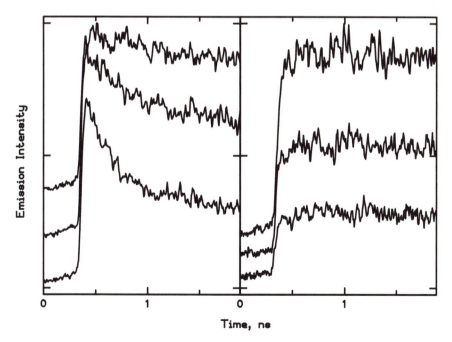

Figure 2. Wavelength dependence of time-resolved emission profiles of Ru(bpy)$_2$(CN)$_2$ in BuOH at –20 °C. Left, lower to upper: 614, 631, and 648 nm. Right, upper to lower: 665, 682, and 699 nm.

in emission intensity follows excitation. Intermediate behavior is found at wavelengths between these two extremes. This behavior is consistent with an emission profile that shifts to the red following excitation. Similar behavior was found for Ru(bpy)$_2$(CN)$_2$ in MeOH, EtOH, and PrOH (at −20 °C), though on different time scales.

To extract solvent relaxation dynamics from the time-resolved emission decays, smoothing was effected by first fitting the single-wavelength decay kinetics to multiexponential functions. A sum of three exponential functions, convoluted with the instrument response function, provided adequate fits to the data. The reconstructed time-resolved spectra generated from the exponential decay parameters were then fit to symmetric Gaussian distribution functions. The dynamics of solvent relaxation are given by the correlation function, $C(t)$ (eq 1) (31–54)

$$C(t) = \frac{\langle \nu(t) \rangle - \langle \nu(\infty) \rangle}{\langle \nu(0) \rangle - \langle \nu(\infty) \rangle} \tag{1}$$

where $\langle \nu(t) \rangle$ is the mean value of the spectral distribution function at time t. Because the time-resolved emission spectra of Ru(bpy)$_2$(CN)$_2$ were fit to symmetric functions, $\langle \nu(t) \rangle$ is simply given by the peak maxima of these spectra. The resulting $C(t)$ functions are biphasic for Ru(bpy)$_2$(CN)$_2$ in EtOH, PrOH, and BuOH at −20 °C. A single exponential function describes $C(t)$ in MeOH at −20 °C, but it is likely that faster components are lost because of the limited time resolution of the TRES apparatus. A plot of $C(t)$ for Ru(bpy)$_2$(CN)$_2$ in BuOH appears in Figure 3, and the $C(t)$ exponential fitting parameters appear in Table I.

Mo$_2$Cl$_4$[P(CH$_3$)$_3$]$_4$. *Steady-State Emission Spectra.*

The binuclear metal complex Mo$_2$Cl$_4$[P(CH$_3$)$_3$]$_4$ fluoresces from its lowest-lying excited singlet state (58). This excited state is described as a δδ* metal-localized excited state in molecular orbital models (58, 63) and as a metal-to-metal charge-transfer (MMCT) excited state in valence bond models (64). The excited-state lifetime is ∼140 ns. In contrast to its absorption spectrum, the emission spectrum of Mo$_2$Cl$_4$[P(CH$_3$)$_3$]$_4$ at room temperature in fluid solution is very sensitive to the solvent (Figure 4). The emission maxima appear at lower energies, and the bandwidths increase with increasing solvent polarity.

The reorganization energy associated with the luminescent transition was determined from the room-temperature emission spectra of Mo$_2$Cl$_4$[P(CH$_3$)$_3$]$_4$. If solvent reorientation is treated as a single classical nuclear coordinate described by harmonic potential surfaces with equal force constants in the ground and excited states, then every vibronic line in the spectrum can be represented by a Gaussian line shape. Under these circumstances, the first moment of the total emission spectral distribution

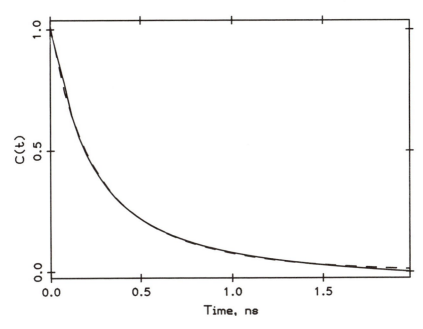

Figure 3. Solvent relaxation correlation function C(t) (eq 1) for Ru(bpy)$_2$(CN)$_2$ in BuOH at –20 °C.

Table I. Kinetic Parameters for Solvent Relaxation in the MLCT Excited State of Ru(bpy)$_2$(CN)$_2$

Solvent	α_1	τ_1	τ_2
MeOH	—	—	34
EtOH	0.74	26	114
PrOH	0.60	94	332
BuOH	0.49	156	513

NOTE: For biexponential relaxation functions, α_1 is the coefficient of the exponential with decay time τ_1, and $(1 - \alpha_1)$ is the coefficient of the exponential with decay time τ_2. Decay times are in picoseconds.

function $I_{ab}(\nu)$ (eq A3) will depend linearly upon λ_{out}. After correcting for the ν^3-dependence of spontaneous emission, the first moments of the Mo$_2$Cl$_4$[P(CH$_3$)$_3$]$_4$ emission profiles were determined by numerical integration. The emission spectrum in hexanes was chosen as a reference: The spectrum was assumed to arise solely from internal-mode distortions, and the solvent reorganization energy was assumed to be zero. Because Mo$_2$Cl$_4$[P(CH$_3$)$_3$]$_4$ absorption spectra are not particularly sensitive to the solvent, half the difference between first moments in the spectrum of Mo$_2$Cl$_4$[P(CH$_3$)$_3$]$_4$ in hexanes and that in a polar solvent provides an estimate

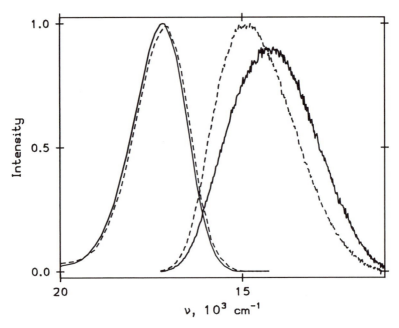

Figure 4. Absorption (left) and emission (right) spectra of $Mo_2Cl_4[P(CH_3)_3]_4$ in hexanes (dashed curve) and benzonitrile (solid curve) at room temperature.

for λ_{out}. The results of these calculations are set out in Table II. Continuum models predict that the solvent reorganization parameter should depend upon the dielectric function F_1 (eq 2) (65)

$$F_1 = \frac{3(\epsilon_s - \epsilon_{op})}{[4(\epsilon_s + 1)(\epsilon_{op} + 1)]} \tag{2}$$

Table II. Solvent Reorganization Energies for $Mo_2Cl_4[P(CH_3)_3]_4$

Solvent	F_1	λ_{out} (eV)
Hexanes	0.0	0.0
Chloroform	0.114	0.024
Ethyl ether	0.121	0.011
Ethyl acetate	0.154	0.017
Tetrahydrofuran	0.165	0.018
Dichloromethane	0.173	0.032
Benzonitrile	0.197	0.027
Dimethyl sulfoxide	0.220	0.032
Butanone	0.221	0.024
Acetonitrile	0.247	0.042

NOTE: Energies are determined from steady-state emission spectra in aprotic solvents at 293 K. The solvent dielectric parameter F_1 is defined in eq 2. Dielectric data are taken from ref. 67.

where ϵ_s and ϵ_{op} are the static and optical dielectric constants of the solvent, respectively. The magnitude of λ_{out} for the $Mo_2Cl_4[P(CH_3)_3]_4$ excited state generally increases with the solvent dielectric parameter F_1, but the relationship is not linear (Figure 5). The solid line in Figure 5 was calculated by using a model that treats the solute as a sphere of low internal dielectric constant ($\epsilon_{int} = 2.5$) embedded in a continuous dielectric medium (66). A partial charge (0.4 electron) was assumed to transfer from one metal to the other (2.1 Å) within a sphere of 3.96-Å radius [the value obtained from an analysis of the rotational correlation time of $Mo_2Cl_4[P(CH_3)_3]_4$ in benzonitrile; vide infra]. The dipole moment corresponding to this charge transfer is 4 D. The partial charge transfer is consistent with the mixed character of the excited state. Although these parameters provide a satisfactory fit to the data, it is important to remember that this treatment is based upon a dielectric continuum description of the solvent. The deviations from the calculated curve can be the result of specific solute–solvent interactions.

Time-Resolved Emission Spectra. No spectral evolution was observed in the time-resolved emission spectra of $Mo_2Cl_4[P(CH_3)_3]_4$ in benzonitrile between 20 and –32 °C (the depressed freezing point of the solutions was near –35 °C). This finding places an upper limit of ~5 ps on the solvation time for benzonitrile at –32 °C. Though solvent reorientation dynamics were beyond the time resolution of the instrument, the rotational dynamics of

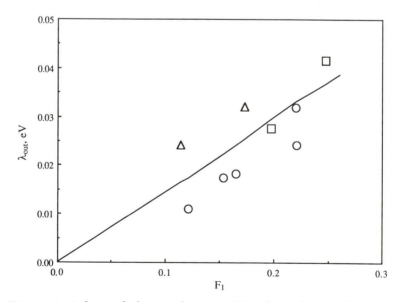

Figure 5. Solvent dielectric function (F_1) dependence of λ_{out} for $Mo_2Cl_4[P(CH_3)_3]_4$. Solid line was calculated from a single-sphere dielectric continuum model. Solvents are oxygen donors (○), nitriles (□), and chlorocarbons (△).

$Mo_2Cl_4[P(CH_3)_3]_4$ were measurable. Solute rotational dynamics were determined in the 20 to −32 °C temperature range by time-resolved fluorescence depolarization (61). The data for parallel and perpendicular polarizations (relative to vertical excitation polarization) were fit to eq A7 (Figure 6). The tumbling dynamics are adequately described by a single exponential rotational time constant τ_r that is strongly temperature dependent. The

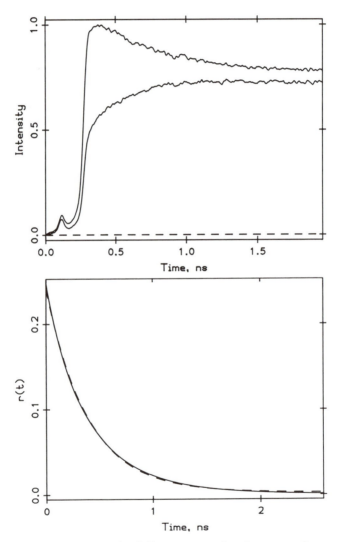

Figure 6. Top: Time-resolved fluorescence depolarization kinetics for $Mo_2Cl_4[P(CH_3)_3]_4$ in benzonitrile at −32 °C. Bottom: Rotational correlation function derived from $Mo_2Cl_4[P(CH_3)_3]_4$ depolarization kinetics. The dashed line is a fit to a single exponential decay function.

magnitude of τ_r increases by a factor of 5.5 between 20 and -32 °C [T (K), τ_r (ps): 293, 86; 273, 106; 258, 174; 248, 396; 241, 470]. The rotational correlation time is expected to depend upon solvent viscosity, η, according to eq 3 *(61)*

$$\frac{1}{\tau_r} = \frac{k_B T}{V\eta}$$ (3)

where k_B is the Boltzmann constant, T is absolute temperature, and V is the effective volume of the solute molecule. Taking $\eta = 1.34$ cP at 20 °C *(67)* (interpolated value) gives a solute volume of 260 Å^3, corresponding to a sphere of radius 3.96 Å. Plots of $\ln(\tau_r T)$ vs. T^{-1} are not linear, perhaps owing to the deviation from Arrhenius behavior of the solvent viscosity near the freezing point.

Discussion

Ru(bpy)$_2$(CN)$_2$.   Continuum models are generally the first recourse for discussions of solvation properties. As was shown in Figure 1, a dielectric continuum model adequately describes the breadths of the Ru(bpy)$_2$(CN)$_2$ steady-state emission profiles. Specific solvation effects, especially hydrogen bonding, can complicate the analysis, but the general trend appears to be described by bulk solvent properties. The dynamics of microscopic solvation can also be described by continuum models. In a Debye-type dielectric, the approach to equilibrium of the dielectric polarization following an instantaneous change in the permanent dipole moment of a spherical solute is exponential, with a time constant τ_L given by eq 4 *(3–11, 68, 69)*

$$\tau_L = \left(\frac{\epsilon_\infty}{\epsilon_s}\right) \tau_D$$ (4)

where τ_D, the Debye time, is the exponential time constant for the approach to equilibrium polarization following an instantaneous change in the external electric field *(70–72)*. The high- and low-frequency dielectric constants, ϵ_∞ and ϵ_s, respectively, arise from the dielectric dispersion of the medium.

Alcohols are not simple Debye solvents and are reported to have three regions of dielectric dispersion. The highest frequency component is attributed to rotation of free hydroxyl groups, the mid-range component to reorientation of free alcohol molecules, and the low-frequency component to disruption of hydrogen bonds in alcohol aggregates (Table III) *(73)*. In MeOH, EtOH, PrOH, and BuOH, the low-frequency process accounts for the major part of the dielectric constant. The definition of τ_L for alcohols is also complicated. A different value of ϵ_∞ is associated with each of the three

Table III. Dielectric Relaxation Properties of Aliphatic Alcohols

Alcohol	τ_{D3}	τ_{D2}	τ_{D1}	ϵ_{0l}	$\epsilon_{\infty l}$	τ_{Ll}
MeOH	—	—	154	43.2	6.40	23
EtOH	—	—	632	31.4	4.85	98
PrOH	4	34	1920	27.4	3.92	274
BuOH	3.5	50	3145	23.6	3.65	486

NOTE: Relaxation times at 253 K are in picoseconds. Data are taken from ref. 73.

regions of dielectric dispersion. These values differ from the optical dielectric constant ϵ_{op}, which is equal to the square of the refractive index of the medium and which is commonly used for ϵ_∞ in nonassociated solvents. The significance of this complication is especially clear in BuOH where, depending upon the choice of ϵ_∞, τ_L can vary by almost a factor of 2 ($\epsilon_{\infty l}$ = 3.65; ϵ_{op} = 1.96). The τ_{Ll} values in Table III were calculated from eq 4 by using the ϵ_{0l}, $\epsilon_{\infty l}$, and τ_{Dl} values in ref. 45 ($\epsilon_{\infty l}$ refers to the lowest frequency region of dielectric dispersion, $\tau_{Dl} > \tau_{D2} > \tau_{D3}$).

In light of the complex dielectric relaxation behavior of the medium, it is not surprising that the solvent relaxation dynamics for the Ru(bpy)$_2$(CN)$_2$ MLCT excited state are biphasic in alcohols. Comparison of the data in Tables I and III reveals that the slower solvent relaxation time (τ_2) is in fair accord with the low-frequency longitudinal dielectric relaxation time (τ_{Ll}) for the four alcohols examined in this work. The contributions of the two relaxation times to their corresponding relaxation functions, however, are quite different. Although the dispersion region with time constant τ_{Ll} corresponds to the largest part of the dielectric constant in these alcohols, τ_2 represents less than half of the observed microscopic solvation dynamics.

These results are similar to other reports of solvent relaxation about excited chromophores in alcohols (32, 33, 36, 41, 42, 50, 53, 54). Biphasic relaxation kinetics are typically observed, and one relaxation time is generally comparable to τ_{Ll}. In some reports the slowest relaxation time is longer than τ_{Ll}, but faster components are also observed. Few of the systems used thus far have been free of potential hydrogen-bonding interactions. It is, therefore, difficult to assess the importance of this specific interaction, and many aspects of solvent relaxation dynamics in alcohols remain unresolved.

$Mo_2Cl_4[P(CH_3)_3]_4$. The steady-state emission spectra of Mo_2Cl_4-$[P(CH_3)_3]_4$ are strongly solvent-dependent, and the variations are in general agreement with a dielectric continuum model. As in the case of Ru(bpy)$_2$(CN)$_2$, specific solute–solvent interactions are also likely to be important in $Mo_2Cl_4[P(CH_3)_3]_4$ because the vacant axial sites at the two metal centers could be occupied by strong donor solvents. In this regard, it is surprising that the absorption spectrum of $Mo_2Cl_4[P(CH_3)_3]_4$ varies so little with solvent; the shift in absorption maximum between hexanes and benzonitrile is only 60 cm^{-1}. In contrast, the shift in emission maximum between these two solvents is 700 cm^{-1}. The simplest explanation of these results is that donor solvents interact much more strongly with electronically excited $Mo_2Cl_4[P(CH_3)_3]_4$ than with the ground-state species.

The dielectric properties of benzonitrile are presented in Table IV (74–76). Unlike alcohols, benzonitrile exhibits simple Debye relaxation properties, although there is some uncertainty in the magnitude of τ_D. The choice between ϵ_{op} and ϵ_∞ (2.24 and 3.80, respectively) adds even more uncertainty to the value of τ_L. The Debye time for benzonitrile has been reported for temperatures between 15 and 40 °C (76), but data are not available for temperatures as low as –32 °C. A rough estimate for the Debye time of benzonitrile at this temperature can be obtained from the $Mo_2Cl_4[P(CH_3)_3]_4$ rotational correlation times. Both τ_D and τ_r are believed to depend approximately linearly on solvent viscosity (61, 70). Assuming that the effective solute volume remains constant, a value of $\eta = 6.0$ cP can be estimated for benzonitrile at –32 °C. Given a room-temperature Debye time of 16 ps, τ_D is estimated to be 87 ps at –32 °C. This gives $\tau_L = 13$ ps, provided that $\epsilon_\infty/\epsilon_s$ remains approximately constant. On the basis of these estimates of dielectric relaxation times, the time-resolved emission spectra of $Mo_2Cl_4[P(CH_3)_3]_4$ indicate that solvent relaxation time about the excited metal complex is shorter than or equal to τ_L. This result differs from some other measurements of solvent relaxation in nitriles, where observed relaxation times are 2 to 5 times longer than τ_L, although none of the observed relaxation times exceed 5 ps (32, 50, 53).

Caveats. Analyses of the time-resolved emission spectra and their interpretation in terms of solvent reorientation dynamics are complicated by several factors. In the first instance, both the $Ru(bpy)_2(CN)_2$ and

Table IV. Measured and Estimated Dielectric Properties of Benzonitrile

T (K)	$\tau_D{}^{a,b}$	$\tau_L{}^{a,c}$
323	9	1.4
313	11	1.7
303	13	2.0
293	16, 38[d]	2.4, 5.7[d]
283	18	2.7
241	87[e]	13[e]

[a]Relaxation times are in picoseconds.
[b]Data are taken from ref. 76.
[c]τ_L was calculated according to eq 4 by using $\epsilon_s = 25.2$ and $\epsilon_\infty = 3.80$. Data are taken from ref. 75.
[d]Data are taken from refs. 74 and 75.
[e]Estimate from fluorescence depolarization determination of solvent viscosity. The (depressed) freezing point of the solution was ~238 K. See text.

$Mo_2Cl_4[P(CH_3)_3]_4$ complexes were excited with laser pulses significantly above the origins of the emission bands ($\sim$1.4 and 0.3 eV, respectively). This excess energy is ultimately deposited into the solvent, where it leads to local heating. The resulting temperature gradient could complicate the solvent relaxation dynamics. Furthermore, in $Ru(bpy)_2(CN)_2$, the above-threshold excitation will likely populate the singlet MLCT state before the triplet. No clear evidence for singlet emission was found in $Ru(bpy)_2(CN)_2$, a finding that suggests that the intersystem crossing time is much shorter than the time resolution of our TRES apparatus. The presence of such fluorescence, however, could confuse the analysis of the solvent relaxation dynamics.

The spectral models described in the Appendix also include a number of assumptions. The most tenuous could be the Condon approximation, which states that the transition dipole, and hence the electronic wave functions, are independent of all nuclear coordinates, including those of the solvent. This may be a poor assumption; the wave functions describing MLCT and MMCT excited states could depend directly upon the solvent orientation because the electric field experienced by the molecule will vary substantially with solvent orientation. It is well known from Stark-effect spectroscopy that excited-state wave functions can be significantly perturbed by external electric fields (77). Fleming and co-workers have proposed a similar explanation for the rapid fluorescence depolarization dynamics found with coumarin 153 in alcohol solvents (33). Finally, specific interactions between solute and solvent may vitiate any comparison between solvation dynamics of these excited chromophores and bulk solvent dielectric relaxation properties. Clearly, the systems to be studied must be selected with care, and the observations interpreted with caution.

Summary

Excitation of $Ru(bpy)_2(CN)_2$ in alcohol solvents to a thermally equilibrated MLCT excited state requires, depending upon the alcohol, 0.07–0.11 eV of solvent reorganization. At –20 °C this solvent reorientation proceeds on a subnanosecond time scale and the function describing solvent relaxation is biphasic. Formation of excited $Mo_2Cl_4[P(CH_3)_3]_4$ in aprotic polar solvents is accompanied by 0.01–0.04 eV of solvent reorientation energy. The dynamics of this reorientation in benzonitrile are faster than 5 ps, even at temperatures as low as –32 °C. At this temperature, fluorescence depolarization measurements suggest that the solvent viscosity increases to nearly 5 times its room-temperature value. For both $Ru(bpy)_2(CN)_2$ in alcohols and $Mo_2Cl_4[P(CH_3)_3]_4$ in benzonitrile, solvent relaxation proceeds on a time scale greater than or equal to the longitudinal dielectric relaxation time.

Appendix: Data Analysis

Steady-State Emission Spectra.   The conventional description of radiative transitions predicts that the total spontaneous emission probability per unit time, w, for a molecule in a condensed phase is given by eq A1 (78)

$$w = \int \left[\frac{n^3}{\epsilon} \left(\frac{E_e}{E} \right)^2 \right] \frac{64 \, \pi^4 \nu^3}{c^3} I_{ab}(\nu) \, d\nu \tag{A1}$$

in which n is the refractive index and ϵ is the dielectric constant of the medium, both at frequency ν; E_e is the effective electric field at the chromophore and E is the macroscopic electric field; c is the speed of light; and $I_{ab}(\nu)$ is the emission probability per unit frequency interval at frequency ν for the electronic transition from state a to state b. The function $I_{ab}(\nu)$ is given by eq A2

$$I_{ab}(\nu) = \text{Av}_m \sum_n | \langle \Psi_{am} | M_{ab} | \Psi_{bn} \rangle |^2 \ \delta(E_{am} - E_{bn} - h\nu) \tag{A2}$$

in which Av_m indicates a Boltzmann average over initial vibrational states Ψ_{am}; Σ_n is a sum over all final states Ψ_{bn}; M_{ab} is the dipole transition operator; δ is the Dirac delta function; and h is Planck's constant. Internal-mode distortions can be treated as harmonic oscillators both quantum mechanically and semiclassically, and solvent reorientation is generally described classically. Invoking of the Condon approximation leads to eq A3

$$I_{ab}(\nu) = \langle M_{ab} \rangle^2 \prod_i \left\{ \sum_{m_i} \exp \left(-\frac{m_i \hbar \omega_i}{k_B T} \right) \sum_{n_i} | \langle \chi_{am_i} | \chi_{bn_i} \rangle |^2 \ g \, (n, m, \nu) \right\} \tag{A3}$$

in which the repeated product extends over the quantum-mechanical modes, χ_{am_i} is the vibrational wave function for mode i in vibrational state m of the initial electronic state a, χ_{bn_i} is the vibrational wave function for mode i in vibrational state n of the final electronic state b, and $\hbar$ is $h/2\pi$. The squares of the vibrational overlap integrals (i.e., the Franck–Condon factors) are given by recursion relations for the Hermite polynomials (79), and it has been assumed that the vibrational frequency ω_i is the same in states a and b. The line-shape function $g(n, m, \nu)$ is taken to be Gaussian and to contain broadening contributions from both the internal semiclassical modes and the classical solvent mode (eqs A4–A6) (80).

$$g(n, m, \nu) = \left(\frac{1}{\sigma \sqrt{\pi}} \right) \exp \left\{ \frac{-[E(n, m) - \lambda_{out} - h\nu]^2}{\sigma^2} \right\} \tag{A4}$$

$$\sigma^2 = 4\lambda_{out} RT + 2 \sum_j (\lambda_{in,j} \hbar \omega_j) \coth \left(\frac{\hbar \omega_j}{2k_B T} \right) \tag{A5}$$

$$E(n, m) = E_{00} - \sum_j \lambda_{in,j} + \sum_i (m_i - n_i)\hbar \omega_i \tag{A6}$$

The factor σ is a width function for the Gaussian. The term λ_{out} is the reorganization parameter for the classical solvent mode and the $\lambda_{in,j}$ are the reorganization parameters for the semiclassical internal modes. The sums over j include only modes treated semiclassically, and the sum over i includes only the quantum-mechanical modes. E_{00} is the electronic origin for the a-to-b transition.

It is a simple matter to show that the first moment of the spectral distribution function $I_{ab}(\nu)$ is a linear function of λ_{out}. This result is quite general and assumes only that the shifting and broadening of individual vibronic lines in the spectrum can be described by a Gaussian line shape as in eq A4 (*see* text). Because ground and excited states are not necessarily stabilized to the same extent by the solvent, the value of E_{00} can also be a function of the solvent. This factor can be eliminated by taking the difference between the absorption and emission first moments. This difference is just a solvent-independent constant plus 2λ. Relative values of λ_{out} can be extracted from spectra by numerical integration of absorption and emission profiles to determine the first moment. If spectra in a nonpolar reference solvent are available, the solvent-independent constant can be determined, and absolute values of λ_{out} can be estimated.

Time-Resolved Fluorescence Depolarization.   Fluorescence decay curves for parallel ($I_{\parallel}(t)$) and perpendicular ($I_{\perp}(t)$) analyzing polarizer orientations were fit, with deconvolution, to multiexponential decay functions by keeping one rate constant equal to the total excited-state decay rate ($1.0 \times 10^7 \ s^{-1}$). The two decay functions were normalized on the coefficient of this fixed exponential, and a rotational correlation function was generated according to eq A7 (*62, 81*).

$$r(t) = \frac{I_{\parallel}(t) - I_{\perp}(t)}{I_{\parallel}(t) + 2I_{\perp}(t)} \tag{A7}$$

Acknowledgment

We thank Bruce Brunschwig for many helpful discussions during the course of this work. This research was carried out at Brookhaven National Laboratory under Contract DE–AC02–76CH00016 with the U.S. Department of Energy and supported by its Division of Chemical Sciences, Office of Basic Energy Sciences.

References

1. Marcus, R. A. *J. Chem. Phys.* **1956**, *24*, 966–978.
2. Hush, N. S. *Trans. Faraday Soc.* **1961**, *57*, 557.
3. van der Zwan, G.; Hynes, J. T. *Physica A (Amsterdam)***1983**, *121*, 227–252.

4. van der Zwan, G.; Hynes, J. T. *Chem. Phys. Lett.* **1983**, *101*, 367–371.
5. van der Zwan, G.; Hynes, J. T. *J. Phys. Chem.* **1985**, *89*, 4181–4188.
6. Hynes, J. T. *J. Phys. Chem.* **1986**, *90*, 3701–3706.
7. Calef, D. F.; Wolynes, P. G. *J. Phys. Chem.* **1983**, *87*, 3387–3400.
8. Calef, D. F.; Wolynes, P. G. *J. Chem. Phys.* **1983**, *78*, 470–482.
9. Sumi, H.; Marcus, R. A. *J. Chem. Phys.* **1986**, *84*, 4272–4276.
10. Sumi, H.; Marcus, R. A. *J. Chem. Phys.* **1986**, *84*, 4894–4914.
11. Nadler, W.; Marcus, R. A. *J. Chem. Phys.* **1987**, *86*, 3906–3924.
12. Bayliss, N. S.; McRae, E. G. *J. Phys. Chem.* **1954**, *58*, 1813–1822.
13. Ooshika, Y. *J. Phys. Soc. Jpn.* **1954**, *9*, 594–602.
14. Lippert, E. *Z. Naturforsch.* **1955**, *10a*, 541–545.
15. Mataga, N.; Kaifu, Y.; Koizumi, M. *Bull. Chem. Soc. Jpn.* **1956**, *29*, 465–470.
16. McRae, E. G. *J. Phys. Chem.* **1957**, *61*, 562–572.
17. Lippert, E. *Z. Elektrochem.* **1957**, *61*, 962–975.
18. Mataga, N. *Bull. Chem. Soc. Jpn.* **1963**, *36*, 654–662.
19. Bakhshiev, N. G. *Opt. Spect. (Engl. Transl.)* **1964**, *16*, 446–451.
20. Bakhshiev, N. G.; Piterskaya, I. V. *Opt. Spect. (Engl. Transl.)* **1965**, *19*, 390–395.
21. Taft, R. W.; Pienta, N. J.; Kamlet, M. J.; Arnett, E. M. *J. Org. Chem.* **1981**, *46*, 661–667.
22. Brady, J. E.; Carr, P. W. *J. Phys. Chem.* **1985**, *89*, 5759–5766.
23. Ware, W. R.; Lee, S. K.; Brant, G. J.; Chow, P. P. *J. Chem. Phys.* **1971**, *54*, 4729–4737.
24. Chakrabat, S. K.; Ware, W. R. *J. Chem. Phys.* **1971**, *55*, 5494–5498.
25. Ware, W. R.; Chow, P. P.; Lee, S. K. *Chem. Phys. Lett.* **1986**, *2*, 356–358.
26. Mazurenko, Y. T. *Opt. Spect. (Engl. Transl.)* **1974**, *36*, 283–286.
27. Mazurenko, Y. T.; Uldaltsov, V. S. *Opt. Spect. (Engl. Transl.)* **1978**, *44*, 417–420.
28. Mazurenko, Y. T.; Uldaltsov, V. S. *Opt. Spect. (Engl. Transl.)* **1978**, *45*, 141–145.
29. Mazurenko, Y. T.; Uldaltsov, V. S. *Opt. Spect. (Engl. Transl.)* **1978**, *45*, 765–767.
30. Mazurenko, Y. T. *Opt. Spect. (Engl. Transl.)* **1980**, *48*, 388–392.
31. Bagchi, B.; Oxtoby, D. W.; Fleming, G. R. *Chem. Phys.* **1984**, *86*, 257–267.
32. Castner, E. W.; Maroncelli, M.; Fleming, G. R. *J. Chem. Phys.* **1987**, *86*, 1090–1097.
33. Maroncelli, M.; Fleming, G. R. *J. Chem. Phys.* **1987**, *86*, 6221–6239.
34. Castner, E. W.; Fleming, G. R.; Bagchi, B. *Chem. Phys. Lett.* **1988**, *143*, 270–276.
35. Maroncelli, M.; Fleming, G. R. *J. Chem. Phys.* **1988**, *89*, 875–881.
36. Castner, E. W., Jr.; Bagchi, B.; Maroncelli, M.; Webb, S. P.; Ruggiero, A. J.; Fleming, G. R. *Ber. Bunsen-Ges. Phys. Chem.* **1988**, *92*, 363–372.
37. Castner, E. W.; Fleming, G. R.; Bagchi, B.; Maroncelli, M. *J. Chem. Phys.* **1988**, *89*, 3519–3534.
38. Maroncelli, M.; MacInnis, J.; Fleming, G. R. *Science (Washington, D.C.)* **1989**, *243*, 1674–1681.
39. Kitamura, N.; Kim, H.-B.; Kawanishi, Y.; Obata, R.; Tazuke, S. *J. Phys. Chem.* **1986**, *90*, 1488–1491.
40. Su, S.-G.; Simon, J. D. *J. Phys. Chem.* **1986**, *90*, 6475–6479.
41. Su, S.-G.; Simon, J. D. *J. Phys. Chem.* **1987**, *91*, 2693–2696.
42. Simon, J. D.; Su, S.-G. *J. Chem. Phys.* **1987**, *87*, 7016–7023.
43. Simon, J. D.; Su, S.-G. *J. Phys. Chem.* **1988**, *92*, 2395–2397.
44. Karim, O.; Haymet, A. D. J.; Banet, M. J.; Simon, J. D. *J. Phys. Chem.* **1988**, *92*, 3391–3394.
45. Su, S.-G.; Simon, J. D. *J. Chem. Phys.* **1988**, *89*, 908–919.
46. Simon, J. D. *Acc. Chem. Res.* **1988**, *21*, 128–134.
47. Su, S.-G.; Simon, J. D. *Chem. Phys. Lett.* **1989**, *158*, 423–428.

basic questions about the electron-transfer process are still numerous and substantial. Long-term goals for understanding electron transfer could lead to the design of truly useful molecular devices for photochemical energy conversion or other energy-channeling processes. Those goals will require us to understand a number of questions in great detail. However, this chapter will not focus on details because the puzzles about very basic questions in electron transfer are so numerous.

The basic "golden rule" expression for nonadiabatic electron-transfer rate processes (1–15) can be used to categorize these puzzles.

$$k_{et} = 2 \left(\frac{\pi}{\hbar^2} \right) H_{rp}^2(r) \text{ FCWD} \tag{1}$$

In this expression, the rate (k_{et}) is a product of an electronic coupling between reactants and products (H_{rp}), which depends strongly on distance (r), on orientation of the donor and acceptor molecules, and on the Franck–Condon weighted density of states (FCWD). In eq 1, $\hbar$ is Planck's constant divided by 2π. The FCWD gives the probability of finding the donor and acceptor groups and the surrounding medium in a nuclear configuration such that an energy level of the product matches that of the reactants. When such a match of energy levels occurs there is a chance, proportional to H_{rp}^2, for the electron to jump to the acceptor. The electronic coupling H_{rp} decays rapidly and exponentially with distance when material between the electron donor and the electron acceptor is classically forbidden to the electron. The exponential attenuation of the rate with distance occurs because the electron transfer is by nature an electron-tunneling process.

The Franck–Condon weighted density of states contains most of the dependence of the rate on free energy change (ΔG^0), temperature, solvent polarity, and changes in the structure of the donor and acceptor molecules when they add or release an electron. Achieving a nuclear configuration in which electron transfer is allowed by energy conservation often involves thermal activation, particularly of the solvent molecules in a polar solvent. It may also involve quantum mechanical motions of nuclei within the donor and acceptor groups themselves. Such an "activated" nuclear configuration achieved by quantum mechanical motions is sometimes referred to as nuclear tunneling.

This chapter will divide the puzzles into those involving electronic coupling, those involving the Franck–Condon part of the problem, and those dependent on both to such an extent that they cannot be classified.

Puzzles of Electronic Coupling

Distance Dependence.   Electron-transfer rates have been measured as a function of distance in rigid glasses (16–21), in difunctional molecules

Puzzles of Electron Transfer

John R. Miller

Chemistry Division, Argonne National Laboratory, Argonne, IL 60439

After much progress several questions remain in our understanding of electron-transfer reactions, including a lack of clear understanding of the factors controlling the dependence of rate on distance. Distance dependence of long-distance electron-transfer rates is similar in a variety of experimental situations, such as rigid glasses, donors and acceptors bound to rigid spacer groups, monolayer assemblies, and proteins. But two puzzling cases in which the distance dependence is dramatically weaker have not been understood and interpreted. An even more serious lack of understanding exists for control of rate by orientation. In some cases rates depend strongly on solvent polarity, but in at least one other case the rates seem to be independent of polarity. Important energetic quantities such as free energy change and solvent reorganization energy are often difficult to obtain and are particularly difficult to predict because of inaccuracies of the dielectric continuum model. This chapter compiles a list of many of these problems, but makes no attempt to suggest the nature of their solutions.

MAJOR PROGRESS HAS DRAMATICALLY CHANGED our concepts of electron-transfer reactions over the last several years by defining the distances, energetics, and even orientations of electron donor and acceptor groups. This chapter will emphasize the aspects of electron transfer (ET) that we understand poorly. Most of the many recent successes of electron transfer are discussed elsewhere in this volume. Other chapters, particularly that of Bolton and Archer, offer an introduction to electron transfer.

Experiment and Theory

Although the past few years have seen explosive growth in our knowledge about electron-transfer processes, the problems or puzzles involving very

0065–2393/91/0228–0265$06.00/0

basic questions about the electron-transfer process are still numerous and substantial. Long-term goals for understanding electron transfer could lead to the design of truly useful molecular devices for photochemical energy conversion or other energy-channeling processes. Those goals will require us to understand a number of questions in great detail. However, this chapter will not focus on details because the puzzles about very basic questions in electron transfer are so numerous.

The basic "golden rule" expression for nonadiabatic electron-transfer rate processes (1–15) can be used to categorize these puzzles.

$$k_{et} = 2 \left(\frac{\pi}{\hbar^2} \right) H_{rp}^2(r) \text{ FCWD} \tag{1}$$

In this expression, the rate (k_{et}) is a product of an electronic coupling between reactants and products (H_{rp}), which depends strongly on distance (r), on orientation of the donor and acceptor molecules, and on the Franck–Condon weighted density of states (FCWD). In eq 1, $\hbar$ is Planck's constant divided by 2π. The FCWD gives the probability of finding the donor and acceptor groups and the surrounding medium in a nuclear configuration such that an energy level of the product matches that of the reactants. When such a match of energy levels occurs there is a chance, proportional to H_{rp}^2, for the electron to jump to the acceptor. The electronic coupling H_{rp} decays rapidly and exponentially with distance when material between the electron donor and the electron acceptor is classically forbidden to the electron. The exponential attenuation of the rate with distance occurs because the electron transfer is by nature an electron-tunneling process.

The Franck–Condon weighted density of states contains most of the dependence of the rate on free energy change (ΔG^0), temperature, solvent polarity, and changes in the structure of the donor and acceptor molecules when they add or release an electron. Achieving a nuclear configuration in which electron transfer is allowed by energy conservation often involves thermal activation, particularly of the solvent molecules in a polar solvent. It may also involve quantum mechanical motions of nuclei within the donor and acceptor groups themselves. Such an "activated" nuclear configuration achieved by quantum mechanical motions is sometimes referred to as nuclear tunneling.

This chapter will divide the puzzles into those involving electronic coupling, those involving the Franck–Condon part of the problem, and those dependent on both to such an extent that they cannot be classified.

Puzzles of Electronic Coupling

Distance Dependence.   Electron-transfer rates have been measured as a function of distance in rigid glasses (16–21), in difunctional molecules

4. van der Zwan, G.; Hynes, J. T. *Chem. Phys. Lett.* **1983**, *101*, 367–371.
5. van der Zwan, G.; Hynes, J. T. *J. Phys. Chem.* **1985**, *89*, 4181–4188.
6. Hynes, J. T. *J. Phys. Chem.* **1986**, *90*, 3701–3706.
7. Calef, D. F.; Wolynes, P. G. *J. Phys. Chem.* **1983**, *87*, 3387–3400.
8. Calef, D. F.; Wolynes, P. G. *J. Chem. Phys.* **1983**, *78*, 470–482.
9. Sumi, H.; Marcus, R. A. *J. Chem. Phys.* **1986**, *84*, 4272–4276.
10. Sumi, H.; Marcus, R. A. *J. Chem. Phys.* **1986**, *84*, 4894–4914.
11. Nadler, W.; Marcus, R. A. *J. Chem. Phys.* **1987**, *86*, 3906–3924.
12. Bayliss, N. S.; McRae, E. G. *J. Phys. Chem.* **1954**, *58*, 1813–1822.
13. Ooshika, Y. *J. Phys. Soc. Jpn.* **1954**, *9*, 594–602.
14. Lippert, E. *Z. Naturforsch.* **1955**, *10a*, 541–545.
15. Mataga, N.; Kaifu, Y.; Koizumi, M. *Bull. Chem. Soc. Jpn.* **1956**, *29*, 465–470.
16. McRae, E. G. *J. Phys. Chem.* **1957**, *61*, 562–572.
17. Lippert, E. *Z. Elektrochem.* **1957**, *61*, 962–975.
18. Mataga, N. *Bull. Chem. Soc. Jpn.* **1963**, *36*, 654–662.
19. Bakhshiev, N. G. *Opt. Spect. (Engl. Transl.)* **1964**, *16*, 446–451.
20. Bakhshiev, N. G.; Piterskaya, I. V. *Opt. Spect. (Engl. Transl.)* **1965**, *19*, 390–395.
21. Taft, R. W.; Pienta, N. J.; Kamlet, M. J.; Arnett, E. M. *J. Org. Chem.* **1981**, *46*, 661–667.
22. Brady, J. E.; Carr, P. W. *J. Phys. Chem.* **1985**, *89*, 5759–5766.
23. Ware, W. R.; Lee, S. K.; Brant, G. J.; Chow, P. P. *J. Chem. Phys.* **1971**, *54*, 4729–4737.
24. Chakrabat, S. K.; Ware, W. R. *J. Chem. Phys.* **1971**, *55*, 5494–5498.
25. Ware, W. R.; Chow, P. P.; Lee, S. K. *Chem. Phys. Lett.* **1986**, *2*, 356–358.
26. Mazurenko, Y. T. *Opt. Spect. (Engl. Transl.)* **1974**, *36*, 283–286.
27. Mazurenko, Y. T.; Uldaltsov, V. S. *Opt. Spect. (Engl. Transl.)* **1978**, *44*, 417–420.
28. Mazurenko, Y. T.; Uldaltsov, V. S. *Opt. Spect. (Engl. Transl.)* **1978**, *45*, 141–145.
29. Mazurenko, Y. T.; Uldaltsov, V. S. *Opt. Spect. (Engl. Transl.)* **1978**, *45*, 765–767.
30. Mazurenko, Y. T. *Opt. Spect. (Engl. Transl.)* **1980**, *48*, 388–392.
31. Bagchi, B.; Oxtoby, D. W.; Fleming, G. R. *Chem. Phys.* **1984**, *86*, 257–267.
32. Castner, E. W.; Maroncelli, M.; Fleming, G. R. *J. Chem. Phys.* **1987**, *86*, 1090–1097.
33. Maroncelli, M.; Fleming, G. R. *J. Chem. Phys.* **1987**, *86*, 6221–6239.
34. Castner, E. W.; Fleming, G. R.; Bagchi, B. *Chem. Phys. Lett.* **1988**, *143*, 270–276.
35. Maroncelli, M.; Fleming, G. R. *J. Chem. Phys.* **1988**, *89*, 875–881.
36. Castner, E. W., Jr.; Bagchi, B.; Maroncelli, M.; Webb, S. P.; Ruggiero, A. J.; Fleming, G. R. *Ber. Bunsen-Ges. Phys. Chem.* **1988**, *92*, 363–372.
37. Castner, E. W.; Fleming, G. R.; Bagchi, B.; Maroncelli, M. *J. Chem. Phys.* **1988**, *89*, 3519–3534.
38. Maroncelli, M.; MacInnis, J.; Fleming, G. R. *Science (Washington, D.C.)* **1989**, *243*, 1674–1681.
39. Kitamura, N.; Kim, H.-B.; Kawanishi, Y.; Obata, R.; Tazuke, S. *J. Phys. Chem.* **1986**, *90*, 1488–1491.
40. Su, S.-G.; Simon, J. D. *J. Phys. Chem.* **1986**, *90*, 6475–6479.
41. Su, S.-G.; Simon, J. D. *J. Phys. Chem.* **1987**, *91*, 2693–2696.
42. Simon, J. D.; Su, S.-G. *J. Chem. Phys.* **1987**, *87*, 7016–7023.
43. Simon, J. D.; Su, S.-G. *J. Phys. Chem.* **1988**, *92*, 2395–2397.
44. Karim, O.; Haymet, A. D. J.; Banet, M. J.; Simon, J. D. *J. Phys. Chem.* **1988**, *92*, 3391–3394.
45. Su, S.-G.; Simon, J. D. *J. Chem. Phys.* **1988**, *89*, 908–919.
46. Simon, J. D. *Acc. Chem. Res.* **1988**, *21*, 128–134.
47. Su, S.-G.; Simon, J. D. *Chem. Phys. Lett.* **1989**, *158*, 423–428.

48. Kahlow, M. A.; Kang, T. J.; Barbara, P. F. *J. Phys. Chem.* **1987**, *91*, 6452–6455.
49. Nagarajan, V.; Brearly, A. M.; Kang, T. J.; Barbara, P. F. *J. Phys. Chem.* **1987**, *86*, 3183–3196.
50. Kahlow, M. A.; Kang, T. J.; Barbara, P. F. *J. Chem. Phys.* **1988**, *88*, 2372–2378.
51. Kang, T. J.; Kahlow, M. A.; Giser, D.; Swallen, S.; Nagarajan, V.; Wlodzimierz, J.; Barbara, P. F. *J. Phys. Chem.* **1988**, *92*, 6800–6807.
52. Wlodzimierz, J.; Walker, G. C.; Johnson, A. E.; Kahlow, M. A.; Barbara, P. F. *J. Phys. Chem.* **1988**, *92*, 7039–7041.
53. Kahlow, M. A.; Wlodzimierz, J.; Kang, T. J.; Barbara, P. F. *J. Chem. Phys.* **1989**, *90*, 151–158.
54. Kinoshita, S.; Nishi, N.; Kushida, T. *Chem. Phys. Lett.* **1987**, *134*, 605–609.
55. Kinoshita, S.; Nishi, N. *J. Chem. Phys.* **1988**, *89*, 6612–6622.
56. Demas, J. N.; Turner, T. F.; Crosby, G. A. *Inorg. Chem.* **1969**, *8*, 674–675.
57. Schilt, A. S. *Inorg. Chem.* **1964**, *3*, 1323–1325.
58. Hopkins, M. D.; Gray, H. B. *J. Am. Chem. Soc.* **1984**, *106*, 2468–2469.
59. Kozik, M.; Sutin, N.; Winkler, J. R., unpublished results.
60. Mielenz, K. D.; Cehelnik, E. D.; McKenzie, R. L. *J. Chem. Phys.* **1976**, *64*, 370–374.
61. O'Connor, D. V.; Phillips, D. *Time-Correlated Single Photon Counting*; Academic Press: Orlando, FL, 1984; pp 252–283.
62. Cooper, J. B.; Wertz, D. W. *Inorg. Chem.* **1989**, *28*, 3108–3113.
63. Cowman, C. D.; Gray, H. B. *J. Am. Chem. Soc.* **1977**, *95*, 8177–8178.
64. Hay, P. J. *J. Am. Chem. Soc.* **1982**, *104*, 7007–7017.
65. Brunschwig, B. S.; Ehrenson, S.; Sutin, N. *J. Phys. Chem.* **1987**, *91*, 4714–4723.
66. Brunschwig, B. S.; Ehrenson, S.; Sutin, N. *J. Phys. Chem.* **1986**, *90*, 3657–3668.
67. Riddick, J. A.; Bunger, W. B.; Sakano, T. K. *Organic Solvents. Physical Properties and Methods of Purification*; Wiley: New York, 1986.
68. Bagchi, B.; Chandra, A. *Chem. Phys. Lett.* **1989**, *155*, 533–538.
69. Bagchi, B.; Chandra, A. *J. Chem. Phys.* **1989**, *90*, 7338–7345.
70. Frölich, H. *Theory of Dielectrics. Dielectric Constant and Dielectric Loss*; Oxford University Press: London, 1958; pp 71–72.
71. Friedman, H. L. *J. Chem. Soc., Faraday Trans. 2* **1983**, *79*, 1465.
72. Kivelson, D.; Friedman, H. L. *J. Phys. Chem.* **1989**, *93*, 7026–7031.
73. Garg, S. K.; Smyth, C. P. *J. Phys. Chem.* **1965**, *69*, 1294–1301.
74. Poley, J. P. *Appl. Sci. Res., Sect. B* **1955**, *4*, 337.
75. Davies, G. J.; Evans, G. J.; Evans, M. W. *J. Chem. Soc., Faraday Trans. 2* **1979**, *75*, 1428–1441.
76. Madan, M. P. *Z. Phys. Chem. (Frankfurt am Main)* **1974**, *89*, 259–264.
77. Oh, D. H.; Boxer, S. *J. Am. Chem. Soc.* **1989**, *111*, 1130–1131.
78. Lax, M. *J. Chem. Phys.* **1952**, *20*, 1752–1760.
79. Ansbacher, F. *Z. Naturforsch.* **1959**, *14a*, 889–892.
80. Ballhausen, C. J. *Molecular Electronic Structures of Transition Metal Complexes*; McGraw-Hill: New York, 1979; pp 125–128.
81. Belford, G. G.; Belford, R. L.; Weber, G. *Proc. Natl. Acad. Sci. U.S.A.* **1972**, *69*, 1392–1393.

Received for review April 27, 1990. Accepted revised manuscript October 30, 1990.

in which a relatively rigid spacer group holds the donor and acceptor apart (*22–34*), in monolayer assemblies (*34–39*), and in proteins (*40–46*). Where it has been possible to accurately measure dependence on distance, it has always been found that electron-transfer rates decayed exponentially because of the exponential decrease of electronic coupling with distance.

$$H_{rp}^2(r) \; = \; H_{rp}^2(r \, = \, 0) \, \exp \, (-\alpha(r \, - \, r_0)) \tag{2}$$

Major questions about the distance dependence remain. Does the attenuation parameter change α substantially as the nature of the material between the donor and acceptor group changes? Are there other important factors influencing α? What, precisely, are the values of α, which can be quite important at long distances because of the exponential nature of the process? Great care is required in the measurement of α because distance dependence of electron-transfer rates can also result from distance-dependent factors in FCWD. Solvent reorganization energy is distance-dependent, and the FCWD tends to be more strongly distance-dependent for weakly exoergic reactions, which are the easiest to measure (*17*). Apparent very large values of α have been quoted incorrectly (*47*), when the actual dependence of the rate predominantly reflected the Franck–Condon terms. This error occurred in some work on glasses and in early work on proteins, where it seemed that the rates were decaying unusually rapidly with distance. This dependence was probably due mainly to Franck–Condon effects.

Values of α near 1.0 Å^{-1} have usually been found in rigid glasses, with values slightly smaller for intramolecular electron transfer between groups connected by rigid saturated spacer groups. The more efficient transmission of intramolecular electronic coupling appears to result from the slightly greater efficiency of through-bond interactions, as compared to interactions that must pass from one molecule to another through regions in which there are no chemical bonds. Electronic coupling appears to pass much more efficiently through conjugated π systems between the donor and acceptor groups (*28–31, 48–50*), although it has not yet been possible to characterize these with a value for α. A wide range of data for glasses, intramolecular electron transfer, electron transfer through proteins, and some data for monolayer assemblies appear to be accommodated by values of α between 0.6 and 1.2 Å^{-1} when the material between the donor and acceptor is mainly saturated.

There are, however, two impressive anomalies in which very efficient transfer of electronic coupling has been reported to occur over long distances (small α). One of these cases involves electron transfer from photoexcited molecules to electron acceptors across a monolayer assembly. Moebius (*39*) reported data that can be interpreted to give an α approximately 3 times smaller than that for most other experiments with saturated spacer material. Although this experiment is approximately 10 years old, it has never become completely clear whether all sources of experimental artifacts have been removed or whether some special feature of these assemblies leads to a very weak attenuation of the electronic coupling with distance.

Another extremely interesting example is electron-transfer fluorescence quenching of metal complexes bound to DNA (51). These experiments are at an early stage; their interpretation is complex and appears to be fraught with a number of difficulties. It is possible, however, to interpret the experiments as providing evidence for long-distance electron transfer through the DNA backbone with a very slow decay of the electronic coupling.

Another curious question involves "hole tunneling". Experiments in rigid glasses have shown that positive charge can be transferred over distances similar to those in which electrons can be transferred (19, 52–54). In our laboratory we measured distance dependence for such processes and concluded that these hole-transfer processes provided excellent evidence for the superexchange picture first advanced for ET by McConnell (55), but that consideration of the energetics also required an additional type of superexchange in which a "hole" rather than an electron was the tunneling quasiparticle (19). More recent experiments on intramolecular ET have demonstrated that these two processes can have almost exactly the same dependence on distance (25). It is not clear whether this was an accident or whether the factors affecting the two can indeed be almost identical, so that in general, at least for saturated hydrocarbon spacers, we can expect hole tunneling and electron tunneling to have similar distance dependence. Superexchange can also involve π mediators (28–31, 49, 50, 56).

Orientation Dependence.   For intramolecular electron transfer in difunctional molecules involving rigid spacers, it has recently become possible to measure the effects of the spatial orientation of donor, spacer, and acceptor groups on electron-transfer rates. Orientation effects are of great theoretical, and possibly practical, interest; the orientation could become a powerful control tool for directing electron-transfer processes and discriminating against undesirable ET paths. For many ET reactions, the electronic coupling, H_{rp}, can be either positive or negative, depending on angles. Therefore there are, in principle, angles at which the coupling can be zero. At such angles the electron-transfer rate could become exceedingly small. The most striking example of orientation dependence has come from experiments in McLendon's laboratory, in which electron transfer occurs between two porphyrin molecules held at a series of angles with respect to one another (57, 58). Although there may be more than one possible interpretation, a satisfactory one has been advanced by Closs (59).

In two other cases, however, interpretations of the effects of orientation are not available. One of these is a porphyrin–quinone (P–Q) molecule separated by two spacers giving different orientations between the P and Q groups (60). In the other case, biphenyl and naphthalene groups attached to decalins and cyclohexane spacers show sizable (about a factor of 4) orientation effects on the rates (*61*). At present no theory appears able to explain these results.

Puzzles About Franck–Condon Effects

Rehm and Weller (*62*) focused attention on the question of an "inverted region" predicted by the Marcus theory (*63, 64*) and other electron-transfer theories. There was a decade of confusion and controversy about the nature of the dependence of electron-transfer rates on the free energy change (ΔG^0). More recently, in a series of stunning successes, the inverted region has been clearly demonstrated by studies in a number of different kinds of systems (*16, 17, 23, 26, 27, 45, 61, 65–77*). Recent work has added an exclamation point to the success in this area by showing that the theory and reorganization parameters that describe the dependence of electron-transfer rates on ΔG^0 can provide a quantitative description of the temperature dependence of electron-transfer reactions (*78*). For this aspect of electron transfer, the results of one kind of experiment can be used to predict the results of a different and seemingly quite unrelated kind of experiment. We might conclude that the FCWD is well understood.

But where there are exclamation points, there are also question marks. One very large question mark is raised by results for dependence of rates on ΔG^0 in Wasielewski's laboratory (*65*). Photoexcited charge-separation electron-transfer rates in a series of porphyrin–quinone compounds showed evidence for both a normal and an inverted region. However, when the relationship between rate and ΔG^0 was investigated (*65*) in two solvents, one quite polar (butyronitrile) and the other very nonpolar (toluene), there was no noticeable difference in the relationship between rate and ΔG^0! The absence of a solvent-polarity effect stands strongly in contrast to theory and to the observations of large effects of solvent polarity in charge-shift reactions.

Solvent dependence was also examined in porphyrin–quinone compounds by Bolton and co-workers (*33, 34*) and Mauzerall and co-workers (*79–81*). In those cases charge-separation reactions were studied for which solvent dependence is expected to be weak because the free energy change and the solvent reorganization energy are in parallel by increasing solvent polarity. Therefore, according to dielectric continuum theory, only changes in the refractive indices of the solvents are expected to lead to substantial changes in the rates. The data from the Bolton group illustrate that this theory is not exact.

This question of whether electron-transfer rates depend upon solvent polarity is connected to another one of our major problems: We have a very poor understanding of the thermodynamics of creating ions in solution. In charge-separation or recombination reactions $(D + A) \rightleftarrows (D^+ + A^-)$ such as those of Wasielewski previously cited, to know the free energy change for the process we must know the solvation free energies of the ions on the right side of the the equation. Those solvation energies can be estimated as a function of solvent polarity and size of the ions by use of the Born equation

$$\Delta G_s(D^{\pm}) - \Delta G_s(D) = \frac{-e^2}{r_D}\left(1 - \frac{1}{\epsilon_s}\right) \tag{3}$$

so that the energetics of charge separation and recombination as a function of solvent polarity can be estimated by use of the Weller equation (82), which adds corrections to the donor's oxidation potential ($E_{D/D^+}{}^0$) and the acceptor's reduction potential ($E_{A/A^-}{}^0$).

$$\Delta G_s^0 = e(E_{D/D^+}{}^0 - E_{A/A^-}{}^0) + \frac{e^2}{\epsilon_s}\left(\frac{1}{2r_D} + \frac{1}{2r_A}\right) - \frac{e^2}{\epsilon_s R_{DA}} \tag{4}$$

Here R_{DA} is the donor–acceptor distance (center to center) and r_D and r_A are mean radii of the donor and acceptor groups, respectively; e is the electronic charge; and ϵ_s is the static dielectric constant. Unfortunately, almost no tests exist that really quantitatively assess the validity of these equations, and there are, indeed, doubts about their accuracy. Consequently, the energetics of charge-separation and recombination reactions are not accurately known in most cases. Furthermore, even in charge-shift reactions, if the donor and acceptor groups have substantially different sizes or charge distributions, it may not be possible to accurately assess the change in energetics. For charge-shift reactions, the energetics can often be measured precisely by the measurement of the electron-transfer equilibria, but in many cases it is not practical. Measuring electron-transfer equilibria is rarely possible for charge-separation and recombination reactions, although there are examples in the special case of very stable ions, which can equilibrate with the corresponding neutrals ($D + A \rightleftarrows D^+ + A^-$) (83, 84). Paddon-Row and co-workers (85, 86) also observed equilibration of ions with excited neutrals [$(D + A)^* \rightleftarrows D^+ + A^-$]. The energetics are very sensitive to solvent polarity and are sensitive to distance, so that the equilibrium is seen only at certain distances.

A related problem is the estimation of solvent reorganization energies, λ_s. This estimation can be done by treating the solvent as a dielectric continuum and making the approximation that the donor and acceptor groups are spherical, which leads to eq 5, where ϵ_{op} is the optical dielectric constant.

$$\lambda_s = \left(\frac{e^2}{4\pi\epsilon_0}\right)\left(\frac{1}{2r_D} + \frac{1}{2r_A} - \frac{1}{R_{DA}}\right)\left(\frac{1}{\epsilon_{op}} - \frac{1}{\epsilon_s}\right) \tag{5}$$

More sophisticated elliptical models can also be developed (87–89). There is reason, however, to have serious doubts about these dielectric continuum models. For intramolecular electron transfer in difunctional steroids, the solvent reorganization energies can be obtained from experiments on rate as a function of ΔG^0 and from dependence of rate on temperature. The two systems are in quantitative agreement. However, the reorganization energies obtained from the measurements of rate differ by a factor of almost 2 from the estimates of eq 5 with reasonable crystallographic sizes used for the donor and acceptor groups. There is also grave doubt about the utility of eq 5 in predicting the distance dependence of solvent reorganization energies. This doubt casts uncertainty on the measurement of the distance dependence of electronic couplings α because of the necessity to correct for the distance dependence of reorganization energies.

One of the main events in the study of electron-transfer processes was the measurement as a function of temperature from cytochrome c to chlorophyll dimer cation in photosynthetic reaction centers by Chance et al. (*90*, *91*). That reaction showed a substantial activation energy at room temperature, but became temperature-independent below about 100 K. Theorists pointed out that this interesting temperature dependence could be explained by nuclear tunneling accompanying the reorganization of a molecular vibration (*6*, *14*, *92*). However, this explanation has foundered because a very large reorganization energy is needed in a mode of frequency near 400 cm^{-1}. Such a large reorganization does not occur in either the cytochrome or the chlorophyll dimer. A number of other electron-transfer processes in protein show a similar kind of temperature dependence. Such similarity suggests that this temperature dependence is not specific to the structural features of those reactants. Although all of these puzzles can be fit to theories that require huge reorganization energies in modes, which must be described quantum mechanically, it is not clear that the reorganization parameters assumed are reasonable in any of the cases.

Another major area of confusion in the Franck–Condon control of electron-transfer rates is the reason for the absence of "inverted region" behavior in charge-separation reactions. Although evidence for the inverted region is substantial in charge-shift reactions and charge-recombination reactions, it is almost nonexistent for charge-separation reactions. The difficulty of observing the inverted region for charge-separation reactions may result from the formation of excited states in these reactions. Excited states will be a pervasive problem because two radical ions are created, and radical ions in general have low-lying excited states.

Another intriguing explanation of general importance in electron transfer has been advanced by Kakitani and co-workers (*93–105*). These authors suggest that there are large frequency changes for the motion of solvent molecules around radical ions, as compared to around neutrals. There is wide agreement that this suggestion must be qualitatively correct, but many workers doubt whether the very large effects suggested occur in reality. The area is controversial in the point of view of both theory and experiment.

Evidence recently obtained suggests that the effects Kakitani and co-workers invoke are insignificant. This area is in flux, a substantial open question.

Puzzles at the Adiabatic–Nonadiabatic Borderline

Equation 1 predicts that electron-transfer rates increase toward infinity as the electronic coupling matrix element increases with decreasing distance between the donor and acceptor groups. Equation 1 is, however, a nonadiabatic equation. The increase of rate with increasing electronic coupling is expected to saturate at a large H_{rp}, typically somewhere near k_bT (the Boltzmann constant times absolute temperature), as the electron-transfer process becomes adiabatic (the chapter by Bolton and Archer presents a discussion of terms, adiabatic and nonadiabatic). Some evidence for the saturation was obtained years ago by experiments of Richardson and Taube (106, 107), who compared slow electron-transfer rates for reactions that had extremely large reorganization energies with the intensities of optical electron-transfer bands. Their data remain the best example of the transition to adiabatic behavior. However, the data do not thoroughly and quantitatively characterize the transition from nonadiabatic to adiabatic behavior.

One interesting question about this saturation is as follows: What is the maximum possible value of an electron-transfer rate? Classical transition-state approaches (63, 64, 108, 109) would predict that electron-transfer rates would saturate at about 10^{13} s^{-1}. However, inclusion of reorganization of high-frequency skeletal vibrations of the donor and acceptor groups and/or fast librational solvent realization mechanisms may lead to the prediction that substantially higher rates are possible because of the high frequencies of these quantum vibrations.

An important, interesting, and closely related question is control of electron-transfer rates by dynamics of the solvent molecules when the rate of the electron transfer becomes competitive with solvent motions. Experimental evidence has been obtained for such solvent dynamic control, but mainly in a case where substantial structural change of the electron donor–acceptor molecules is required in order for the electron transfer to occur (110–119). Experimental examinations of solvent dynamic control (110–122) have not yet fully examined its relationship to high-frequency vibrations or examined the role of free energy, but there are indications that rates may be obtained that are unexpectedly fast according to the classical theory (122).

Acknowledgment

This work was performed under the auspices of the Office of Basic Energy Sciences, Division of Chemical Science, U.S. Department of Energy under contract number W–31–109–ENG–38.

References

1. Ulstrup, J.; Jortner, J. *J. Chem. Phys.* **1975,** *63,* 4358.
2. Van Duyne, R. P.; Fischer, S. F. *Chem. Phys.* **1974,** 5, 183.
3. Ulstrup, J. *Charge Transfer Processes in Condensed Media*; Springer-Verlag: Berlin, 1979; p 419.
4. Levich, V. O. *Adv. Electrochem. Electrochem. Eng.* **1966,** *4,* 249.
5. Kuznetsov, A. M.; Ulstrup, J. *J. Chem. Phys.* **1981,** *75,* 2047.
6. Jortner, J. *J. Chem. Phys.* **1976,** *64,* 4860.
7. Grigorov, L. N.; Chernavskii, D. S. *Biophysics (Engl. Transl.)* **1972,** *17,* 202.
8. Dogonadze, R. R. In *Reactions of Molecules at Electrodes*; Hush, N. S., Ed.; Wiley-Interscience: New York, 1971; p 135.
9. Dogonadze, R. R.; Kuznetsov, A. M.; Vorotyntsev, M. A. *Z. Phys. Chem. (Frankfurt am Main)* **1976,** *100,* 1.
10. Dogonadze, R. R.; Kuznetsov, A. M.; Vorotyntsev, M. A.; Zakaroya, M. G. *J. Electroanal. Chem. Interfacial Electrochem.* **1977,** *75,* 315.
11. Newton, M. D. *Int. J. Quantum Chem., Quantum Chem. Symp.* **1980,** *14,* 363.
12. Newton, M. D.; Sutin, N. *Rev. Phys. Chem.* **1984,** *35,* 437.
13. Hush, N. S. *Trans. Faraday Soc.* **1961,** *57,* 577.
14. Hopfield, J. J. *Proc. Natl. Acad. Sci. U.S.A.* **1974,** *71,* 3640.
15. Marcus, R. A.; Sutin, N. *Biochim. Biophys. Acta* **1985,** *811,* 265.
16. Beitz, J. V.; Miller, J. R. *J. Chem. Phys.* **1979,** *71,* 4579.
17. Miller, J. R.; Beitz, J. V.; Huddleston, R. K. *J. Am. Chem. Soc.* **1984,** *106,* 5057–5068.
18. Miller, J. R.; Hartman, K. W.; Abrash, S. *J. Am. Chem. Soc.* **1982,** *104,* 4296–4298.
19. Miller, J. R.; Beitz, J. V. *J. Chem. Phys.* **1981,** *74,* 6746–6757.
20. Kira, A.; Nosaka, Y.; Imamura, M.; Ichikawa, T. *J. Phys. Chem.* **1982,** *86,* 1866–1868.
21. Kira, A.; Imamura, M.; Tagawa, S.; Tabata, Y. *Bull. Chem. Soc. Jpn.* **1986,** *59,* 593–597.
22. Calcaterra, L. T.; Closs, G. L.; Miller, J. R. *J. Am. Chem. Soc.* **1983,** *105,* 670–671.
23. Miller, J. R.; Calcaterra, L. T.; Closs, G. L. *J. Am. Chem. Soc.* **1984,** *106,* 3047.
24. Huddleston, R. K.; Miller, J. R. *J. Chem. Phys.* **1983,** *79,* 5337–5344.
25. Johnson, M. D.; Miller, J. R.; Green, N. S.; Closs, G. L. *J. Phys. Chem.* **1989,** *93,* 1173–1176.
26. Harrison, R. J.; Pearce, B.; Beddard, G. S.; Cowan, J. A.; Sanders, J. K. M. *Chem. Phys.* **1987,** *116,* 429.
27. Irvine, M. B.; Harrison, R. J.; Beddard, G. S.; Leighton, P.; Sanders, J. K. M. *Chem. Phys.* **1986,** *104,* 315.
28. Heitele, H.; Michel-Beyerle, M. E. *J. Am. Chem. Soc.* **1985,** *107,* 8286–8288.
29. Heitele, H.; Michel-Beyerle, M. E.; Finckh, P. *Chem. Phys. Lett.* **1987,** *138,* 237–243.
30. Heitele, H.; Michel-Beyerle, M. E.; Finckh, P. *Chem. Phys. Lett.* **1987,** *134,* 273–278.
31. Heitele, H.; Finckh, P.; Weeren, S.; Poellinger, F.; Michel-Beyerle, M. E. *J. Phys. Chem.* **1989,** *93,* 5173–5179.
32. Connolly, J. S.; Bolton, J. R. In *Photoinduced Electron Transfer. Part D: Inorganic Substrates and Applications*; Fox, M. A.; Chanon, M., Eds.; Elsevier: Amsterdam, 1988; pp 303–393.

33. Archer, M. D.; Gadzekpo, V. P. Y.; Schmidt, J. A.; Liu, J.-Y.; Bolton, J. R. *Proc. Electrochem. Soc.* **1988**, *88*, 417–427.
34. Schmidt, J. A.; Liu, J.-Y.; Bolton, J. R.; Archer, M. D.; Gadzekpo, P. Y. *J. Chem. Soc., Faraday Trans. 1* **1989**, *85*, 1027–1041.
35. Seefeld, K.-P.; Moebius, D.; Kuhn, H. *Helv. Chim. Acta* **1977**, *60*, 2608–2632.
36. Kuhn, H. *Pure Appl. Chem.* **1979**, *51*, 341–352.
37. Kuhn, H. *Pure Appl. Chem.* **1981**, *53*, 2105–2122.
38. Moebius, D. *Ber. Bunsen-Ges. Phys. Chem.* **1978**, *82*, 867.
39. Moebius, D. *Acc. Chem. Res.* **1981**, *14*, 63–68.
40. Meade, T. J.; Gray, H. B.; Winkler, J. R. *J. Am. Chem. Soc.* **1989**, *111*, 4353.
41. Petersen-Kennedy, S. E.; McGourty, J. L.; Ho, P. S.; Sutoris, C. J.; Liang, N.; Zemel, H.; Blough, N. V.; Margoliash, E.; Hoffman, B. M. *Coord. Chem. Rev.* **1985**, *64*, 125–133.
42. Liang, N.; Kang, C. H.; Ho, P. S.; Margoliash, E.; Hoffman, B. M. *J. Am. Chem. Soc.* **1986**, *108*, 6468–6470.
43. McLendon, G.; Miller, J. R. *J. Am. Chem. Soc.* **1985**, *107*, 7811.
44. McLendon, G.; Miller, J. R.; Simolo, K.; Taylor, K.; Mauk, A. G.; English, A. M. In *Excited States and Reactive Intermediates: Photochemistry, Photophysics, and Electrochemistry*; Lever, A. B. P., Ed.; ACS Symposium Series 307; American Chemical Society: Washington, DC, 1986; pp 150–165.
45. McLendon, G.; Miller, J. R. *J. Am. Chem. Soc.* **1985**, *107*, 7811–7816.
46. Cheung, E.; Taylor, K.; Kornblatt, J. A.; English, A. M.; McLendon, G.; Miller, J. R. *Proc. Natl. Acad. Sci. U.S.A.* **1986**, *83*, 1330–1333.
47. Aleksandrov, I. V.; Khairutdinov, R. F.; Zamaraev, K. I. *Chem. Phys.* **1978**, *32*, 123.
48. Aleksandrov, I. V.; Khairutdinov, R. F.; Zamaraev, K. I. *Dokl. Akad. Nauk SSSR* **1978**, *241*, 119.
49. Finckh, P.; Heitele, H.; Volk, M.; Michel-Beyerle, M. E. *J. Phys. Chem.* **1988**, *92*, 6584–6590.
50. Finckh, P.; Heitele, H.; Michel-Beyerle, M. E. *Chem. Phys.* **1989**, *138*, 1–10.
51. Barton, J., Columbia University, private communication, 1989.
52. Kira, A.; Nakamura, T.; Imamura, M. *J. Phys. Chem.* **1978**, *82*, 1961–1965.
53. Kira, A.; Nosaka, Y.; Imamura, M. *J. Phys. Chem.* **1980**, *84*, 1882–1886.
54. Kira, A.; Imamura, M. *J. Phys. Chem.* **1984**, *88*, 1865–1871.
55. McConnell, H. M. *J. Chem. Phys.* **1961**, *35*, 508 and 515.
56. Wasielewski, M. R.; Niemczyk, M. P.; Johnson, D. G.; Svec, W. A.; Minsek, D. W. *Tetrahedron* **1989**, *45*, 4785–4806.
57. Heiler, D.; McLendon, G.; Rogalsky, P. *J. Am. Chem. Soc.* **1987**, *109*, 604–606.
58. McLendon, G., University of Rochester, private communication, 1989.
59. Closs, G. L., University of Chicago, private communication, 1989.
60. Sakata, Y.; Nakashima, S.; Goto, Y.; Tatemitsu, H.; Misumi, S. *J. Am. Chem. Soc.* **1989**, *111*, 8978.
61. Closs, G. L.; Calcaterra, L. T.; Green, N. J.; Penfield, K. W.; Miller, J. R. *J. Phys. Chem.* **1986**, *90*, 3673–3683.
62. Rehm, D.; Weller, A. *Isr. J. Chem.* **1970**, *8*, 259–271.
63. Marcus, R. A. *J. Chem. Phys.* **1956**, *24*, 966.
64. Marcus, R. A. *J. Chem. Phys.* **1965**, *43*, 58.
65. Wasielewski, M. R.; Niemczyk, M. P.; Svec, W. A.; Pewitt, E. B. *J. Am. Chem. Soc.* **1985**, *107*, 1080.
66. Johnson, S. G.; Small, G. J.; Johnson, D. G.; Svec, W. A.; Wasielewski, M. R. *J. Phys. Chem.* **1989**, *93*, 5437–5444.
67. Gould, I. R.; Ege, D.; Mattes, S. L.; Farid, S. *J. Am. Chem. Soc.* **1987**, *109*, 3794.

68. Gould, I. R.; Moody, R.; Farid, S. *J. Am. Chem. Soc.* **1988**, *110*, 7242.
69. Gould, I. R.; Moser, J. E.; Armitage, B.; Farid, S.; Goodman, J. L.; Herman, M. S. *J. Am. Chem. Soc.* **1989**, *111*, 1917.
70. Yoshimori, A.; Kakitani, T.; Enomoto, Y.; Mataga, N. *J. Phys. Chem.* **1989**, *93*, 8316–8323.
71. Asahi, T.; Mataga, N. *J. Phys. Chem.* **1989**, *93*, 6575–6578.
72. Hirata, Y.; Mataga, N. *J. Phys. Chem.* **1989**, *93*, 7539–7542.
73. Kobashi, H.; Funabashi, M.; Shizuka, H.; Okada, T.; Mataga, N. *Chem. Phys. Lett.* **1989**, *160*, 261–266.
74. Mataga, N. *Denki Kagaku* **1989**, *57*, 1025–1030.
75. Mataga, N. In *Photochemical Energy Conversion*; Proceedings of the 7th International Conference on Photochemical Conversion and Storage of Solar Energy, Evanston, IL, July 31–Aug 5, 1988; Norris, J. R., Jr.; Meisel, D., Eds.; Elsevier: New York, 1989; pp 32–46.
76. Nakatani, K.; Okada, T.; Mataga, N.; De Schryver, F. C. *Chem. Phys.* **1988**, *121*, 87–92.
77. Ohno, T.; Yoshimura, A.; Mataga, N.; Tazuke, S.; Kawanishi, Y.; Kitamura, N. *J. Phys. Chem.* **1989**, *93*, 3546–3551.
78. Liang, N.; Miller, J. R.; Closs, G. L. *J. Am. Chem. Soc.* **1989**, *111*, 8740–8741.
79. Lindsey, J. S.; Delaney, J. K.; Mauzerall, D. C.; Linschitz, H. *J. Am. Chem. Soc.* **1988**, *110*, 3610–3621.
80. Mauzerall, D.; Weiser, J.; Staab, H. *Tetrahedron* **1989**, *45*, 4807–4814.
81. Delaney, J. K.; Mauzerall, D. C.; Lindsey, J. S. *J. Am. Chem. Soc.* **1990**, *112*, 957–963.
82. Weller, A. *Z. Phys. Chem.* **1982**, *133*, 93.
83. Yamagishi, A.; Watanabe, F.; Masui, T. *J. Chem. Soc., Chem. Commun.* **1977**, *1977*, 273.
84. Iida, Y. *Bull. Chem. Soc. Jpn.* **1980**, *53*, 2673–2674.
85. Paddon-Row, M. N.; Oliver, A. M.; Warman, J. M.; Smit, K. J.; de Haas, M. P.; Oevering, H.; Verhoeven, J. W. *J. Phys. Chem.* **1988**, *92*, 6958–6962.
86. Smit, K. J.; Warman, J. M.; de Haas, M. P.; Paddon-Row, M. N.; Oliver, A. M. *Chem. Phys. Lett.* **1988**, *152*, 177–182.
87. Cannon, R. D. *Chem. Phys. Lett.* **1977**, *49*, 299–304.
88. Brunschwig, B. S.; Ehrenson, S.; Sutin, N. *J. Phys. Chem.* **1986**, *90*, 3657–3668.
89. Brunschwig, B. S.; Ehrenson, S.; Sutin, N. *J. Phys. Chem.* **1987**, *91*, 4714–4723.
90. Chance, B.; DeVault, D.; Legallais, V.; Yonetani, T. In *Nobel Symposium No. 5; Fast Reactions and Primary Processes in Chemical Kinetics*; Claesson, S., Ed.; Interscience: New York, 1967; pp 437–468.
91. Chance, B.; De Vault, D.; Frauenfelder, H.; Marcus, R. A.; Schrieffer, J.; Sutin, N. In *Tunneling in Biological Systems*; Academic Press: New York, 1979.
92. Blumenfeld, L. A.; Chernavskii, D. S. *J. Theor. Biol.* **1973**, *39*, 1.
93. Kakitani, T.; Mataga, N. *J. Phys. Chem.* **1985**, *89*, 4752–4757.
94. Kakitani, T.; Mataga, N. *Chem. Phys.* **1985**, *93*, 381–397.
95. Kakitani, T.; Mataga, N. *J. Phys. Chem.* **1985**, *89*, 8–10.
96. Kakitani, T.; Mataga, N. *J. Phys. Chem.* **1986**, *90*, 993–995.
97. Kakitani, T.; Mataga, N. *J. Phys. Chem.* **1987**, *91*, 6277–6285.
98. Kakitani, T. *Primary Processes in Photobiology*; Springer Proc. Phys. 20; Springer-Verlag: Berlin, New York, 1987; pp 14–22.
99. Kakitani, T.; Mataga, N. *J. Phys. Chem.* **1988**, *92*, 5059–5068.
100. Kakitani, T. *J. Lumin.* **1988**, *40–41*, 43–46.
101. Mataga, N.; Asahi, T.; Kanda, Y.; Okada, T.; Kakitani, T. *Chem. Phys.* **1988**, *127*, 249–261.

102. Mataga, N.; Kanda, Y.; Asahi, T.; Miyasaka, H.; Okada, T.; Kakitani, T. *Chem. Phys.* **1988**, *127*, 239–248.
103. Yoshimori, A.; Kakitani, T.; Enomoto, Y.; Mataga, N. *J. Phys. Chem.* **1989**, *93*, 8316–8323.
104. Sarai, A.; Kakitani, T. *Chem. Phys. Lett.* **1981**, *77*, 427–432.
105. Kakitani, T. *Bunko Kenkyu* **1986**, *35*, 365–384.
106. Richardson, D. E.; Taube, H. *J. Am. Chem. Soc.* **1983**, *105*, 40–51.
107. Richardson, D. E.; Taube, H. *Coord. Chem. Rev.* **1984**, *60*, 107–129.
108. Marcus, R. A. *Can. J. Chem.* **1959**, *37*, 155.
109. Marcus, R. A. In *Special Topics in Electrochemistry*; Rock, P. A., Ed.; Elsevier: New York, 1970; p 180.
110. Huppert, D.; Kanety, H.; Kosower, E. M. *Faraday Discuss. Chem. Soc.* **1982**, *74*, 161–175.
111. Kosower, E. M.; Kanety, H.; Dodiuk, H.; Striker, G.; Jovin, T.; Boni, H.; Huppert, D. *J. Phys. Chem.* **1983**, *87*, 2479–2484.
112. Kosower, E. M.; Huppert, D. *Chem. Phys. Lett.* **1983**, *96*, 433–435.
113. Huppert, D.; Ittah, V.; Kosower, E. M. *Chem. Phys. Lett.* **1988**, *144*, 15–23.
114. Huppert, D.; Ittah, V.; Masad, A.; Kosower, E. M. *Chem. Phys. Lett.* **1988**, *150*, 349–356.
115. Huppert, D.; Ittah, V.; Kosower, E. M. *Chem. Phys. Lett.* **1989**, *159*, 267–275.
116. Masad, A.; Huppert, D.; Kosower, E. M. *Chem. Phys.* **1990**, *144*, 391–400.
117. Simon, J. D.; Su, S.-G. *J. Chem. Phys.* **1987**, *87*, 7016–7023.
118. Simon, J. D.; Su, S.-G. *J. Phys. Chem.* **1988**, *92*, 2395–2397.
119. Simon, J. D.; Su, S.-G. *J. Phys. Chem.* **1990**, *94*, 3656–3660.
120. Kahlow, M. A.; Kang, T. J.; Barbara, P. F. *J. Phys. Chem.* **1987**, *91*, 6452–6455.
121. Kang, T. J.; Kahlow, M. A.; Giser, D.; Swallen, S.; Nagarajan, V.; Jarzeba, W.; Barbara, P. F. *J. Phys. Chem.* **1988**, *92*, 6800–6807.
122. Weaver, M. J.; McManis, G. E.; Jarzeba, W.; Barbara, P. F. *J. Phys. Chem.* **1990**, *94*, 1715–1719.

Received for review April 27, 1990. Accepted revised manuscript September 17, 1990.

Epilogue

R. A. Marcus

Arthur Amos Noyes Laboratory of Chemical Physics, California Institute
of Technology, Pasadena, CA 91125

THE FIELD OF ELECTRON-TRANSFER REACTIONS has expanded dramatically
since the early days in the late 1940s and the 1950s when the rates of many
such reactions (isotopic-exchange reactions) were studied by using isotopic
labeling techniques. The chapters in this volume demonstrate some of the
more recent developments. They speak eloquently for themselves, and my
summary will be relatively brief.

The excellent prefactory chapters by Bolton et al. and by Bolton and
Archer introduce the electron-transfer field. The latter surveys key concepts
underlying the theory for these reactions, together with relevant equations
used in comparisons with experimental results.

Among the topics treated in some detail in the symposium and in this
volume are

1. factors influencing the effect of donor–acceptor separation distance on
   electron transfer (ET) reaction rates, the distance influencing the rate
   constant via both electronic and reorganizational factors,

2. electron transfers in proteins, including photosynthetic systems,

3. the "inverted effect" for ET rates,

4. comparisons of photoinduced charge separation and charge recombination,

5. solvent and/or temperature and molecular bridge effects on ET rates,

6. charge-transfer states in porphyrin–chlorophyll systems and their enhancement of the quenching of locally excited states in polar solvent by
   mixing with CT states, the role of perpendicularity in favoring the CT
   state,

0065–2393/91/0228–0277$06.00/0
© 1991 American Chemical Society

7. solvent dynamics and intramolecular ET,

8. the effect of applied electric fields on the long-range $BChl_2^+ \ Q^-$ recombination in a photosynthetic system, and

9. intramolecular transfers involving triplets.

This volume thus embraces a broad range of topics in the electron-transfer field. It is, of course, not possible to be all-inclusive in a relatively small publication. There are a number of other active areas in the electron-transfer field: electron transfers at metal–liquid, semiconductor–liquid, and liquid–liquid interfaces; ET on semiconductor colloidal particles, micelles, and at modified metal electrodes; ET with bond rupture; salt effects on ET rates; computer simulations of reorganizational and dynamical aspects of electron transfers; and the detailed relation to charge transfer and photo-electric emission spectra. Several recent results from our laboratory on electron transfer in liquid–liquid and semiconductor–liquid systems and on electronic matrix elements in donor–acceptor rigid molecular bridge and protein systems are described elsewhere (1, 2).

The widespread growth into new areas may strike many observers of the electron-transfer field. New problems, new questions, and challenges continue to arise and are being addressed, both experimentally and theoretically. Systems of increasing complexity, or systems at a greater level of molecular detail, are being examined.

In the early days of modern electron-transfer study (namely, in the few decades after 1945) a primary focus was delineation of the main features of simple inorganic electron transfers, the impact of the standard free energy of reaction on their rate, and the effect of an "intrinsic reorganizational parameter" related to bond-length changes and to molecular size. Such concepts as the extent of adiabaticity vs. nonadiabaticity were also important, as was the relationship between homogeneous and heterogeneous electron-transfer rates. Now, with the aid of the information and methodology derived in these earlier studies, many areas are being explored, including those at various other interfaces and in biological systems. Computer simulations provide a useful added supplement to the earlier analytical-type (equations) development, and recent calculations of electronic effects have supplemented current experimental work on long-range electron transfer.

Further communication among researchers working in rather different aspects of the ET field continues to be desirable. Some may not be familiar, for example, with the relatively recent work by Iwasita et al. (3) showing adiabaticity for electron transfer between $Ru(NH_3)_6^{+2, +3}$ and metal electrodes. They showed that the electron-transfer rate between ion and electrode (the "exchange current") was constant for metals having widely different density of electronic states. This example may also serve to show

how information can be readily obtained in one area, whereas comparable information for homogeneous bimolecular ETs is less readily derived.

References

1. Marcus, R. A. *J. Phys. Chem.* **1990,** *94,* 1050, 4152.
2. Siddarth, P.; Marcus, R. A. *J. Phys. Chem.* **1990,** *94,* 2985, 8430.
   Iwasita, T.; Schmickler, W.; Schultz, J. W. *Ber. Bunsen-Ges. Phys. Chem.* **1985,** *89,* 138.

RECEIVED May 17, 1990.

INDEXES

Author Index

Affiliation Index

Subject Index

A

Copy editor and indexer: Colleen Stamm
Production editor: Paula M. Bérard
Acquisitions editor: Cheryl Shanks

Typesetting by Techna Type, Inc., York, PA
Printing and binding by Maple Press Company, York, PA

Paper meets minimum requirements of American National Standard
for Information Sciences—Permanence of Paper for Printed Library
Materials, ANSI Z39.48–1984 ∞